1 MONTH OF
FREE
READING

at

www.ForgottenBooks.com

By purchasing this book you are eligible for one month membership to ForgottenBooks.com, giving you unlimited access to our entire collection of over 1,000,000 titles via our web site and mobile apps.

To claim your free month visit:

www.forgottenbooks.com/free838066

ISBN 978-0-364-68205-0
PIBN 10838066

THE UNIVERSITY OF CHICAGO PRESS
CHICAGO, ILLINOIS

Agents

THE CAMBRIDGE UNIVERSITY PRESS
LONDON AND EDINBURGH

THE MARUZEN-KABUSHIKI-KAISHA
TOKYO, OSAKA, KYOTO

KARL W. HIERSEMANN
LEIPZIG

THE BAKER & TAYLOR COMPANY

THE

BOTANICAL GAZETTE

EDITOR
JOHN MERLE COULTER

VOLUME LX

JULY–DECEMBER 1915

WITH TWENTY-ONE PLATES AND TWO HUNDRED AND NINETY-FIVE FIGURES

THE UNIVERSITY OF CHICAGO PRESS
CHICAGO, ILLINOIS

Published
July, August, September, October, November, December, 1915

Composed and Printed By
The University of Chicago Press
Chicago, Illinois, U S.A

TABLE OF CONTENTS

DATES OF PUBLICATION

No. 1, July 16; No. 2, August 14; No. 3, September 16; No. 4, October 15; No. 5, November 15; No. 6, December 16.

ERRATA

Vol. LIX

P. 436, line 6 from bottom, for 23 read 13

P. 442, line 12 from bottom, for germination read germination of

Vol. LX

P. 150, line 18 from top, after species read of

P. 367, line 9 from bottom, for nucleus read nucleolus

P. 389, line 14 from top, for eclipse read ellipse

P. 391, line 9 from top, for third read second

P. 408, insert the following legend for chart:

0, *Lathyrus maritimus;* 1, *Ammophila arenaria;* 2, *Atriplex arenaria;* 3, *Salsola Kali;* 4, *Chenopodium rubrum;* 5, *Arenaria peploides;* 6, *Cakile edentula;* 7, *Artemisia Stelleriana;* 8, *Solidago sempervirens;* 9, *Euphorbia polygonifolia;* A, *Suaeda maritima;* B, *Salicornia europaea;* C, *Limonium carolinianum;* D, *Ligusticum scoticum;* X, *Zostera marina;* Y, *Spartina glabra pilosa.*

ICAL GAZ

Editor: JOHN M. COULTER

JULY 1915

VOLUME LX   NUMBER 1

THE
BOTANICAL GAZETTE

JULY 1915

THE ORIGIN AND RELATIONSHIPS OF THE ARAUCARIANS. I

L. LANCELOT BURLINGAME

Perhaps I can make the purpose of this paper clear in no better way than by quoting a sentence or two from a recent paper by JEFFREY (42):

> The comparative, developmental, paleobotanical, and experimental investigation of the Coniferales is likely to throw more light on the stable and sound general principles of biology than that of any other large group of animals or plants, on account of their great geological age and remarkably continuous display, both as regards external form and internal structure in the strata of the earth. . . . .
>
> It is finally clear that morphologists will find it necessary in the future more and more to adopt certain general working principles, as for example in the sister sciences of chemistry and physics. If there prove on trial to be no generally applicable fundamental principles in morphology, that branch of biological science cannot be too soon cast into the outer darkness, which prevails outside the scientific view of nature.

It is doubtful whether anyone is likely to dispute the truth of the first sentence quoted above. At any rate, it is sufficiently accurate to make an examination of the various opinions concerning the origin and relationships of the araucarians well worth while. It is a conspicuous fact that among those most familiar with the facts now known regarding conifers, the most diverse opinions are strongly held. It is further to be observed that these various opinions are grounded on the different values set upon different classes of facts. The object of this paper is to attempt to set forth the principal views and to show on what sort of evidence each rests. Having gotten at the facts, we may then inquire into the relative merits of each sort of evidence and attempt to evaluate them. If such an evaluation could be made, we might rescue the subject of morphology from the fate of being "cast into outer darkness." I am very little inclined to think that it will be possible for any one investigator to set forth the evidence in a manner to satisfy the adherents of all views, but my purpose will be served if I shall induce those more competent to take up the subject with the serious intention of determining just what principles are generally applicable, if there are any, and what are the limitations to which various sorts of evidence are subject.

In a broad way the opinions regarding the origin of the arau-carians may be grouped under three heads: (1) the lycopod theory, (2) the cordaitean theory, (3) the abietinean theory. In setting forth these various theories I shall attempt to present the salient facts and opinions of a number of those who have written on the subject, but no attempt will be made to present exhaustively the facts or even the opinions of any one. It will undoubtedly happen that in choosing what seem to me the strongest evidences for the different views and the strongest objections to them, I shall be unable to escape the effects of personal bias, and still less able to choose just as the authors themselves would choose. These are objections that cannot be wholly avoided. The best I can do is to be as impartial as possible and to beg indulgence in advance for the slips that are sure to occur.

The lycopod theory

The external resemblance of certain conifers to the lycopods has attracted notice from early times. The first serious attempt that I have found to set forth the evidence for believing that Coniferales, or at least certain of them, have originated from Lycopodiales is that of POTONIÉ (47) in 1899. A rather free condensation of his statement of this theory and the facts on which it is based is as follows: (1) conifers, except possibly Taxaceae, are derived from "Lepidophyten"; (2) the ancestor of a cone scale was a leaf that was at once foliar and spore bearing; (3) this foliar organ became differentiated into two parts, (a) a reduced spore-bearing part and (b) a foliar part, somewhat after the fashion of an *Ophioglossum* leaf; (4) these compound organs were aggregated into cones; (5) in the process of cone formation the foliar portion was gradually reduced and the ovuliferous (spore-bearing) portion was increased; (6) in the pine cone the spore-bearing portion is the ovuliferous scale, in *Araucaria* it is the ligule, and in many other plants ligule-like outgrowths are to be explained in the same manner (47, pp. 320–323). He offers the following arguments in its support: (1) it harmonizes a great variety of facts; (2) this gradual development of the ovuliferous scale and reduction of the bract is in accord with geological history, because it is only in the Abietineae that the distinction between scale and bract is comprehensible, and they are later and more specialized forms than the earlier Araucariae and Taxodiae; (3) the brachyblast theory requires us to think of the ancestral forms as having less compact strobili and with cone scales more like a leafy shoot, and this is directly opposite to the historical fact. We shall return in a later paragraph to the question of how much of this evidence is still pertinent after the presentation of the more recent views.

In 1905 CAMPBELL (8) stated his opinion that "as both the pollen tube and seed formation are but further developments of heterospory, it is quite conceivable that these might have arisen independently more than once. The close resemblance between the conifers and the lycopods, especially *Selaginella*, probably points to a real relationship. The strobiloid arrangement of the sporophylls, as well as the development of the prothallium and embryo, are extraordinarily similar, and it is not unreasonable to suppose that this is something more than accidental. The strong resemblance between the method of the secondary thickening of the stem in the arborescent fossil Lycopodineae, and that of the conifers, as well as the anatomy of the leaves, suggests a real affinity. It is known that some of these bore seeds, which in structure and position may very well be compared to those of typical conifers." He reaffirmed this opinion in less precise terms in 1911 (9).

In 1905 SEWARD and FORD (54) surveyed the work upon araucarians up to that time, added a considerable body.of observations
of their own, and discussed thoroughly the bearing of all the available facts. They came to the conclusion that the Araucariae
occupy a place so remote from other conifers that they should be
set off as a separate order under the name of the Araucariales.
It is difficult to summarize so extensive a paper, in itself more or
less a review, within the limits necessarily set on such a paper as
this. However, we may summarize briefly their conclusions and
indicate more or less inadequately the nature of the evidence on
which they are based. So far as their discussion of affinities is
concerned, the argument may be divided into three parts: (1)
Araucarineae are primitive plants; (2) there are numerous grave
objections to the assumption that they have originated from the
Cordaitales; (3) there are significant resemblances to various
living and extinct lycopods.

1. *The Araucarineae are primitive plants.*—This thesis these
authors attempt to prove in two ways. They show from a review
of the fossils that have been assigned to this family that it is
extremely probable that fossil stems are known as far back as the
Permian, and possibly the Carboniferous, both as impressions and
as petrifactions, that find their closest resemblances and affinities
with the present-day araucarians. Cone scales resembling those
of the EUTACTA section of the genus *Araucaria* are known far back
in the Jurassic, farther back in fact than those of the Abietineae.
Historically, then, they argue the Araucarineae are primitive
plants. To the support of their argument from the geological
record they bring the testimony of comparative anatomy and
morphology. They think the primitive character of the group is
indicated by the gradual transition of foliage leaves to cone scales
or sporophylls, by the resemblance of the two cones in some species,
by the simplicity (as they attempt to prove) of the ovulate cone,
by the persistent leaf traces, by the anatomical character of the
leaves, by the homogeneous character of the wood without resin
canals or wood parenchyma, by the multicellular male gametophyte, by the lateral distribution of the numerous archegonia,
and by the similarity of the embryo cap to a root cap.

2. *A cordaitean connection does not seem probable.*—The chief
arguments for this view they summarize under three heads: (*a*) the
presence of hexagonal contiguous pits on the tracheal walls,
(*b*) a comparatively wide transition zone from protoxylem to
secondary wood, (*c*) a resemblance in the form of the leaf and in
the general habit of the vegetative shoots. They are of the opinion
that the first two resemblances are of little significance because
of their common occurrence among paleozoic plants. They are
primitive characters, but do not necessarily indicate a relationship
to Cordaitales. The external form of the leaves is not specially
significant, for the internal structure does not indicate a close
affinity with Cordaitales, and furthermore the more ancient
araucarian leaves are less like those of Cordaitales than those of
certain modern species.

Having refuted the supposed arguments for the cordaitean
connection, they offer certain other objections for good measure.
They point out that the leaf trace of Cordaitales is double when it
leaves the primary wood, while that of the Araucarineae is single.
They cite several other investigators in support of their opinion
that the ovulate cone is simple in structure. They recognize that
a comparison can be made between the simple appearing cone
scales of *Araucaria* and the apparently double structures of the
Abietineae. The evidences for the duplicity of *Araucaria* must
be derived second hand from the latter. The evidence for their
double nature rests on the assumption that certain abnormalities,
in which an ovulate scale is replaced by a foliar shoot, indicate that
the cone has been derived by condensation of a branching leafy
shoot whose leaves bore ovules abaxially. While this may be probable enough (they do not even commit themselves to so much) for
this group, it is not considered valid evidence in the interpretation
of the apparently simple cone of *Araucaria*, which they believe to be
an older type and more likely to exhibit primitive structures than
the Abietineae. Their contention is that in the Araucarineae there
is nothing to explain. The cone is just what it appears to be.
The inverted vascular supply of the ovule is a normal feature of
such bundles, and the duplicity is no more than a sort of ligular
excrescence, such as is common among the Lycopodiales.

The resemblance of the male gametophyte to that of Cordaitales is no more than a parallel carried over by both from the pteridophytes. In short, there are not very many resemblances between *Araucaria* and *Cordaites*, and even these are no more than an expression of a common heritage from the pteridophytic ancestors of both. From all this SEWARD and FORD conclude that the Araucarineae are too unlike Cordaitales in too many respects to make the supposition of actual relationship probable. This conclusion leads logically up to the presentation of what they believe to be evidences of real affinity with the Lycopodiales.

3. *The significant resemblances of araucarians and lycopods* appear to these authors to be numerous and important. They recognize, however, that the application of JEFFREY'S (27, 28, 29) well known and widely accepted division of the vascular plants into two great divisions, the Lycopsida and Pteropsida, on the ground of the presence in the latter of leaf gaps and their absence in the former, would constitute a serious objection to a derivation of any conifers from a lycopsid ancestry. They prefer rather to question the validity of his generalization, and express the opinion that the characters on which it is founded "have been estimated at too high a value as indices of affinity."

Having thus cleared the ground by attempting to show that the Araucarineae are primitive, that it is unlikely that they have been derived from the Cordaitales, and that their possession of leaf gaps does not preclude a lycopsid ancestry, they reach the really critical portion of the argument. Direct comparisons are instituted with a number of living and fossil lycopods.

In the main body of the paper they have shown to their own satisfaction that the araucarian cone is simple in structure and so "poles asunder" from that of the Abietineae or Cordaitales. This suggests the direct comparison with the ovulate cone of *Lepidocarpon*. In opposition to the view of the discoverer (51), they are of the opinion that it may "constitute a (possible) connecting link between the Araucariae and lycopods" and approvingly cite SCOTT'S (53) admission that it furnishes an excellent argument for their view. The organization of the cone, the simple sporophylls, the single median ovule, and the ligule appear to present so close a homology as to afford strong indications of real affinity. The method of attachment of the single megasporangium by a stalk and the similar cone scales serve to form the basis of a comparison with *Spencerites* (49).

The resemblance of the microsporophylls to those of *Equisetum* is noticed, but is not held to indicate a direct relationship. Though they admit that the typical lycopod did not have its microsporangia on the dorsal side of the microsporophyll, they think they see a certain significance in the fact that the *Araucaria* type is found in *Cheirostrobus* (50), a genus that has been thought to be one of those generalized types which serve as finger-posts to the paths which evolution has followed, and which is considered to be intermediate between club mosses and horsetails (54).

These investigators assert that not only are the ovulate cones simpler and more primitive in structure than those of other conifers and more like those of paleozoic lycopods, but they are more nearly like those fossils from the Mesozoic than are those of any other group. This accords with their view; whereas, if they had been derived directly from Cordaitales or indirectly through the Abietineae, they should show some approach to the supposed leafy shoot predicated by the brachyblast theory, as the fossil history is followed backward toward the ancestral forms. They are of the opinion that no such transitional forms are known. On the contrary, they are of the opinion that the older forms show an approach to the lycopod situation in having smaller leaves and cone scales, with a gradual transition between the two organs.

They point out that the stem apex more closely resembles that of lycopods than that of ferns, though they do not attach much value to this fact. The exarch veins of the leaf may be regarded, they think, as a possible "ancestral feature which has disappeared from the vegetative stems." The leaf traces are accompanied by a group of cells in the cortex which the authors compare to the "parichnos" in lepidodendroid stems.

While they admit that there are many points of dissimilarity between araucarian stems and those of the lycopods, they do not think that any of them constitute an insuperable barrier to the derivation of the one from the other. They point out that the

scalariform tracheids of the lepidodendroid stems is paralleled
by the same structures in the transitional primary wood of *Araucaria*. They do not urge this point as a strong argument, but
merely point out that the presence of the two distinct types of
pitting in the mature secondary wood of the two groups does not
"necessarily imply separate lines of descent." The resin canals
of the araucarians can be derived as well from the mucilage cells
and canals of the Lepidodendreae as from those of the Cycadofilicales.

They are strongly of the opinion that the multicellular pollen
grains of the araucarian alliance are very different from those of
Cordaitales or any other recent seed plants. They differ from the
former in the arrangement of the cells, and from the latter in the
much greater development of the vegetative cells. A comparison
with the microspore of *Selaginella* or *Isoetes* appears more convincing. The reduction they "connect with the substitution of sipho-
nogamous for zoidogamous fertilization, which would demand as
much space and material as possible for the production of the
pollen tube."

STILES (**61**) argues that the conifers can be derived more readily
from the lycopods than from the Cordaitales. His argument is
divisible into two parts: (1) an attempt to show that Podocarpeae
(and hence other conifers, for he holds that all have had a common
origin) cannot have been derived from Cordaitales and must,
therefore, have been derived from the only other(?) available
source, the lycopods; and (2) a direct comparison of conifers and
lycopods to show the possibility of deriving the former from the
latter.

Under the first head he adduces much excellent evidence to show
that the podocarps are closely related to the araucarians. He
also attempts to show that the conifers are monophyletic. The
next step is to show that podocarps cannot have been derived from
Cordaitales, in consequence of which the other conifers are likewise
excluded from such an origin. He enumerates four points which
he considers sufficient to preclude the possibility of the primitive
podocarps having originated from Cordaiteans: (*a*) the stem of
these podocarps is no more like that of the Cordaitales than it is like

that of conifers in general; they may have derived certain resemblances from a common ancestry but are not on that account closely
related; (b) the roots of the podocarps are not "particularly
reminiscent" of those of Cordaitales; (c) the primitive type of
leaf among the araucarian-podocarp alliance was small and narrow
and provided with a single midvein and unlike the broad parallel
veined leaves of the Cordaitales; (d) the structure of the microsporophyll of the podocarps "no more favors this view [cordaitean
origin] than the three preceding pieces of evidence." It is scarcely
necessary to mention that most of these objections would be
equally valid as arguments against a relationship between
podocarps and araucarians, a relationship which he champions
vigorously, nevertheless. It seems rather unfortunate that
so many of the facts known about gymnosperms may be used
almost equally well to prove a variety of quite antagonistic
views.

Under the direct argument, he places first the similarity of the
ovulate cone of the simpler and more primitive podocarps and of
the araucarians to the lycopod cone. They are, he thinks, alike
in their general structure, being in both cases composed of simple
sporophylls with a gradual transition to the leaves. In both each
simple sporophyll bears a single median megasporangium. In
both the sporangial organ is at first erect in the axil of the scale. In
certain of the podocarps he sees a tendency to the development of
a double structure of the cone scale analogous to that of the
Abietineae.

Secondly, "the microsporophylls are also easily comparable
with those of the lycopods." The presence of more than one
sporangium in the conifers "is not a serious" difficulty, since
septation is well known·in the sporangia of lycopods. Moreover,
"the shifting of the sporangia to the under side presents little
difficulty to the view under consideration," since it "has certainly
taken place in other cases." The other cases cited are the possibility of its having occurred among the ferns, and the fact that
"among the Equisetales in Palaeostachya the sporangiophores are
found on the upper side of the sporophyll, while in Cingularia the
sporangia are below the sporophylls" (italics mine). In any case,

whether conifers have sprung from ferns or lycopods, one sort of sporangium must have migrated to the other side of the sporophyll. This is true, of course, only if one accept the author's view that the cone scales of conifers are really simple.

The presence of seeds in *Lepidocarpon* and *Miadesmia* is held to prove that "there is thus abundant evidence that the potentiality of seed production existed in this phylum as well as in the fern phylum." Any differences that exist in the vascular supply of the sporophylls between podocarps and lycopods "is to be accounted for by the greater relative importance of the ovule as compared with the sporophyll." Small, narrow, uninerved leaves are characteristic of both conifers and lycopods, but are unknown among the Cordaitales. Those podocarps and araucarians with broad parallel veined leaves are not primitive, but have derived their leaves from narrow-leaved ancestral forms.

One of the most interesting points in the argument is the attempt to show that, while a siphonostele with leaf gaps is certainly characteristic of the fern alliances, it is not necessarily limited to them. It merely represents a goal toward which vascular plants of all sorts have tended. The ferns reached it early, while the paleozoic lycopods did not quite reach it. They did actually attain the seed habit, another one of the milestones of plant evolution, but attained only to a medullated siphonostele in which nearly all the metaxylem had been obliterated and which had become broken up in some forms into separate strands. These bundles were still exarch, however, and the leaf traces did not produce leaf gaps in the stele. These would have been the next logical steps in the evolution of the lycopod stele. The inference probably is that they were actually attained by the yet unknown lycopodialean ancestors of the conifers. While admitting that the presence of bordered pits in the secondary wood of conifers is a point against the lycopod theory, he thinks that the presence of a modified sort of pit in *Sigillariopsis Decaisnei* (**48, 53**) shows the possibility of their development in this phylum. The double leaf trace of the Abietineae (and Araucarineae), which has been used as an argument for the cordaitean origin, he thinks is offset by the single trace of the primitive podocarps.

The conclusion is indicated by the following quotations: "To the writer the evidence seems to point . . . . to the primitiveness in the Coniferales of a type bearing female cones composed of aggregations of simple sporophylls, each sporophyll bearing a single erect axillary ovule." "This supposed primitive conifer is very suggestive of the Lycopodiales, but is not reminiscent of the Cordaitales."

It is evident from the preceding that, aside from what support may be gained by discrediting rival theories, the lycopod theory derives its greatest strength from the three following sources.

The first and strongest argument comes from the very close resemblance in form and structure of the ovulate araucarian cone to the strobilus of the lycopods. If there were no other reasons for suspecting a filicinean origin of araucarians, and there were no Abietineae with their perplexing structures, no one would, I think, even suspect that the ovulate cone is other than what it appears to be. Notwithstanding these influences most (though not all, 20, 42, 59) investigators (44, 61, 69, 76) who have studied the araucarian cone have concluded that it is really simple in structure and its cone scales simple sporophylls.

Next in importance and even more difficult to dispose of is the structure of the seed and pollen tube. SEWARD and FORD have pointed out the close resemblance of the seed structure (54) and the writer has elsewhere (7) shown that these structures could easily have arisen from the condition found in *Miadesmia* and *Lepidocarpon*, but that it is exceedingly difficult to see how, and still more difficult to see why, they should have arisen from the known types of cordaitean seeds. It is easy to see how pollination of the scale instead of the nucellus would be the most probable type in plants which developed the strobiloid habit before the pollination and seed habit. But it seems hardly probable that having been in possession of the habit of depositing pollen grains in a specially prepared and protected pollen chamber in the nucellus, any group of plants would pass through a course of evolution requiring them to give up all the advantages comprised in these arrangements and to acquire an entirely new and certainly less efficient method of pollination. Considering how very little we know of the structure of paleozoic seeds of any sort, it would be rash indeed to suppose that the·known types of seeds were the only ones found among Cordaitales, and even more rash to generalize more than provisionally on the assumption that they did not possess a particular type.   We do not know whether the seed habit was developed in the phylum once or more than once, much less whether it was developed before or after the organization of cones.   In short, this is today a strong argument for the lycopod theory, that the discoveries of tomorrow may become an equally valid argument for the cordaitean theory.

A third group of resemblances between lycopods and conifers is presented by the leaves.   The small, narrow, uninerved type of leaf so characteristic of lycopods is very common among the conifers.   The arrangement in many cases is also similar.   The gradual transition from leaves to sporophylls in the lycopods presents a very close resemblance to certain araucarians.   Most, or perhaps all, of these resemblances, indeed, can be explained away, but that is just where their strength as evidence for this theory lies, they do have to be explained away.

On most other points the theory appears to be on the defensive. It can, to be sure, offer more or less plausible explanations and possibilities for some of the evidence that appears to be against it, but still it must explain them in some other than the obvious way to bring them into harmony with itself.   The first and most serious objection in the opinion of most of its opponents appears to be the structure of the stem, more particularly the stele.   Notwithstanding STILES's ingenious and convincing exposition of the evolutionary tendencies of the lycopod stele, it yet remains true that no known lycopod did attain to the possession of a mesarch or endarch siphonostele with leaf gaps.   That they might have done so appears very probable, but there is yet no evidence that they actually did so, and much less that any one that could be supposed to be a form ancestral to the conifers had even nearly approached it.   The same sort of objections apply with even greater force to the attempts to explain the origin of the staminate cone structures.   It is admitted by most botanists that septation has probably occurred in certain lycopod sporangia.   It may even be admitted that a shifting of sporangia from one surface of the sporophyll has occurred in some pteridophytes. Again, there is no evidence that they have occurred in any lycopod that can by any possibility serve as a starting-point for modern conifers.

I have indicated above what seem to me to be the most fundamental objections to the theory, namely, that the individual comparisons which can be made between conifers and lycopods must be made with plants of the latter phylum which are admittedly very remotely related to one another. No single lycopod is known that combines within itself any very considerable number of resemblances to the conifers. In a later paragraph I shall return to the attempt to evaluate evidence of this sort. It is an interesting fact that practically all the evidence for this theory is derived from the comparative morphology of adult plants. The two conspicuous exceptions are (1) STILES's comparison of the erect axillary mature megasporangium of the lycopods with the position of the very young ovule of the podocarps, which is also erect and axillary, but which may later be inverted and carried out and away from the axis by the growth of the base of the sporophyll; and (2) his argument that the primitive leaf in the conifers was small, narrow, and uninerved, because many conifers have juvenile foliage of this sort.

The cordaitean theory

The majority of writers have held that conifers are ultimately to be derived from the Cordaitales. SCOTT (53), OLIVER (45), WORSDELL (77, 78), COULTER and CHAMBERLAIN (16), JEFFREY (29, 42), THOMSON (70, 73), and many others have brought forward much convincing evidence in support of this view. Although these authors agree in general as to the ultimate origin of all conifers from a common stock, there is considerable diversity of opinion as to the relationship of the tribes. In the present paper we shall consider these divergent views only so far as they pertain to the origin and relationship of the araucarians. There is a prevailing opinion that this tribe is either the primitive basal group of conifers or constitutes an independent line by itself. Opposed to this view is that of JEFFREY and his students, who have presented much evidence to show that the Abietineae are the oldest and most primitive group, and that the araucarians have been derived from them.

The first view finds its support in (1) the many close resemblances between the modern araucarians and the paleozoic cordaiteans; (2) the apparently greater geological age of the Araucarineae; and (3) transitional forms of the Triassic and Cretaceous, which appear to become more like the Abietineae from the earliest to the later ones.

The second view does not deny the similarities pointed out as supporting the previous view, but in view of other sorts of evidence believes them to have been secondarily acquired. It does not necessarily deny that they are indications of relationship, but merely that they do not indicate direct and immediate relationship. It derives most of its positive support from (a) vestigial structures, (b) recapitulation phenomena in seedling and young wood, (c) traumatic responses.

Since the supporters of the lycopod theory have chosen the araucarians as the tribe most favorable to their contention, the views of their opponents can be best set forth, perhaps, by marshaling the evidence that tends to show that the araucarians and, by implication, the other conifers have had a direct and immediate origin from Cordaitales.

1. GYMNOSPERMS ARE A MONOPHYLETIC GROUP.—There has been a general tendency toward the view that the gymnosperms resemble one another so much more closely than they do any other group that they must therefore have had a monophyletic origin. This point of view was apparently prominent in the minds of more than one (45, 52, 77) of the speakers at the Linnaean Society discussion. If the monophyletic origin of gymnosperms be admitted, it follows almost without dispute that they all have had a filicinean origin. Among the known fossil groups of gymnosperms no other can present anything like so strong a claim to be the ancestors of the conifers as the Cordaitales. Just how numerous, striking, and significant are the resemblances between Cordaitales and Coniferales (more particularly Araucarineae) can be best shown by a brief review.

Ever since JEFFREY'S (27, 28, 29) division of vascular plants into Lycopsida and Pteropsida on the basis of the presence or absence of leaf gaps, there has been a very general disposition to accept this distinction as entirely valid so far as it concerns the Pteropsida. Striking as is this fact in the other groups of the Pteropsida, it is preeminently so among the conifers, a small-leaved group where it is not only present in the mature stem but also in the seedling and reproductive axes (31). It is true (61) that the cladosiphonic exarch stems of the ancient lycopods did occasionally become medullated, and it is possible that in the course of evolution some member might have lost all of its centripetal wood, and have developed centrifugal wood and leaf gaps, but there is no evidence as yet that any of them ever actually did either.

The histological structure of the stem is only less strikingly uniform among gymnosperms than the general organization. In fact, the wood of araucarians and cordaiteans is so nearly identical that no absolutely trustworthy tests have yet been discovered for distinguishing them. Although the other gymnosperms do not all have exactly the same arrangement of the bordered pits, they do all have such pits on the radial walls of the tracheids, and they are, on the contrary, with a single exception (48, 53), unknown among the lycopods. While there is greater diversity in the phloem, perhaps that of lycopods differs still more widely.

Aside from the Araucarineae, the structure of the ovulate cones is more readily brought in line with a filicinean than a lycopodinean ancestry. Though the ovulate cone readily lends itself to the derivation of araucarians from lycopods, it can nevertheless be explained in terms of the Cordaitales; while, on the contrary, it is very difficult to explain the cone of a pine in terms of a lycopod ancestry, and next to impossible to so explain those of cycads.

The structure of the seed is remarkably uniform through the entire phylum, from the oldest to the living representatives. Very few lycopods (3, 51, 53) are known to have borne seeds of any kind and even those are much simpler than those of any gymnosperms. I have elsewhere (7) pointed out that these seeds do offer us an analogy of the way in which the peculiar pollination processes of the araucarians may have originated. It may be objected, however, and I think rightly, that if lycopods had developed high grade seeds, they would have been likely to parallel the structures present in the Pteropsida. Still it remains true that they are not known actually to have done so.

There are no microsporangiate structures known among lycopods that are at all comparable with the pollen cones of the gymnosperms. Notwithstanding the puzzling diversity within the group, it is still far easier to derive them from filicinean ancestors than from club mosses. In the latter the sporangia are uniformly single and adaxial instead of multiple and abaxial, as they are in ferns and gymnosperms.

With the exception of the Gnetales, the female gametophyte of the gymnosperms is so uniform in mature structure and in development as almost of itself to preclude any question of its diverse origin. The deep-seated megaspore, the vacuolated free nucleate embryo sac, the centripetal growth of walled tissue, the origin and development of remarkably uniform archegonia are common to all known members of the group, and form a unique and characteristic series unknown outside of it.

The development and mature form of the embryo, with its free nuclear phase, organization into a walled proembryo, elongating suspensors, and terminal embryo, are no less striking and equally without analogy outside the group. Nor is there any sufficient diversity in the mature structure or in the course of development of the male gametophytes to cast serious suspicion on their common origin. The differences are strictly of degree, and find a ready explanation in the changes incident to a long course of evolution.

As a result, it seems to the reviewer that all gymnosperms resemble one another in very many and very significant ways. On the contrary, it is the araucarians alone that present anything more than very slight resemblances to the lycopods, and even here the significant points of resemblance are few and less exact than the numerous ones that relate them to other gymnosperms.

2. THE ARAUCARINEAE RESEMBLE THE CORDAITALES MORE CLOSELY THAN ANY OTHER CONIFERS.—Among those who hold this view no one has expressed himself more clearly or strongly than SCOTT, who says: "The Araucarineae present a close agreement with the Cordaiteae in the structure of the stem, and particularly in that of the wood, which, as universally admitted, is often indis-

tinguishable in the two families. The essential feature is that the mass of the wood, apart from the medullary rays, is composed of tracheids with multiseriate bordered pits on their radial walls" (53, p. 654).

The more recent and comprehensive argument for this view is that of THOMSON (70). His arguments may be summarized under four heads: (1) the Araucarineae closely resemble the Cordaitales in the anatomy of the stem, root, and leaf; (2) they are the oldest known conifers; (3) certain mesozoic forms show a transition from Araucarineae toward Abietineae; (4) vestigial structures in leaf, stem, root, and reproductive axes, some of which indicate (a) the origin of araucarians from cordaiteans, and (b) others of which indicate the origin of Abietineae from Araucarineae.

Although I have not seen the papers by GOTHAN (21, 22), the references to them by other writers, particularly JEFFREY and THOMSON, would indicate that he holds similar views respecting the relationship between the Araucarineae and Abietineae.

Speaking of the pith, THOMSON (70) says: "In the variability of the size of the pith, and in the magnitude which it may attain, the Araucarineae are the only forms of the conifers at all comparable to those of the cordaitean alliance."

The root is usually diarch, and the protoxylem points are separated into two forks by the presence of a resin duct, as in the Pineae (70); nor "is there any indication of a resin duct in the center of the metaxylem, as in the Abieteae."

There is a very broad transitional zone from the primary to the secondary wood in the stem and particularly so in the cone axis. "In no other group of the conifers is there an approach to this cordaitean condition" (70).

There are many resemblances between araucarian and cordaitean leaves. The araucarian-podocarp alliance includes the only conifers with leaves at all comparable in size to those of the cordaiteans. In both, the leaves persist for many years; in both, the leaves are parallel and dichotomously veined, with mesarch collateral bundles and remarkably persistent leaf traces.

Although the pitting of the more ancient araucarians was so nearly identical with that of cordaiteans, it "is much reduced in the mature (stem) wood, and occurs mainly at the ends of the tracheids" (70) in the living forms. Some cordaiteans show a notable tendency in the same direction. The characteristic paleozoic type of pitting is found only in the primitive regions of the living forms (cones and roots).

A torus is present in the bordered pits of all conifers except araucarians and some podocarps (70). It is very poorly developed, when present at all, in the Araucarineae, and entirely absent in Cycadales, Ginkgoales, Cordaitales, Cycadofilicales, and Filicales.

Miss GERRY (20) has proposed to separate araucarians from other conifers on the ground that they lack bars of Sanio in the radial walls of the tracheids, which are possessed by all others. JEFFREY has gone even farther and held that it is the most certain distinguishing feature in separating fossil araucarians from the abietineans (35). On the contrary, THOMSON holds "that a rudimentary bar of Sanio is present in all Araucarineae" and "that the araucarians are not to be separated from the other conifers because of the lack of a bar of Sanio, but rather that they are to be regarded as the basic forms from which this structure in the other conifers has been derived" (70). I have found no reference to its presence in cordaitean wood, so that its absence or feeble development is another point of similarity.

The absence of resin canals in the wood of both Araucarineae and Cordaitales is a well known and striking resemblance. THOMSON holds it to be primitive in both cases (70). He brings forward much argument to show that resin canals in the pines are primitively solid, and that they have been derived from resin parenchyma, which has in its turn replaced the resinous tracheids characteristic of cordaiteans and araucarians. He concludes that "the origin of the resin tissue of the pine alliance from tracheary elements as in the Araucarineae, and the retention of similar stages in its development, forms what the writer regards as one of the fundamental features of relationship between the two groups."

So far as living araucarians are concerned, the cells of the medullary rays are characteristically thin-walled and unpitted, just as they are in the Cordaitales. There are known several mesozoic forms in which the rays approach the abietinean type.

It is obvious that forms intermediate in character may be interpreted as araucarians that are being modified in the direction of the
abietineans (70), or as pines that are about to be transformed into
araucarians (32, 34, 35). It appears (70) that the older forms are
more like the araucarians, while the later ones resemble the pines
more closely. The geological sequence thus appears to be in favor
of the origin of pinelike conifers from araucarian ancestors.

Recently evidence (75) has been adduced to show that marginal
ray tracheids have arisen through a modification of the tracheids
of the wood, and not by a transformation of parenchymatous cells.
The oldest known forms with ray tracheids do not antedate the
Cretaceous. Since the Araucarineae are known with great certainty from the Jurassic and probably from earlier strata, the geological evidence appears to favor the view that thin-walled unpitted
ray cells are the primitive type.

The albuminous cells of the phloem have been considered (13,
65, 75) homologous with the ray tracheids. Their absence from
the Araucarineae, accordingly, has been interpreted (70) in the
same way.

Annual rings are absent or feebly developed in the Cordaitales
and in most Araucarineae, though STOPES (64) has recently reported
a cretaceous *Araucarioxylon* from New Zealand with very definite
growth rings.

Aside from the structure of the seed itself, the ovulate cone of
Araucaria is nearer to that of the Cordaitales than are those of the
Abietineae. The essential feature of the cordaitean cone is that
the single seed is borne on an axis standing in the axil of a bract.
The seed-bearing axis is not always axillary if one may trust the
illustrations (16). In some cases the bract appears to be borne
on the seed stalk. The seed itself is terminal and erect. The cone
of *Araucaria* differs in that the bract and axis are much more intimately associated and in that the ovule is inverted and not terminal.
There are at least three obvious interpretations of the cordaitean
cone. First, one may suppose the cone to be simple, consisting
of an axis covered with branched sporophylls, some of which are
sterile and some fertile. Secondly, one may suppose that the cone
is compound and that the ovule is borne directly on the branch

axis, which stands in the axil of a bract or leaf, or merely arises directly from the main axis among the bracts but without a fixed relation to any of them. Thirdly, one may apply the theory of CELAKOVSKI (16) of the pine cone to it, and suppose that the seed-bearing axis really represents a branch, in the axil of a bract, so intimately united with a sporophyll, which itself bears the seed, that no traces are left of its complex nature.

Without attempting to review the extensive and well known literature relating to this third theory, the writer is disposed to admit that it offers a reasonable explanation of the cones of the Abietineae. It seems much less probable when applied to the araucarian cone or to the ovulate structures of the podocarps and taxads. The attempt to explain the cordaitean cone according to it would appear to be beset with very many grave difficulties. In the first place, most of the evidence used to support it for the modern forms is here unavailable. In the second place, there is no indication in the cordaitean cone itself of such a union of branch and sporophyll. In the third place, it is hardly to be supposed that if such a process had taken place in the ancestors of the pines, there would be still in the present geological age clear indications of it, and that the paleozoic ancestors would have apparently gone so much farther than their modern representatives as to have made their cone appear even simpler than any of them, including even the apparently simple araucarians. Evolution plays strange tricks, it is true, but it really puts a considerable strain on one's credulity to believe, as I think we must if we accept the theory that araucarians are derived from Cordaiteans *through the Abietineae,* that a complex branch system was reduced to a cone in the paleozoic cordaiteans, showing practically no trace of its complexity, then reverted in the abietinean descendants to a stage where the evidence of complexity is again clear, and finally passed on into the araucarians, where the evidence of complexity is again at least doubtful.

It seems to the writer far simpler to make no such difficult assumptions, but to consider that the cones of *Cordaites* and *Araucaria* are no more complex than they appear to be. In any case, the cones of the two resemble one another closely in apparent structure, and will probably both be eventually satisfactorily explained in the same way.

The seed of *Araucaria* resembles that of *Cordaites* in having the nucellus free from the integument to a zone below the female gametophyte. It is doubtful whether this is a character of any great consequence, inasmuch as there were seeds of both types known among paleozoic plants. Nevertheless, it remains true that this ancient type is not known in any other modern plants. Cordaitean seeds, so far as yet definitely known, appear to have had pollen chambers in the nucellus. Since the pollen has been found in these chambers, in some cases apparently sealed in, one can only infer that the method of pollination was essentially the same as that of modern Abietineae. In this respect the Araucarineae differ very markedly. To the writer the difference appears so great, and the method of the Cordaitales so much superior, that it is difficult to believe that having been once attained it would ever have been given up (36). If there is any dependence to be placed on the facts that appear to indicate that podocarps have been derived from an araucarian ancestry, it would appear that the tendency of evolutionary selection had been in this case in the other. direction. This objection does not appear to be very formidable at present, for we know many more seed genera from impressions than we have plants to assign them to. Moreover, we know the internal structure of very few of them. It is not unreasonable to suppose that the Cordaitales may have borne more than one sort of seed, and that among them may have been some which were pollinated in the araucarian fashion. ·

When the pollen cones are considered, it is at once evident that the closest resemblances to those of the Cordaitales are found in three rather widely unrelated groups: *Ginkgo*, araucarians, and taxads. Doubtless they have all inherited their resemblances from a common source, though along little related lines. So far as the evidence goes, it constitutes a notable resemblance between the araucarians and cordaiteans, in which the abietineans do not share. The araucarian type, with its free pendent sporangia, has apparently been transformed into the more common conifer type with two imbedded sporangia in the podocarps. It would appear more probable that this transformation had taken place in several lines of descent than that it had taken place in the supposed *Cordaites-Pinus* line, and had then reversed itself in the supposed *Pinus-Araucaria* line. The only very obvious difference between the microsporophyll of *Araucaria* and *Cordaites* is that the pollen sacs are erect and terminal in the latter and reversed in the former. This is precisely the difference in the ovulate cones.

But little evidence can be gathered from the gametophytes, owing to our ignorance of those of the Cordaitales. In both araucarians and cordaiteans the male gametophyte (4, 5, 36) is larger than in other modern conifers. It is uncertain whether the gametophyte of the cordaiteans had a more extensive prothallial tissue, like that of the araucarians (5, 7), or a more extensive spermatogenous tissue, like that of certain modern cycads (36, 42). If I correctly apprehend the abietinean theory of the descent of araucarians, it involves the assumptions that the original male gametophyte possessed a more or less extensive prothallial tissue and at least one antheridium; that in the course of evolution it lost its prothallial tissue with the exception of two primary cells, but retained the spermatogenous tissue of the antheridium (this would perhaps represent the cordaitean stage); that it further lost all of its spermatogenous tissue during its evolution into the Abietineae, except that part giving rise to two male cells; and, finally, that in the course of the evolution of an abietinean into an araucarian the place of pollen deposition became shifted (for reasons not stated) to a point much farther away from the female gametophyte, thereby necessitating the production of a more extensive prothallial tissue (36) to supply the needs of the larger amount of cytoplasm required to fill the more extensive pollen tube. Such a course of evolution is presumably possible, though I am inclined to think that the evidence favoring it is yet very inadequate. In the present state of our knowledge the large size of the gametophytes is a point of resemblance between araucarians and cordaiteans, while the pine type of male gametophyte can be easily derived in the same manner from either by the reduction of either or both the prothallial or spermatogenous tissue.

Not enough is known about female gametophytes to make a comparison of much value. The cordaiteans apparently possessed apical archegonia in the manner of all modern conifers. A comparison of embryos of course is at present impossible.

3. THE ARAUCARINEAE ARE VERY ANCIENT PLANTS.—Although wood of the *Araucarioxylon* type is known from the Paleozoic to the present, it is not yet possible to say with certainty just how old are plants corresponding in other essential points with modern araucarians. SEWARD and FORD (54) have given us a very careful review of the fossils that have been assigned to the araucarians, to which SCOTT (53) has given general agreement. It appears probable, though not beyond question, that such genera of the Permo-Carboniferous as *Walchia* and *Voltzia* were more nearly allied to araucarians than to any other known conifers (54). *Voltzia* and *Ullmannia* appear very probable triassic representatives (24). There is abundant evidence of impressions, cones, and wood of araucarians in the Jurassic and Cretaceous (53, 54).

The Abietineae have been said to extend to the Paleozoic (33), and this assertion has been vigorously disputed. The carboniferous form has been discredited on the ground that its source is not known to be from rocks of that age (21, 22). The form from the Permian is said by THOMSON and ALLIN not to be a *Pityoxylon* at all, but a cordaitean or *Araucarioxylon*. PENHALLOW appears to have originally regarded it as a *Pityoxylon* on account of what he supposed were horizontal resin canals (46). These are now (71) said, on a reexamination, not to be resin canals at all, but leaf traces. If these forms are rejected, no true Abietineae are known that can be compared in age with the araucarians.

4. TRANSITIONAL FOSSIL FORMS.—Of late there have been described (particularly by JEFFREY) a number of mesozoic plants with wood more or less intermediate between the true *Araucarioxylon* and abietinean wood. As will be shown in a subsequent section, the JEFFREY school interprets these as evidence of the origin of araucarians from the Abietineae. THOMSON (70), however, points out that the earliest of these transitional forms, *Woodworthia* (38), is much more like true *Araucarioxyla* than the later ones (as *Araucariopitys*), while the latter are much more

abietinean.   He refers particularly to the absence of resin canals, even of the revival type, in the former, and their presence in the latter.   The rays of the latter are thick and pitted in the fashion of the Abietineae, while those of the former are thin and resemble those of modern araucarians as well as cordaiteans.   In like manner the pitting of the former is more extensive, more crowded and flattened, and with the pits mostly alternately arranged; whereas in the latter they are less numerous, more restricted to the ends of the tracheids, less crowded, and more frequently opposite.   He contends that this is consistent either with a cordaitean or an araucarian ancestry for the pines, but difficult to reconcile with an abietinean ancestry of araucarians.

5.   VESTIGIAL STRUCTURES, RECAPITULATION, TRAUMATIC REACTIONS.—The broad transitional zone between primary and secondary wood in the araucarian cone has already been mentioned as a remarkable parallel to the condition found in the cordaitean stem. The pitting in the cone is also more extensive.   The pits cover the whole radial surface of the tracheids, are crowded and mutually flattened, and there may sometimes be as many as five rows to a tracheid.   THOMSON remarks (70) that not only does the pitting in the araucarian cone resemble cordaiteans, but that "instead of the opposite pitting, the pitting of the cone axis and early wood of the Abietineae has characteristically either scattered uniseriate pits or biseriate ones which are alternately arranged."

A torus, characteristically present in mature wood of Abietineae and feebly developed in mature wood of Araucarineae, is entirely absent (70) in such primitive regions of the latter as cone axis, first-year stem wood, primary and young secondary root wood. They should be expected in some or all of these places if araucarians had descended from abietinean ancestors which possessed them.

Bars of Sanio are well developed in the Abietineae and feebly so in the araucarians (42, 70).   They are also poorly developed in the primitive regions of Abietineae and in the mesozoic *Pityoxyla*. From this THOMSON infers that well developed bars of Sanio were not characteristic of the ancestors of the pine alliance.

He also argues that, since resin canals are ontogenetically developed from solid parenchyma and are frequently solid in the abietinean cone and in certain fossil forms, resin canals are not actually a character, as Jeffrey maintains (42), of the ancestors of modern Abietineae, and much less so of the ancestors of modern araucarians.

The absence of ray tracheids from the seed cone of *Pinus* and of the erect cells of the phloem from the cone and first few years' growth of the stem and root (13, 70) is interpreted to mean that these structures have been acquired in the comparatively recent geological history of the group, very probably long since the time at which Araucarineae are supposed to have originated from it.

In the outer extremities of the vascular bundles of the leaf of the Araucarineae there is a considerable amount of centripetal xylem. It has been interpreted in various ways. Thomson holds that, though much of it is of the transfusion type, there is yet always a certain number of elongated elements that are true centripetal wood lying next the protoxylem (70).

Attention has often been called to the fact that seedling pines have only primary needle leaves and only later develop spur shoots. If they are ancestral to araucarians, the latter might be expected to develop spur shoots on the seedling.

Traumatic reactions play little part in Thomson's argument, though he does invoke its aid in the attempt to show that the ancestral type of resin canal (70) in the pines was solid. He points out that the resin canals produced by wounding modern pines are much more numerous than can reasonably be expected to have been the case in the ancestral forms, and that they are frequently solid. The argument would appear to cut both ways. In a later paper (73) he has studied the normal variability of the spur shoot of *Pinus* and has made free use of the effects of wounding to substantiate his conclusions. His general conclusions are that the larger numbers of leaves on the less definite spurs of fossil forms find their counterparts in the normal variations in the number of leaves to a fascicle of living pines, particularly such wide variations as occur frequently on very vigorous branches, on reproductive axes, and on vigorous seedlings. Variation in the number of needle leaves, branching of the spur shoot, the production of primordial leaves on branches, and even on proliferating·spur shoots can all be produced by wounding. He interprets these reactions as a reversion to the ancestral condition as indicated by their occurrence in the fossil forms and in the primitive regions of living forms just mentioned. LLOYD in a recent paper (43) has touched the same subject and given his approval to the general conclusion as stated by THOMSON. The latter calls attention to the fact that the conifers which do not have spur shoots show no evidences, either in their primitive regions or as the, result of wounding, that they have descended from ancestors possessing spur shoots.

STANFORD UNIVERSITY, CALIFORNIA

THE EFFECTS OF ILLUMINATING GAS ON ROOT SYSTEMS

CONTRIBUTIONS FROM THE HULL BOTANICAL LABORATORY 205

EDWARD M. HARVEY AND R. CATLIN ROSE

(WITH NINE FIGURES)

Introduction

The injurious effects of illuminating gas upon trees and shrubs have been testified to by numerous observers. In 1864 GIRARDIN (5) reported severe injuries to trees in several cities of Germany which he attributed to escaping illuminating gas. Similar observations have since been recorded by many writers, among whom are VIRCHOW (17), KNY (8), SPÄTH and MEYER (14), EULENBERG (4), WEHMER (18), SHONNARD (13), and others.

The shade tree commissions of every city find themselves face to face with this serious problem. The trees of our city streets and parks are unusually subject to the various tree-injuring agencies, the chief of which are insects, fungi, and atmospheric and soil impurities. The two latter factors are particularly characteristic of the troubles of city trees. The problem of gas injuries, therefore, is one of considerable economic importance. City foresters should know the exact cause of any tree death, not only to enable them to provide means for future protection, but also in order to determine with whom responsibility rests for the present financial losses. They should therefore be able to say with certainty whether or not a tree has been killed by gas in the soil. At present this is no small undertaking, because there are few, if any, reliable symptoms known by which one may differentiate with certainty gas injuries from those due to several other causes. For example, fungi sometimes quickly become prominent in a tree injured by gas, as purely a secondary effect (see KNY 8 and STONE 15); but in a case like this the primary injury might easily be attributed to the fungi. It is claimed that characteristic odors often accompany gas poisonings; sometimes in the

27]

roots as well as in the soil; in fact, these odors form the chief diagnostic characters. STONE (15) claims much for them, and it is
probably true that in many cases they furnish some evidence.
However, many instances are known where illuminating gas
odors have not given sufficient evidence for fixing responsibility.
A rational system of diagnosis of injuries in city trees would be
of value and would be welcomed by all parties concerned. No
such system, from the gas injury standpoint, is possible, owing
to the lack of definite knowledge concerning the effects of the
constituents of illuminating gas under various controlled conditions.

The experimental work previously done on gas injuries to root
systems will be briefly summarized. KNY (8) passed known
amounts of illuminating gas through the soil at the roots of a maple
and two linden trees. Among the symptoms of injuries recorded
were the bleaching and final fall of leaves and the appearance of a
blue coloration in the xylem of the roots. Finally the trees died.
SPÄTH and MAYER (14) passed small amounts of gas into soil in
which grew a number of species of trees and shrubs. General
death resulted, but otherwise the only symptom recorded was the
yellowing of the leaves. BÖHM (1) grew willow slips in water into
which he had passed a stream of illuminating gas. The short
roots produced soon died, but the slips themselves lived for three
months. Potted plants of *Fuchsia* and *Salvia* died after gas had
flowed to their roots for four months. Again, he found that when
gas had been passed through a soil for a long period of time, this
soil became very toxic to seedlings germinated in it. Also a
Dracaena transplanted to this soil became dry and died in ten days.
MOLISCH (10) found that roots of corn increased in diameter and
were bent in certain concentrations of illuminating gas. SHON-
NARD (13) noted exudation of sap from the trunk and branches
of a lemon tree treated with gas in the soil. RICHARDS and MAC-
DOUGAL (11) found that carbon monoxide and illuminating gas
retarded the rate of elongation of roots of *Vicia Faba*, sunflower,
wheat, rice, etc. Swelling also appeared in the leaf sheaths of
wheat, being somewhat more pronounced in illuminating gas than
in carbon monoxide. Such increases in thickness were largely due
to the enlargement of the cortical cells. In some cases, however,

the cambium seemed to have become more active. Recently STONE (16) has reported proliferations of tissue at the lenticals of willow slips growing in water which had been charged with illuminating gas, He also noted a rapid proliferation at the cambium in stems of *Populus deltoides* due to the influence of gas.

Another important phase of the gas injury problem is that after trees have been killed by gas, a question arises regarding the safety of resetting trees where the dead ones have been removed, assuming, of course, that the gas leak has been located and stopped. It seems to be the general opinion that resetting should only be done either after a considerable time has elapsed, or after large amounts of the old soil have been removed and replaced by fresh soil. Neither of these methods of procedure is entirely satisfactory; the first involves great loss of time, the second is expensive. The practicability of resetting trees in any given case is often determined only by the crude method of smelling a handful of soil taken from the place of injury, and if the odor of gas is still present, resetting is deemed unsafe. One is thus led to ask whether the odor itself is a true index of the toxicity of the soil to the roots of plants.

The investigation reported below was undertaken with the two problems in mind: (1) that of determining some of the effects of illuminating gas on root systems, having in mind the securing of further diagnostic characters of gas poisoning; and (2) whether the chief causes of injury are those constituents of illuminating gas which are readily absorbed by the water film of the soil particles, or those which remain mainly in the soil interstices (not being so readily soluble).

Methods and materials

The illuminating gas used was the so-called "water gas" from the Chicago Gas Light and Çoke Company's system. Along with the illuminating gas experiments, many parallel ones were carried out with an ethylene-air mixture. The Chicago illuminating gas contains 2–6 per cent ethylene; therefore, to facilitate comparison between the ethylene alone and the ethylene of the illuminating gas, the ethylene of the mixture above was made to constitute 4 per cent (by volume). Thus, volume for volume, the ethylene-air mixture and the illuminating gas contained approximately the same amount of ethylene. In the following experiments, where the term "ethylene in corresponding concentrations to the illuminating gas" is used, it refers to the ethylene content of the latter gas. For example, the ethylene concentration corresponding to an illuminating gas concentration of one part gas to four parts of air ("1:4") would be one part ethylene to 100 of air ("1:100"). Such parallel experiments were considered important owing to the fact that ethylene has been found to be by far the most toxic constituent of illuminating gas for the aerial organs of several plants (see CROCKER and KNIGHT 2, KNIGHT and CROCKER 7, and HARVEY 6).

The experiments were primarily arranged so as to yield evidence with regard to the two phases of the second problem above. The methods will be described separately and in the order named. Investigation of the second problem should yield data with regard to the first.

A. THE TOXICITY OF THE CONSTITUENT OF ILLUMINATING GAS
ABSORBED BY THE SOIL

Good potting soil in 10–20-liter cans was treated with illuminating gas by allowing the gas to flow through at a definite rate, at room temperature, for varying lengths of time. The rate of flow was approximately 2 liters per hour. The time periods were from 30 hours to 20 days, hence the lots of soil received 40–1000 liters of gas. The moisture content of the soil was kept as near the "optimum" as possible. In one experiment 10 liters of soil received gas at the rate of about 0.2 liter per hour for 68 days. In another experiment 8 liters of soil received gas at the rate of 2 liters per hour for 53 days, and was kept at a temperature of 1–5° C. throughout the period. The purpose of this soil treatment was to allow the soil particles to absorb as much of the gas constituents as possible. In the case of treatment at low temperature, it was the intention to allow still better opportunity for condensation of substances on the soil particles. After stopping the flow of gas, the soils were removed from the cans and thoroughly stirred in pure air to free them from the gas in the interstices. They were then taken to the greenhouse, placed in shallow boxes, and planted to 41 different species of plants,

representing 18 families. Controls were maintained throughout.
The seeds were allowed to germinate and grow for periods of 25–60
days.

**B. THE EFFECT OF ILLUMINATING GAS ON ROOTS WITH NO SOIL
PRESENT**

1. *With roots alone exposed to gas*

In order to expose the roots without exposing the shoots, the
following method was employed. Moist-air chambers were made
from 8-liter, wide-mouthed bottles; 6–8 short glass tubes were
inserted in the corks of each of these bottles so that the tap roots
of young seedlings of *Vicia Faba* could be admitted to the chamber.
The space between a root and the sides of a tube was sealed by
means of a short length of pure gum tubing which had been previously fitted to the outer end of the glass tube. Definite amounts
of illuminating gas or ethylene were admitted through a small
glass tube reaching to the bottom of the bottle. The pressure
resulting from the addition of the small volume of gas was relieved
through a second short tube.

2. *With entire plant exposed to gas*

Seeds of tomato, radish, and mustard were allowed to germinate,
and when the hypocotyls had reached a length of 0. 5 cm. they were
transferred to flower pots under bell jars provided with water seals.
The seedlings were fixed to the rims outside of the flower pots, so
that the roots in growing would hang free in the air. Definite
amounts of illuminating gas could be easily added.

**C. EFFECT OF ILLUMINATING GAS ON ROOTS GROWING IN SOIL
MEDIUM**

1. *Quantitative tests*

In these tests young seedlings (two or three months old) of
Catalpa speciosa, Ailanthus glandulosa, and *Gleditschia triacanthos*
were used. A few days before the beginning of the experiments,
the seedlings were transplanted from the pots, in which they had
germinated, to large battery jars filled with coarse quartz sand.
Two glass tubes for admitting gas were thrust into the sand and

then the roots and glass tubes were sealed in with a vaseline-paraffin mixture by the BRIGGS and SHANTZ method. The total volume of the interstices (allowing for a definite water content) was previously determined, and upon the basis of this volume, known concentrations of gas could be secured. The duration of experiments was 5–21 days. Gases were changed every three days by drawing pure air through the jars for 10–15 minutes by means of an aspirator, and new gas added. Parallel experiments with ethylene were also run in many cases.

2. *Qualitative tests*

In this series of experiments it was desired to subject a number of plants to illuminating gas under conditions met with in the field where leaking gas mains are involved. Potted woody plants, including 36 individuals of 11 species, were used. Illuminating gas was allowed to pass slowly into the soil around the roots. Meanwhile symptoms of injury were noted. When the plant had died, or become seriously injured, or after a certain time had elapsed, the roots were washed clear of soil, and careful examinations made. The rate of gas flow was often less than one-eighth liter per hour.

Results and discussion

The results will be presented in the order in which the methods were described.

A. EFFECT OF ILLUMINATING GAS ON SEEDS PLANTED IN SOIL PREVIOUSLY TREATED WITH GAS AND THEN AERATED

In all cases, at the time of the planting of the seeds, the treated soil gave an exceedingly strong odor of illuminating gas. Of the 41 species planted, 5 failed to germinate, but the failure was in both the controlled and the treated soils. Throughout the 25–60 day period, the aerial portions of the seedlings were watched for signs of injury. However, it was found that the plants in all the treated soils gave no superficial evidence of injury. All seemed perfectly normal. At the end of the period the plants were taken up, their roots washed free of soil, and examinations made. Only two species gave any evidence of injury; these were cotton and lupine. In them the root systems were perhaps somewhat less developed in the treated soils, and in the cotton there appeared to be a greater development of anthocyanin in the treated soils. At the close of the experiments, the treated soils always gave a noticeable gas odor and in most cases the odor was very strong.

The results of these experiments would indicate that of the constituents absorbed from illuminating gas by soil, those which give the odor to the gas are very prominent, and when once absorbed are held for extended periods, even after the soil is freely exposed to the air. But the most important fact from the standpoint of the question in hand seems to be that these odorous constituents are evidently not extremely toxic to plant roots growing in the same soil. The plants tried in such soils included a wide range in regard to relationship, and also they were taken at what is considered the critical stage, that is to say, the germinating and young seedling stages. Just what these odorous compounds are is an interesting question. The odor of any illuminating gas is probably the combined odor of a number of substances, for example, pyridine, thiophene, picoline, quinoline, cumene, cymene, and others. Very little is known concerning the effect of these on vegetation. CROCKER, KNIGHT, and ROSE (3) found cumene, thiophene, and pyridine were many hundred times less effective in reducing growth in the etiolated sweet pea seedlings than was ethylene.

The results of the experiments described indicate that the presence of a gas odor in soil is not an index of its toxicity for vegetation, and that odors would be valuable merely as a means of determining whether or not illuminating gas had been in the soil. With regard to using odors in diagnosis, CROCKER has suggested to us the possibility of distilling (at high temperature *in vacuo*) soil from places where gas injuries have been suspected, but where the odor even at the time is not discernible, thereby drawing off some of the odors previously held too firmly by the soil particles.

B. THE EFFECT ON ROOTS WITH NO SOIL MEDIUM

1. *With roots alone exposed*

Material, seedlings of *Vicia Faba*. Exposure period five days.

With illuminating gas.—(1) Concentration 1:4000; no effects were noted, evidently grew as well as controls; (2) concentration 1:400; growth in length somewhat retarded and two other strongly marked effects entered: (*a*) swelling back of the root tip, and (*b*) turning and coiling of root (fig. 1); (3) concentration 1:40; growth in length greatly retarded, also considerable swelling; coiling present, but the coils smaller and tighter.

With ethylene.—(1) Concentration 1:100,000 (that is, corresponding to ethylene of illuminating gas with (1) above); apparently no effect; normal growth ensued; (2) concentration 1:10,000; responses similar to those shown in fig. 1 but somewhat more pronounced; (3) concentration 1:1000; little or no growth in length; strongly swollen, no coils.

The parallel experiments with ethylene gave some evidence that the toxic effect recorded for the illuminating gas may be due to ethylene present in that gas because corresponding concentrations of illuminating gas and the ethylene-air mixture gave quite parallel results.

FIG. 1.—Outline of *Vicia Faba* roots, showing the effects of illuminating gas on growth; *A*, control.

2. *With entire plants*

Material, radish, mustard, and tomato. The roots and hypocotyls gave responses as follows:

Radish.—Exposure period 48 hours.

With illuminating gas.—(1) Concentration 1:500; bending of root evident (similar to those of fig. 1); no coiling; no swelling of hypocotyl or root; (2) concentration 1:5000; coiling and bending of root evident; no enlargement of hypocotyl or root.

Mustard.—Exposure period 48 hours.

With illuminating gas.—(1) Concentration 1:100; coiling and bending of root evident (similar to those of fig. 1); swelling of hypocotyl evident; no swelling of root; (2) concentration 1:11,000; bending of root slight; no other effect; (3) concentration 1:20,000; bending of root slight; no other effect.

Tomato.—Exposure period 3 days.

With illuminating gas.—(1) Concentration 1:500; swelling between stem and root (fig. 2); growth considerably retarded; (2) concentration 1:10,000; swelling as above, but less marked;

swollen zone longer and not so thick; more growth of hypocotyl;
(3) concentration 1 : 100,000; little if any effect.

With ethylene.—(1) Concentration 1 : 12,500; short swollen
knob between stem and root; growth greatly retarded; (2) concentration 1 : 250,000; swollen knob longer and not so thick as in
(1); (3) concentration 1 : 2,500,000; little if any effect.

The response shown by the tomato seedling differs very markedly from that shown by the radish and mustard seedlings. While
the roots of the radish and mustard seedlings show a coiling and

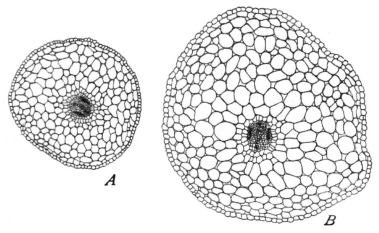

FIG. 2.—Sections of the root of tomato seedlings: A, control; B, treated with
illuminating gas; ×18.

bending similar to those of the *Vicia Faba*, in addition to a slight
swelling of the hypocotyl, the only response shown by the tomato
seedling is a decided swelling of the hypocotyl and root at the point
where the two join. With the tomato seedling the parallel experiments with ethylene again give some evidence that the toxic effect
recorded for illuminating gas is due to the ethylene constituent
of that gas.

The results recorded above for *Vicia Faba*, radish, and mustard
show that injuries to roots may readily be brought about by placing
them in an atmosphere containing small amounts of illuminating gas

or ethylene. If we consider here also the results of the first series of experiments, it would seem that the chief causes of injury to root systems are those constituents of illuminating gas which are present mainly in the interstices of the soil, rather than those dissolved in the soil water.

The greatest danger in replanting where other plants have been killed by gas, seems to lie in the constituents remaining between the soil particles. If, therefore, a method was devised for quickly aerating this soil, trees might be safely planted at once without removing large amounts of soil.[1]

C. THE EFFECTS OF ILLUMINATING GAS ON ROOTS IN SOIL MEDIUM

1. *Quantitative tests*

a) Catalpa speciosa seedlings; illuminating gas; exposure period 8 days; concentrations 1:2000, 1:400, 1:200, 1:40, 1:20, and controls.

Stems and leaves showed no modifications; neither were there any strongly marked effects on the root systems. In concentrations 1:40 several roots gave indications of swelling 1–2 cm. back of the tips, while in concentrations 1:20 these swellings were very evident.

b) Same as above, with the exception of ethylene in place of illuminating gas but in corresponding concentrations (that is, 1:50,000, 1:5000, 1:1000, and 1:500).

No effect on stems or leaves. The responses with concentrations 1:1000 and 1:500 were like those above, but in addition, the tendency of the roots to coil at the tips (as noted in *Vicia Faba*, fig. 1).

c) Catalpa seedlings; illuminating gas; duration of exposure 21 days; concentration 1:4 and controls.

Stem and leaves gave no response. Roots of 1:4 gave swelling of main root extending from near the surface of the sand 4–7 cm. downward (fig. 3). This increase in amount was 2–3 times that of the normal thickness. The epidermis was often cracked and sloughed off in places. The drawing of fig. 4 shows that the swelling

---

[1] Perhaps this could be accomplished by passing pure air through the soil by means of a pipe thrust below the surface.

is brought about through an increase in size of the cortical cells, and particularly by rapid proliferation at the phellogen layer.

d) Ailanthus seedlings; illuminating gas; exposure period 15 days; concentrations 1:400, 1:10, and controls.

1:400 gave slight swelling in tap root near the surface of the soil; while in the 1:10 the leaves began falling after 5 days, and had all fallen before the end of the exposure. The tap roots were much swollen for 3–4 cm. below the surface of the soil.

e) Same as (*d*) but with ethylene instead of illuminating gas; concentrations 1:10,000 and 1:250.

Swollen zone of tap root was somewhat more pronounced in the 1:10,000 than in the 1:400 illuminating gas above. In the 1:250 all plants had lost their leaves within 8 days, and the tap roots were much swollen. The morphological nature of these swellings may be seen in figs. 5 and 6. In fig. 6 are two outlines of trans-

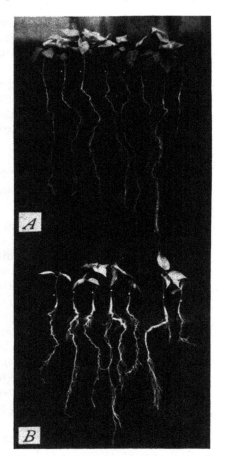

FIG. 3.—*Catalpa* seedlings· *A*, controls, *B*, treated with illuminating gas.

verse sections of the tap roots of *Ailanthus* seedlings in the region where swellings take place; *A*, the control plant; *B*, a treated plant. Fig. 5 shows the detailed structures of the regions outlined in *A* and *B* respectively of fig. 6. Through an examination of figs. 5 and 6 it becomes evident that the stelar region has remained un-

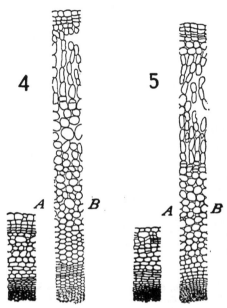

changed, while the cortex, extending to the phellogen layer, has increased in thickness, partly through increase in the diameter of the cells, and partly through cell division. At the phellogen layer of the treated root, cell division has been rapid, resulting in the production of a loose tissue not present at all in the normal. The tissue lying outside the phellogen layer in the beginning has been only slightly modified in the abnormal root.

f) A number of tests were carried out with *Gleditschia* seedlings, used in illuminating gas in various concentrations up to 1:3 (concentrations higher than 1:3–1:4 were

FIGS. 4, 5.—Fig 4., section through the cortex of tap root of *Catalpa: A*, control; *B*, treated with illuminating gas; fig. 5, same for *Ailanthus* roots, showing detail of the tissue regions outlined in fig. 6, ×48.

not used for fear of oxygen becoming a limiting factor). Concentration 1:3 gave leaf fall, but no definite injuries were detected in the root system.

2. *Qualitative tests with potted woody plants*

When illuminating gas was passed rapidly through the soil of a potted plant, injuries might be observed the first day and death in all cases in the course of a few days. This is not at all surprising, because one should expect injury and final death to result as an effect of the shutting off of the oxygen supply, even though an inert gas (that is, nitrogen or hydrogen) be passed through the soil in this manner. Accordingly, illuminating gas could do great damage to vegetation independent of any direct toxic property. This view is advanced by KOSAROFF (9), who found that wilting took place if a stream of carbon dioxide or of hydrogen was passed through soil in which roots were growing. In all experiments where illuminating gas was passed into the soil rapidly enough to cause the death of the plant, the symptoms manifest in the aerial parts were of a type which would indicate that the injury might be due simply to the cutting off of the water supply as a result of injury to the root system, and not, necessarily, due to a conduction of toxic substances to them.

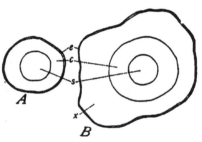

With high evaporating conditions, the symptoms of gas injury (that is, wilting, yellowing, and falling of leaves in the aerial portions) always became manifest very much sooner than under conditions favoring a low water loss. It was found that when roots were killed quickly by using high concentrations of gas, few if any symptoms other than odors appeared which might enable one to diagnose the death as specifically gas poisoning.

FIG. 6.—Diagrams of transverse sections of *Ailanthus* roots; *A*, control; *B*, treated with illuminating gas; *s*, stele; *c*, cortex; *e*, epidermis and tissue lying outside of phellogen; *x*, new tissue developed by the phellogen; ×6.

With regard to the injury to parts above soil, STONE (16) believes that "poisonous principles" are absorbed by the roots and conducted upward to the leaves, hence the yellowing and wilting would result as a direct poisoning of them.

When illuminating gas was so regulated that it passed very slowly to the roots (sometimes as little as 40 cc. per day), a stimulation of the roots often took place, resulting in the development of new tissue resembling those reported from the quantitative experiments made with young tree seedlings. Stems were likewise affected when covered by soil or when slightly above the surface.

In stems this proliferation of cells became first evident at the lenticels by their increase in size and the protruding of white tissue, as shown in the lower portion of the stem in fig. 7. RICHTER (12) has noted a similar proliferation of tissue at the lenticels of *Vicia villosa* under the influence of tobacco smoke. STONE's similar results have been cited. In the roots this tissue was abundantly developed in some cases, particularly in *Hibiscus*, which is shown in fig. 7, where *B* and *C* are plants which were subjected to a slow stream of illuminating gas for a period of 30 days. Fig. 8 shows detailed structure of these proliferations in the roots of *Hibiscus*. The cork

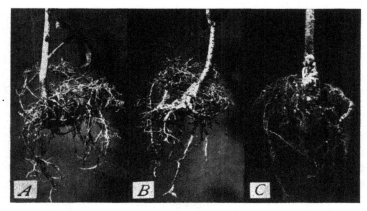

FIG. 7.—*Hibiscus: A*, control; *B, C*, treated with illuminating gas

layer of the abnormal root has been sloughed off. Practically the same morphological situation appears in this case as in *Ailanthus* previously figured.

A similar response by lilac is shown by drawings of fig. 9. These abnormalities were recorded for *Hibiscus*, lilac, *Croton*, *Diervilla*, *Ricinus*, *Ulmus*, and pear.

An experiment similar to the foregoing was carried out with an *Ailanthus* growing on the campus near the Hull Botanical Laboratory. The tree had a diameter of about 8 cm. and a height of 3.5 m. At a short distance were other trees of the same species which served as controls. Illuminating gas was admitted to the roots of the

experimental tree through a glass tube thrust 0.7 m. into the soil,
0.6 m. from the base of the tree.   The rate of flow of gas was 1.5
liters per hour.   The experiment began July 3, and the gas was
stopped flowing September 2.   Except for two or three short periods,
amounting in all to less than three days, the flow of gas was
continuous during the two
months.   Thus the soil
near the tree must have
received 1.5–2 cubic
meters of gas.

The first symptoms of
injury were manifest July
14.   Leaves of some of the
young shoots, growing on
the same side of the tree
from which the gas entered
the soil, showed signs of
wilting.   Three days later
these leaves and others
had shriveled and died,
but remained attached to
the shoots.   In some cases
only a portion of a leaf
was injured.   A few leaves
of older branches also
wilted, but there was no
general effect evident
throughout the tree at that
time.   In the middle of
September the apparently
unaffected leaves began to

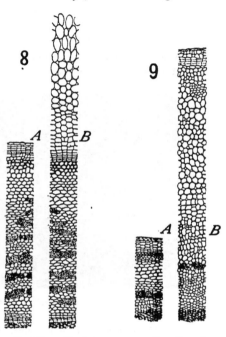

FIGS. 8, 9.—Fig. 8, transverse sections through
cortex of *Hibiscus* at base of stem; ×40; *A*, control, *B*, treated with illuminating gas; fig. 9,
same for lilac; ×35

fall, and finally the tree was free from leaves much before those
of the controls.   In October the tree looked as though it were
entirely dead, and when an examination was made, after it had
been removed from the soil, such was found to be the case.   A
general dryness of the tissues was noted, but neither anatomical
changes nor gas odors were detected in them.

During the gas flow, the odor of gas was evident only at the point where the tube entered the soil. In order to determine whether the gas might be detected by a more delicate means, the etiolated sweet pea seedling test, as described by KNIGHT and CROCKER (7) was tried. When the seedlings were 1.5–2 cm. high, the petri dishes containing the seedlings were taken to the *Ailanthus* tree, placed directly on the soil, and the 10-liter cans inverted over them. The cans, each with a dish of seedlings beneath, were numbered and placed, with reference to distance from the point of entrance of gas, as follows: no. 1, 5 cm., toward tree; no. 2, 0.6 m., at base of tree; no. 3, 0.8 m., also near tree but on side opposite no. 2; no. 4, 1.2 m., on side directly opposite point of entrance of gas; no. 5, 3 m., as no. 4. The cans were placed August 27 and observations were made three days later. The results were as follows: no. 1 gave no growth (observations refer to epicotyls only); no. 2, some growth, with a little swelling; no. 3, growth slightly reduced and diageotropic; no. 4, slender and straight, 9–10 cm. high (normal); no. 5, as no. 4. These results indicate that the pea seedling probably offers a rather delicate test for the presence of illuminating gas in the soil. In this case the injurious effect on the seedlings was very evident in those placed near the base of the tree, where, as stated before, no odor of gas could be detected in the customary manner.

In consideration of the great difference in behavior of a given plant when exposed to low and to high concentrations of illuminating gas, it seems appropriate to make the following suggestions with regard to the diagnosis of gas injuries. One reason why the foregoing abnormal tissue developments have not been recorded in trees killed by gas is probably the fact that examination was made for pathological symptoms in trees which have died or have become seriously injured. Such conditions would mean that the gas had been at their roots in too high concentration to allow the stimulating effects to enter. Sometimes the proliferation reported above was found in roots when no suggestion of injury could be observed in the aerial portions. Therefore, when one is attempting to diagnose with certainty a serious injury suspected to be due to gas, he ought also to make an examination of roots of other trees in the vicinity which have not yet shown injuries in the leaves, thereby perhaps adding one more line of evidence. Another change found to take place under the influence of low concentrations of illuminating gas was the disappearance of starch from the cortex of the roots, an observation in agreement with that of Richter (12) and others.

Summary

1. When illuminating gas is passed through soil, the odor-giving constituents of the gas are readily absorbed by the soil particles and strongly held.

2. These odorous substances are very slightly, if at all, toxic to roots of plants growing in a soil containing them.

3. The constituents of illuminating gas which remain in a gaseous state in the soil interstices are the chief cause of injury to root systems.

4. Among these constituents, ethylene is probably the most harmful,· except in extremely high concentrations of illuminating gas, where the toxicity of other substances, together with other factors, would be expected to play a part.

5. Low concentrations of gas induce abnormal development of tissue.

a) Illuminating gas.—These abnormalities appear in certain tree seedlings within 8–21 days, with concentration one part illuminating gas to four parts air (air of the soil), or as low as one part illuminating gas to forty of air.

b) Ethylene.—This gas alone, when used in concentrations corresponding to the ethylene content of the illuminating gas used in the tests, gives abnormalities similar in type and degree.

6. High concentrations of illuminating gas result in the rapid killing of the roots, and the only symptom of injury to be observed is death.

7. If illuminating gas is allowed to flow very slowly through a soil in which woody plants are growing, abnormal tissue development in the root will very often ensue.

8. In low concentrations of illuminating gas, hydrolysis of starch and some other related chemical reactions are accelerated.

9. It was found that, by use of the etiolated sweet pea seedling. small amounts of illuminating gas in the soil could be detected where the odor of gas was indistinguishable by the usual methods.

We wish to acknowledge our indebtedness to Dr. WILLIAM CROCKER of this laboratory for many valuable suggestions during the course of this investigation.

UNIVERSITY OF CHICAGO

LITERATURE CITED

1. BÖHM, J., Über die Einwirkung des Leucht gases auf die Pflanzen. Bot. Zeit. 32:74–75. 1874.
2. CROCKER, WM., and KNIGHT, L. I., Effect of illuminating gas and ethylene upon flowering carnations. BOT. GAZ. 46:259–276. 1908.
3. CROCKER, WM., KNIGHT, L. I., and ROSE, R. C., A delicate seedling test. Science 37:380. 1913.
4. EULENBERG, H., Handbuch des Gewerbe-Hygiene. 1876 (p. 601).
5. GIRARDIN, J. P. L., Einfluss des Leuchtgases auf die Promenaden- und Strassen-Bäume. Jahresb. Agrikultur. 7:199–200. 1864.
6. HARVEY, E M., The castor bean plant and laboratory air. BOT. GAZ. 56:439–442. 1913.
7. KNIGHT, L. I., and CROCKER, WM., Toxicity of smoke. BOT. GAZ. 55:337–371. 1913.
8. KNY, L., Um den Einfluss des Leucht gases auf Baumvegetation zu prüfen. Bot. Zeit. 29:852–854, 867–869. 1871.
9. KOSAROFF, P., Einfluss äusserlich-Faktoren auf der Wasseraufnahme. Inaug. diss. Leipzig. 1897.
10. MOLISCH, H., Über die Ablenkung der Wurzeln von ihrer normalen Wachstumsrichtung durch Gaze (Aërotropismus). Sitzungsber. Kaiserl. Akad. Wiss. Wien 90:111–196. 1884.
11. RICHARDS, H. M., and MACDOUGAL, D. T., The influence of carbon monoxide and other gases upon plants. Bull. Torr. Bot. Club 31:57–66. 1904.
12. RICHTER, O., Über den Einfluss der Narkotika auf die Anatomie und der chemischen Zusammensetzung von Keimlingen. Verh. Gesells. Deutsch. Naruff. und Aerzte 80:189–190. 1908.
13. SHONNARD, F., Effect of illuminating gas on trees. Dept. Public Works, Yonkers, N.Y. pp. 48. 1903.
14. SPÄTH, J. L., and MEYER, Beobachtungen über den Einfluss des Leuchtgases auf die Vegetation von Bäumen. Landwirtsch. Versuchstat. 16:336–341. 1873.
15. STONE, G. E., Effect of escaping illuminating gas on trees. Mass. Exp. Sta. Report. pp. 180–185. 1906.
16. ———, Effects of illuminating gas on vegetation. Mass. Exp. Sta. Report. pp. 13–29. 1913.
17. VIRCHOW, R., Einfluss des Leuchtgases auf die Baumvegetation. 13–15: 1–237. 1870–1872.
18. WELMER, C., Über einen Fall intensiver Schädigung einer Alee durch ausströmendes Leuchtgas. Zeitschr. Pflanzenkr. 10:267–269. 1900.

NOTES ON NORTH AMERICAN WILLOWS. II[1]

CARLETON R. BALL

(WITH THREE FIGURES)

The first paper[2] in this series dealt with four western willows, including two new species, one new variety, and one new combination. The present paper contains one new species (§ PHYLICIFOLIAE) and descriptions of typical *Salix sessilifolia* Nutt. and of the plant which is usually described in manuals of botany under ·that name.

The writer wishes to acknowledge here the courteous assistance of Professor C. V. PIPER in communicating his notes on the types of NUTTALL's *Salix sessilifolia* and other species in § LONGIFOLIAE, studied by him at the British Museum. The same acknowledgement is gladly given to Dr. C. F. MILLSPAUGH of the Field Museum and to Dr. B. L. ROBINSON of the Gray Herbarium, and to their associates, for the loan of material representing the species discussed.

Salix pennata, n. sp. (fig. 1).—Low shrub with dark, divaricate, stoutish, glabrous branchlets and large chestnut buds; leaves obovate or elliptic-obovate, 3–6 cm. long, acute, narrowed at base, entire, very dark green above, glaucous beneath, the raised midrib and parallel primary veins conspicuous beneath, glabrous; aments

---

[1] Since the above was in press the writer has collected true *S. sessilifolia* Nutt. on the Umpqua River at Roseburg, Oregon, and on the Willamette River at Corvallis, Oregon. In both localities it was very abundant, and associated with *S. bolanderiana* Rowlee. It was not found around Portland nor along the Willamette River and adjacent sloughs for a distance of about 15 miles below Portland, including part of the famous Sauvie's Island. The species here treated as *S. fluviatilis* Nutt. is a common plant about Portland, from Ross Island, 2 miles above the city, to Sauvie's Island, some 15 miles below. It is the dominant willow at the Dalles, Oregon, just east of the Cascade Mountains. Both here and on the lower Willamette it apparently is the only species of the *Longifoliae*.

[2] BALL, CARLETON R., Notes on North American willows. I. BOT. GAZ. **40**:376–380. *pls. 12, 13.* 1905.

FIG. 1.—*Salix pennata* Ball a portion of the pistillate type (*Suksdorf* 15 from Mt Adams) in Bebb Herb., Field Museum of Natural History, nat. size.

sessile, stout; pistillate 2.5–7 cm. long; capsule subsessile, 6–8 mm. long, densely silvery pubescent; style about 1.4 mm. long, stigmas long; scales obovate, acute, black, densely pilose; stamens 2, filaments glabrous, free.

Low shrub, probably about 0.5–1 m. in height, of spreading habit; bark (probably) dark gray to brownish, branchlets stoutish, divaricate or somewhat virgate, dark brown to black, dull or somewhat shining; smooth or sometimes longitudinally finely wrinkled, glabrous throughout; buds mid-sized to large, 6–10 mm. long, ovate, apiculate, chestnut, glabrous.

Leaves usually exstipulate, petiolate; stipules wanting on fruiting branchlets, on one sterile shoot (*Applegate* 2758) 4–8 mm. long, lanceolate to semicordate, acute; petioles slender to stoutish, 5–10 mm. long, greenish yellow or becoming brown, glabrous; blades on fruiting branchlets narrowly to broadly obovate or obovate-oval, acute to abruptly acute at the apex, cuneate to acute at the base, 3–6 cm. long, 1.5–3.5 cm. wide, common sizes being 4 by 2, 5 by 2.5–3, and 6 by 2.5–3.4 cm., those on luxuriant sterile twigs elliptic-oblanceolate to elliptic-oval or even elliptic-oblong, larger, 4–8 cm. long, 1.5–3.8 mm. wide, the narrowest 7 by 1.5, the widest 8 by 3.8 cm., all entire or distal leaves on sterile twigs crenate or crenulate, dark green and somewhat shining above, strongly glaucous beneath, midvein and primary veins rather prominent above, strongly so beneath, where the secondaries make a strong reticulation, the primaries simple, parallel, glabrous throughout or the under surface sparsely sprinkled with short shining hairs.

Aments precocious, stout, sessile, ascending or spreading, loosely to closely arranged on the branchlet, the staminate naked, the pistillate subtended by 1 or 2 very small, scalelike leaves, the peduncle in fruit becoming 2–5 mm. long.

Pistillate aments 2.5–5 cm. long in flower, becoming 3–7 cm. long in fruit, 1.2–2 cm. wide; capsule 6–8 mm. long, narrowly lanceolate-rostrate, acute, subsessile, densely pubescent with silvery white hairs; pedicel stout, 0.2–0 4 mm. long, pubescent; style 1 2–1.5 mm. long, entire, brown; stigmas 0 4–0 6 mm. long, mostly divided; scales in both sexes narrowly obovate, acute or obtusish,

1.2–1.5 mm. long, black, densely clothed on both surfaces with long, straight, shining, white or yellowish hairs.

Staminate aments stout, sessile, 2.5–3 cm. long; stamens 2, filaments slender, 6–8 mm. long, glabrous, free, anthers short, oval; scales as described or somewhat narrower, hairs white or flavescent.

Salix pennata (fig. 1) is easily one of the most beautiful species of § PHYLICIFOLIAE. In relationship it lies between *S. chlorophylla* and *S. pulchra*. Geographically, also, it occupies a position between these two species. It is abundantly distinguished from both. From *S. chlorophylla* it is separated by the much larger and darker leaves, obovate rather than elliptical, with their conspicuous parallel primary veins, especially on the lower surface, and by the longer aments and capsules. From *S. pulchra* it may be distinguished by much the same series of characters. While *S. pulchra* also has large leaves, they are more rhombic-oblanceolate than obovate in outline, always bright and shining green above and less strikingly veined. The stipules in that species also are linear-lanceolate, glandular, and very persistent. Owing to the cold and wet situations in which it occurs, *S. pennata* is a late-flowering species. All the specimens bearing flowers were collected in July, while fruit may be found well into August.

Our species is also an interesting example of apparently restricted distribution. So far, collections have come only from the two great peaks of the central Cascades, Hood and Adams, which face each other across the gorge of the Columbia. Further search may reveal the species on Jefferson and Rainier. The specimen collected by SUKSDORF in Skamania County, Washington, on July 31, 1883, and the battered, fragmentary specimen of the Wilkes Expedition are somewhat doubtfully referred here. Both have leaves more nearly oblanceolate than obovate, but otherwise they have the characters of our species. *S. chlorophylla* and *S. nelsoni* have never been found in the Cascades, though little *S. monica* is found in the central Sierra Nevada.

SPECIMENS EXAMINED.—OREGON: Marion County: 10 miles west of Olay Butte, in alder swamp at head of a canyon, *E. I. Applegate* 2758, Sept. 4, 1898 (N); north base of Olay Butte, in wet meadow, alt. 4000 ft., *Applegate* 2766, Sept. 6, 1898 (N); Clackamas County: edge of swamp, 10 miles north of Olay Butte, alt. 4000 ft., *Applegate* 2770, Sept. 6, 1898 (N); vicinity of Mt. Hood: in swamp, north base of Mt. Hood, *Thomas Howell*, Oct. 2, 1886 (FBb); Government Camp, Mt. Hood, edge of wet meadow, *E. I. Applegate* 2801, Sept. 12, 1898 (N); *F. A. Walpole* 354, Aug. 28, 1899 (N), growth low, dense, compact, 3–4 ft.; shore of Lost Lake, Mt. Hood, *H. D. Langille* 20, July 6, 1901 (B, N).

WASHINGTON: Mt. Adams and vicinity: Mt. Paddo (Adams), *W. N. Suksdorf*, July 31, 1883 (B, N); mountains of Skamania County, *Suksdorf* 1371, Sept. 6, 1883 (B, N); Mt. Paddo (Adams), *Suksdorf* 15 (pistillate type), 16 (staminate type), 17, July 13, Aug. 12, 1886 (FBb); no locality, U.S. Exploring Expedition under Capt. Wilkes, interior of Washington Territory (N).

SALIX SESSILIFOLIA Nuttall, N.A. Sylva 1:68. 1842; Bebb (in part), Willows of California 85. 1879; not of most authors (fig. 2).

FIG. 2.—*Salix sessilifolia* Nutt.: a portion of the pistillate twig of *Hall* 474 in Gray Herb.; nat. size.

S. *sessilifolia* var. *villosa* Andersson, Monog. Sal. 56. 1867; Anders. in DC. Prod. 16²:214. 1868.

S. macrostachya (in part) as interpreted by Rowlee, Bull. Torr. Bot. Club
27:250. *pl. 9, fig. 5* (named *S. macrostachya cusickii* in the description of the
plate but this variety never published). 1900; not of Andersson.

Shrub 2–3 (or more) m. in height; branchlets slender, the older
dark brown, longitudinally striated, the younger yellowish, more or
less tomentose, those of the season densely white villous-tomentose
with spreading white or gray hairs.

Leaves sessile or subsessile, usually stipulate; stipules ovate to
lanceolate, acute, entire or occasionally minutely denticulate; petioles none or 2–4 mm. long on vigorous sterile shoots; blades narrowly to broadly lanceolate or elliptic-lanceolate; the smaller
rounded, the larger somewhat narrowed and acute at the base,
acute or short-acuminate at the apex, terminating in an extremely
sharp spinulose point, 2.5–5 cm. long, 8–15 mm. wide, on vigorous
sterile shoots larger, 5–7 cm. long, 15–20 mm. wide, all denticulate
to spinulosely denticulate (the spinulose teeth gland-tipped, sometimes 1 mm. in length), green and densely to thinly villous-
tomentose with spreading hairs on both sides, especially on the
midribs and on very young leaves, not obscuring the green color
on fully expanded leaves, only the very youngest silvery white in
color.

Aments appearing after the leaves, solitary in the specimens
seen, terminal on short leafy branches 1–4 or 5 cm. long and bearing several well developed leaves, or old aments apparently lateral
and nearly sessile by the seasonal prolongation of the fruiting
branchlet from the bud subtended by the apical leaf.

Pistillate aments dense, 4–6 cm. long, spreading or drooping;
rachis densely villous; scales mostly deciduous, broadly elliptic-
lanceolate, acute, densely villous-tomentose with gray hairs; capsule lanceolate, acute, 5–6.5 mm. long, villous; style and stigmas
together about 1–1.5 mm. long, style apparently 0.5 mm. long
(difficult to differentiate style from stigma in old dried material),
divided; stigmas about 1 mm. long, divided.

Staminate aments not seen on typical material (on approximately typical specimens they are 3–4 cm. long; scales elliptical,
pilose-pubescent; stamens 2, filaments pubescent, free, anthers
about 1 mm. long).

The types of NUTTALL'S numerous species in § LONGIFOLIAE have recently been studied by Professor C. V. PIPER at the British Museum. As a result of this examination it appears, among other things, that the type of *Salix sessilifolia* is quite different from the plant usually described under that name by American authors. From the illustration (fig. 2) and from the full redescriptions prepared, the identity of *S. sessilifolia* should be established readily. It is seen to be a plant with the leaves on the fruiting branches small, truly lanceolate, and actually sessile! The leaves and the twigs of the season are densely clad with a pilose tomentum of gray hairs which are spreading on the petiole and midrib and more or less appressed on the blade. The capsule also is densely pilose, even in age. True *S. sessilifolia* belongs, therefore, in the group containing *S. argophylla* and *S. macrostachya* (as these are at present understood) rather than with the *S. fluviatilis-melanopsis* aggregation. Just what are the limits of variation in *S. sessilifolia*, as well as the more certain identification of the relatives named above, cannot be settled at this moment. Further study of NUTTALL'S types and of collections from the type localities will be necessary to a final decision. The plant commonly identified and described as *S. sessilifolia* is here treated provisionally as *S. fluviatilis* Nuttall.

SPECIMENS EXAMINED.—OREGON: *Elihu Hall* 474, in 1871 (FBb 6271, Gray [fig. 2] distributed as *S. desertorum* Richardson; the Gray Herbarium specimen with *"desertorum"* elided and *"S. sessilifolia* var. *villosa"* written in; the specimen in the Bebb Herbarium identified by BEBB as *S. sessilifolia* var. *villosa;* both specimens attested as *S. macrostachya* Nuttall by ROWLEE); *Cusick* (said by ROWLEE to be no. 1514), said by CUSICK (fide PIPER) to have been collected in Linn County, Oregon, probably on the Willamette River (FBb 994).

CALIFORNIA: In most of the Californian specimens having leaves densely villous-tomentose with somewhat spreading gray hairs, instead of villous with appressed silvery hairs, the leaves are much narrower than in the typical *S. sessilifolia*. Such material is found under the names *S. sessilifolia*, *S. argophylla*, and *S. hindsiana*. Two Californian specimens which rather closely approximate true *S. sessilifolia* in habit and in shape, size, and vesture of the leaves are cited below. Their leaves, however, are not so distinctly lanceolate, nor are they spinulose-denticulate so far as observed, but they do possess small stipules. Marin County, *W. H. Brewer* 2360, in 1863 (FBb 6172) or 1866 (G); Nasismento R., *W. H. Brewer* 544, May 3, 1860–1862 (G, on same sheet with the preceding).

J. G. Jack's plant from Grant's Pass, Oregon, Aug. 23, 1904 (G) may represent the glabrate autumnal aspect of the species. The leaves, though broad, are blunter and quite entire. The spinulose teeth may be deciduous, however. *Cusick's* 4497 from Roseburg, Oregon, 1914 (B) is the same as *Jack's* plant, but with densely pilose-tomentose, stipulate, sharply apiculate leaves which are spinulose-denticulate. His no. 4457 differs only in the more linear leaves; both bear foliage only.

SALIX FLUVIATILIS (?) Nuttall, N.A. Sylva 1:73. 1842 (fig. 3).

S. sessilifolia of various authors, as Sargent, Silva N.A. 9:127. *pl. 475.* 1896; Rowlee, N.A. Willows. I. Longifoliae Bull. Torr. Bot. Club 27:250. *pl. 9. fig. 8.* 1900; Howell, Fl. N.W. U.S. 618. 1903; not of Nuttall, N.A. Sylva 1:68. 1842.

Shrub, 2–6 and probably 8 m. in height; branchlets slender, brown, longitudinally striated (probably from shrinking in drying), glabrous, those of the season yellowish, often pubescent.

Leaves subsessile or the larger distinctly petiolate, stipulate on young shoots; stipules ovate to lanceolate, 2–6 mm. long, sparingly denticulate; petioles becoming 5–8 mm. long on the large leaves subtending branchlets, mostly pubescent; blades very narrowly elliptical or linear-lanceolate or those on fruiting branchlets narrowly oblanceolate, acute or short acuminate at the very sharp or spinulose apex, acute or short acuminate at the base, narrowing to the very short petiole, 5–7 cm. long, 8–14 mm. wide, the subtending leaves much larger, 8–10 cm. long, 15–20 mm. wide, all closely to remotely and more or less spinulosely denticulate, especially near the apex; thinly pubescent, becoming glabrate, the young leaves often silvery silky-villous, finally almost glabrous, green above, paler or occasionally slightly glaucescent beneath.

Aments appearing after the leaves, usually clustered, one terminal and 2–5 younger lateral at the ends of leafy branchlets 5–10 cm. long; pistillate aments 3–7 cm. long, 10–15 mm. wide; ovary silvery villous; capsule lanceolate, 5–7 mm. long, brown, villous to thinly villous to glabrate or wholly glabrous in age; style and stigmas together about 1–1.5 mm. long (difficult to determine how much is style), style divided usually to the base, stigmas also divided to the base, scales lanceolate or elliptical, 2.5–3.5 mm. long, sometimes shallowly erose-dentate at the apex, yellow, 3–5-striate, thinly pubescent when expanding to glabrate or glabrous in age (obovate before anthesis).

Staminate aments clustered as the pistillate, 4–5 cm. long; scales as in the pistillate ament; stamens 2; filaments densely pubescent below and thinly so toward the apex, anthers about 1 mm. long.

The exact identity of this plant is uncertain. The descriptions given by
NUTTALL of his seven new species in § LONGIFOLIAE are often meager. Frequently, too, there is very little difference in the series of characters assigned

FIG. 3.—*Salix fluviatilis* (?) Nutt.: portions of two fruiting twigs, collected by
Howell in Multnomah County, Oregon, July, 1875 (specimens in Bebb Herb., Field
Museum of Natural History; left FBb 6575; right, FBb 6576); nat. size.

to two or more separate species. There is some confusion, also, in the types
and cotypes of some species. The plant here provisionally referred to *S.*

fluviatilis Nutt. is the species which has been called *S. sessilifolia* by many authors and is so named in most herbaria. It is the species described as *S. sessilifolia* by ROWLEE in his revision of the LONGIFOLIAE. No specimen of NUTTALL's collecting is known to exist. The species is quite different from the true *S. sessilifolia.* It is closely related to *S. melanopsis* Nutt., a species described from old Fort Hall, near Pocatello, Idaho. It resembles that species in the broad green leaves (fig. 3), often slightly glaucous beneath, and in the extremely long and glabrate or glabrous scales. It differs from *S. melanopsis*, however, in the much longer, acute capsule, initially pubescent, and in the elongated style and stigmas. The style and stigmas, indeed, are very similar to those of true *S. sessilifolia.* It agrees with the brief description of *S. fluviatilis* in the spinulose-serrate, finally glabrous leaves. The capsules, however, are not normally glabrous but sometimes become entirely glabrous in age. The stigmas, also, apparently are not sessile, as NUTTALL records for *S. fluviatilis.* The present species seems to be confined to the lower part of the Willamette Valley and adjacent Columbia River. The type locality for *S. fluviatilis* is "the immediate border of the Oregon (Columbia), a little below its confluence with the Wahlamet." The writer plans to visit the type locality during the present season.

SPECIMENS EXAMINED.—OREGON: Multnomah County: On river banks near water; *Joseph Howell* (145?), July 1875 (FBb 6575, 6576); July 1876 (F 206534); "Oregon, *Howell*" (FBb 4858); *T. J. Howell*, July 1877, distributed by *G. C. Woolson*, no. 362 (G); the first three collections cited above are probably all one and the same thing but do not bear the same full label; Corbett, *F. A. Walpole* 1032, April 29, 1900 (N); Columbia River bottoms: *Thomas J. Howell*, July 1880, as "*Salix sessilifolia*" Nutt. (B, F 206860; FBb 4859; N); Sandbars, Columbia River, *Thomas J. Howell*, July 1880 (as *S. sessilifolia* var. *villosa* Nutt., F 206859; FBb 4860; N, 19244); Portland, between Portland and the mouth of the river, *E. P. Sheldon* 10862, July 10, 1902 (N, 2 sheets); Portland, *Sheldon* 12029, July 9, 1903 (F 216993).

U.S. DEPARTMENT OF AGRICULTURE
WASHINGTON, D.C.

OXIDATION IN HEALTHY AND DISEASED APPLE BARK

DEAN H. ROSE

During an investigation of oxidase activity in the bark of trees affected with Illinois canker, caused by *Nummularia discreta* (Schw.) Tul., there appeared an interesting correlation between oxidation, as measured by BUNZEL's simplified apparatus (1), and the acidity of the bark extracts. These were made, in a manner to be described later, from material of three different kinds: (1) green bark from limbs of trees unaffected by canker; (2) green and still seemingly healthy bark from trees badly affected by canker; (3) brown diseased bark from around the edges of cankered areas.

Bark obtained during the winter months did not separate easily from the wood, but care was taken to make the removal as complete as possible without scraping or cutting off any of the wood. While the diseased bark may have contained fungus parasites other than *Nummularia*, the precautions taken make this unlikely. All of the really dead dry bark on the surface of the canker was cut and scraped away, until the moist brown or blackened bark around the edge was reached. A strip of this 3–4 cm. wide, including sometimes a little of the green bark next to it, was then cut off down to the wood and used as "diseased bark." The various samples obtained were dried to constant weight at 65–70° C., ground up almost to a dust with a meat-grinder, and stored in glass-stoppered museum jars or tightly corked flasks. The writer realizes, of course, the need of tests on extracts from undried bark, and expects to carry through a series of them as soon as time and other work will permit.

Extracts for the various experiments were prepared under conditions which made them quantitatively comparable. In all cases distilled water was added at the rate of 8.5 cc. per gram of dried ground bark, and toluene at the rate of 0.5 cc. per 100 cc. of water used. The beaker containing the mixture was set on top of an incubator at a temperature of about 29° C. for one hour, during which time the mixture was thoroughly stirred five or six times.

55]

The extract was then filtered through four thicknesses of cheese-cloth, and finally through S. & S. filter paper no. 588.

As previously stated, all tests were conducted with BUNZEL's simplified oxidase apparatus. In this there is no provision made for absorption of the carbon dioxide produced, hence only compara-tive results can be obtained. It is probable, too, as BUNZEL points out (2, p. 30), that with no separate alkali solution in the apparatus a longer time is required for the reaction to come to an end, because of the slower absorption of the carbon dioxide by the mixture in the apparatus. This may partly account for the time required in the experiments described later, but it seems unlikely that a mercury rise which continued for 21 days in the absence of a means for absorbing carbon dioxide would have ceased within a few hours with such a means present. So far as BUNZEL's (2) published data are concerned, there is no evidence that oxidation would not have continued longer if his experiments had covered a longer period. REED (3) and APPLEMAN (4), using BUNZEL's larger apparatus, though they do not state whether or not they used the alkali basket, set the limit for completion of the reaction at 2–4 days.

Corrections for temperature variations were made by running with each experiment a blank apparatus containing only water, and subtracting from the readings in the others the reading above zero (negative pressure) in the blank. No readings were taken when the mercury in the blank stood below zero (positive pressure). It was feared that discrepancies might be introduced by the temperature variations, but as a matter of fact the differences between the results in any two comparable experiments were found to be no larger than those between duplicate apparatus in the same experiment.

As a preliminary experiment (no. 1) an extract was prepared, as already described, from dried ground bark (sample 17) of a healthy Ben Davis limb 3 inches in diameter, and the apparatus set up as follows: 8, 9, 10, 11, 12, with 1 cc. ext.+1 cc. H_2O+4 cc. 1 per cent pyrogallol; 13, with 6 cc. of distilled water.

The shaking was performed by tipping back and forth six or eight times the wire culture-tube holder in which the oxidase apparatus were fastened. Oxidation, as indicated by rise of mercury (negative pressure really), was not complete even at the end of seven days, as was well demonstrated by later experiments.

TABLE I

MANOMETER READINGS IN EXPERIMENT NO. 1; EXTRACT OF HEALTHY BARK, SAMPLE 17

DAY OF TEST	TIME OF READING	ELAPSED TIME IN HOURS AND MINUTES	TEMPERA-TURE AT TIME OF READING	MANOMETER READINGS CORRECTED AGAINST AN APPARATUS CONTAINING ONLY WATER					
				Extract unchanged					H₂O
				8	9	11	12	10	13
1st. .	3:55 P.M.	o*	36 o	o oo	o oo	o oo	o oo	o oo	
2d...	9:30 A.M.	17:35	35 5	—o 15	—o 20	—o 25	—o 20	—o 20	
	2:30 P.M.	22.35	36 o	—o 20	—o 25	—o 35	—o 30	—o 20	
3d...	5:00 P.M	49.05	35 o	—o 60	—o 60	—o 66	—o 60	—o 60	
4th.	2:37 P.M.	70 42	33 9	—o 80	—o 75	—o 93	—o 82	—o 90	
	5·00 P.M.	97·05	34 o	—o 80	—o 78	—o 92	—o 95	—o 98	
5th.	8:50 A.M.	112.55	32 o	—o 95	—o 90	—1 08	—1 04	—1 08	
6th.	11:25 A.M.	139 30	34 o	—1 25	—1 25	—1 25	—1 15	—1 30	
7th.	9.00 A.M.	161:05	33 o	—1 40	—1 35	—1 35	—1 30	—1 40	8 8

* Put in incubator at 2·55 P M ; closed and shaken at 3 55 P M to mix thoroughly the extract and pyrogallol; shaken again next morning

A similar experiment (no. 2) was set up, using the extract of diseased bark from another Ben Davis tree in the same orchard. The results are given in table II.

TABLE II

MANOMETER READINGS OBTAINED FROM EXPERIMENT NO. 2; EXTRACT OF DISEASED BARK, SAMPLE 15

DAY OF TEST	TIME OF READING	ELAPSED TIME IN HOURS AND MINUTES	TEMPERA-TURE AT TIME OF READING	MANOMETER READINGS CORRECTED AGAINST A TUBE CONTAINING ONLY WATER					
				Extract unchanged					H₂O
				8	9	10	11	12	13
1st .	4:10 P M.	o*	36 3	0.00	o oo	o oo	o oo	o oo	o o
2d...	4:12 P.M.	24 02	35 4	—0.99	—o 95	—1 09	—o 85	—o 85	o o
	4·42 P.M.	24 32	35 6	—1 oo	—o 93	—1 10	—o 95	—o 95	o o
3d...	2:22 P.M.	46:12	36 4	—1.70	—1 45	—1 70	—1 52	—1 50	o o
	5:09 P.M.	48:59	37 o	—1.75	—1 50	—1 77	—1 55	—1 55	o o
4th..	4:45 P.M.	72·35	36 o	—1.75	—1 65	—1 75	—1 80	—1 75	o o
5th...	8:18 A.M.	88·08	36 2	—1 93	—1 75	—1 93	—1 95	—1 90	o o

* Put in incubator at 3:00 P M , all apparatus closed and shaken at 4 10 P M , shaken again next morning.

It is seen from table II that oxidation by the extract of diseased bark is more rapid than by the extract of healthy bark, being greater at the end of five days than that caused by healthy extract at the end of seven days.

These two extracts were titrated immediately after extraction with n/20 NaOH, using phenolphthalein as an indicator. The difficulty of determining the end point, due to darkening of the extract on addition of alkali, was avoided by using the test-plate method described by Miss SCHLEY (5). In this two phenolphthalein solutions are used, one slightly alkaline, the other slightly acid; the end point is reached when a drop of the solution being titrated just fails to decolorize a drop of the alkaline solution and just shows a faint pink tinge in a drop of the acid solution. By this method the following results were obtained, using 20 cc. of extract in 50 cc. of distilled water: to neutralize 1 cc. of extract of healthy bark to phenolphthalein requires 0.86 cc. n/20 NaOH; to neutralize 1 cc. of extract of diseased bark to phenolphthalein requires 0.38 cc. n/20 NaOH.

These titrations, of course, measure only the base-absorbing power of the extracts, not the H^+ concentration, for the extracts are probably mixtures of strong and weak acids and acid salts, all having different degrees of ionization. But since the concentration of free H^+ ions is known to have a marked influence on many of the reactions taking place in living matter (HÖBER, p. 176), it was important to determine just what this concentration is in the bark extracts. In the absence of facilities for doing this accurately by the gas-chain method, tests were made of the effect of diluted and undiluted extracts on the color changes of various indicators. The most clear-cut results were obtained with mauvein, which is known to give the following color changes: yellow at $H^+ = 2 \times 10^{-3}$; green at 10^{-3}; green-blue at 10^{-4}; blue at 10^{-5}; violet at 10^{-6}. In repeated tests on extracts from different samples of diseased bark, mauvein was turned to a definite green, indicating that here H^+ is about 10^{-3}. Extract of healthy bark turned mauvein yellow, indicating that $H^+ = 2 \times 10^{-3}$. It was found, however, that any given quantity of this extract had to be diluted to 2.4 times its original volume to give the same green as the extract of diseased bark; consequently (not allowing for increased ionization with dilution), H^+ concentration here would figure, not 2.0 times, but 2.4 times 10^{-3}, or $10^{-2\ 62}$; that is, both indicator and titration figures show diseased bark to be less acid than healthy bark. With this in mind, a reference to tables I and II above shows that oxidation in the BUNZEL apparatus is in inverse ratio to the acidity of the solution being tested.

This correlation between oxidation and acidity seems also to have obtained in REED's work (3, pp. 56 and 76), though he fails to bring out the point, where juice of bitter rot apples caused greater oxidation than the juice of more acid healthy apples. The results obtained by BUNZEL (2, p. 26) in testing the effect of different concentrations of pyrogallol show that as the concentration decreased from 16 per cent to 1 per cent pyrogallol the negative pressure increased from 2.11 cm. of mercury to 2.86 cm. This may not be due to decrease in acidity, as determined by the pyrogallol concentration (attention is not called to it in BUNZEL's discussion of the table), but such an explanation seems the simplest and most likely.

Titration and indicator tests on a 1 per cent solution of pyrogallol and on a two-thirds of 1 per cent solution, the strength used in the apparatus if the diluent were only water, gave the results shown in table III.

TABLE III

RESULTS OF TITRATION AND INDICATOR TESTS WITH PYROGALLOL SOLUTIONS

PYROGALLOL SOLUTION	NO OF CC OF N/20 NaOH NECESSARY TO NEUTRALIZE 1 CC OF SOLUTION TO PHENOLPHTHALEIN		H^+ CONCENTRATION AS INDICATED BY METHYL ORANGE	
	1 per cent	⅔ of 1 per cent	1 per cent	⅔ of 1 per cent
Fresh . . .	0.13	0.086	$<10^{-3.33}$	$<10^{-3.33}$
6 months old.. ..	0.34	0.23	$3.50\times10^{-3.33}$ $=10^{-2.79}$	$2.33\times10^{-3.33}$ $=10^{-2.96}$

The indicator figures for H^+ concentration of a fresh two-thirds of 1 per cent pyrogallol solution (about 1/20 molecular) agree well with those calculated from the electrical conductivity of such a solution. This is known to be 3 12 μ, whence the percentage dissociation $= 3.12 \div 355$ (μ at infinite dilution) $= 0$ 88 per cent. If all the dissociation in pyrogallol at 1/20 molecular concentration is due to H^+ dissociation and only one H dissociates, the H^+ concentration is less than 0.000616 equivalent. If all three H's dissociate it is less than 0.000616×3, or, in other words, the theoretical H^+ concentration of the pyrogallol solution in the apparatus is between 10^{-2} and 10^{-3} or less than either. Because of BUNZEL'S statement (2, p. 28) that either fresh or old pyrogallol solution may be used it is worthy of note that a 1 per cent solution six months old required dilution to 3.5 and a two-thirds of 1 per cent solution to 2.33 times its original volume before it failed to give the red color with methyl orange, thus indicating for the two-thirds solution an H^+ concentration of $2.33 \times 10^{-3\ 33}$ or $10^{-2\ 96}$ (not allowing for increased dissociation with dilution). That is, by slow spontaneous oxidation the H^+ concentration of a two-thirds of 1 per cent solution of pyrogallol solution is increased in six months from less than $10^{-3\ 33}$ to $10^{-2\ 96}$. It follows that fresh pyrogallol solution should be made up for each experiment. In this connection it seems rather more than a coincidence that in BUNZEL'S experiment 6 (2, table VI, p. 28) fresh 1.0 per cent and 0.1 per cent pyrogallol solutions gave, in a test only one hour long, respectively 0.1 and 0.05 cm. higher negative pressure (greater oxidation) than similar solutions one year old.

The reaction to phenolphthalein of the mixtures in the apparatus may be calculated from the data given in table III. For those containing extract of healthy bark, sample 17, it would be $\dfrac{0.0+0.86+(4\times0.34)}{6}=0.37$; for those containing extract of diseased bark, sample 15, $\dfrac{0.0+0.38+(4\times0\ 34)}{6}=0.29$. The acidity of the once distilled water was found to be negligible. On account of the correspondence already observed between titration and indicator figures for both the pyrogallol solution and the extracts, the reaction of the mixtures in the apparatus may be taken to represent roughly the H^+ concentration also until further work can be done.

To test the effect of varying the acidity of the extract on the rate and total amount of oxidation, experiment 3 was set up, using extract of healthy Ben Davis bark, sample 17. The apparatus contained the following solutions: 8, 9, 10, with 1 cc. extract (acidity reduced one-fourth)+1 cc. H_2O+4 cc. 1 per cent pyrogallol; 13, with 6 cc. distilled water; 11, 12, with 1 cc. extract (acidity reduced one-half)+1 cc. H_2O+4 cc. 1 per cent pyrogallol. The results are given in table IV.

TABLE IV

MANOMETER READINGS IN EXPERIMENT NO. 3; EXTRACT OF HEALTHY BARK, SAMPLE 17

DAY OF TEST	TIME OF READING	ELAPSED TIME IN HOURS AND MINUTES	TEMPERATURE AT TIME OF READING	MANOMETER READINGS CORRECTED AGAINST AN APPARATUS CONTAINING ONLY WATER					
				Three-fourths acid			One-half acid		H_2O
				8	9	10	11	12	13
1st ..	4:40 P.M.	0*	33 0	0 00	0 00	0 00	0 00	0 00	0.0
2d...	9:20 A.M.	16.40	33 0	− 16	− 14	− .89	− .34	− 32	0
	10:00 A.M.	17:20	33 0	− .19	− 7	− 04	− 55	− 49	.0
	4:55 P.M.	24:15	33 5	− 3	− 9	− .14	− 69	− 74	.0
4th..	3:50 P.M.	71:10	34 0	− 0	− .2	− 96	− 59	− 64	.0
5th..	1:40 P.M.	93:00	32 0	− 3	− 7	− 39	− 98	− 92	0
6th..	4:35 P.M.	119:55	32 0	− .5	− 9	− .59	− 3	− 3	0
7th...	4:20 P.M.	143:40	33 8	− .7	− 9	− .84	− .44	− 4	.0
8th. .	5:50 P.M.	169:10	34 5	− 94	− 4	− 99	− 49	− 5	0
9th...	8:12 A.M.	183:32	33 1	− 94	− 4	− 09	− 74	− .7	.0

* Put in incubator at 3:10 P.M.; apparatus closed and shaken at 4:40 P.M; shaken again the next morning.

Experiment 4 was arranged with extract of diseased Ben Davis bark, sample 15, as follows: 12, with 1 cc. extract unchanged (acidity 0.38)+1 cc. H_2O+4 cc. 1 per cent pyrogallol; 11, with 6 cc. distilled water; 8, 9, with 1 cc. extract (acidity 0.29)+1 cc. H_2O+4 cc. 1 per cent pyrogallol. It was thought desirable to run the reactions thus close together (see table V) to test the effect of very small differences in acidity on the rate and total amount of oxidation. The results are given in table V, which summarizes the results from all the experiments described above and from others duplicating them in various ways but not described in detail in this paper.

A study of table V shows that when results are averaged from several different experiments the statement made earlier still holds, that unchanged extract of diseased bark causes greater and more rapid oxidation of pyrogallol than unchanged extract of healthy bark. Moreover, when the acidity of healthy bark extract is reduced the oxidation is increased, but greatly out of proportion to the reduction in acidity.

TABLE V

SUMMARY OF EXPERIMENTS WITH EXTRACTS OF HEALTHY AND DISEASED APPLE BARK, SHOWING EFFECT OF VARIOUS DEGREES OF ACIDITY OF EXTRACT UPON OXIDATION OF PYROGALLOL

DAY OF TEST	Extract healthy bark, sample 17				Extract diseased bark, sample 15			Extract healthy bark, no 11	Extract diseased bark, no. 22a
	1	2	3	4	5	6	7	8	9
	*(3.66)	(0.86)	(0.65)	(0.43)	(0 44)	(0.38)	(0.32)	(0.73)	(0.34)
	† 0.83	0 37	0 33	0 29	0 30	0.29	0.28	0 35	0 28
1st....	0 00	−0 26	−1.24	−1.72	−0 87	−0.98	−0.95	−0 26	−0.57
2d.. .	0.00	−0 61	.....	..	−1 35	−1 60	−1 45	−0 53	−0 94
3d....	0 00	−0 84	−2 04	−2 68	−1 53	−1 73	−1 76	−0.77	−1 20
4th.	−0 05	−1 01	−2.34	−2.92	−1 63	−1.89	−1 85	−0.86	−1 36
5th.....	−0 10	−1.25	−2.49	−3 29	−1.88	−1 95	−2.00	−1 07	−1 50
6th.....	−0.13	−1.38	−2.77	−3 43	.....	.....	.....		
7th.....	−0.17	.....	−2.90	−3 59	−1.89	−2.13	−2.19	.....	.
8th....	.....	.....	.....	.....	.....	.....	.....		
9th..	.....	.....	.....	.....	.....	.....	.....	−1.51	−1.94
	1 test	Average of 6	Average of 3	Average of 3	Average of 2	Average of 6	Average of 2	Average of 2	Average of 2

* The figures in parenthesis indicate reaction of extract
† The figures in this horizontal line indicate reaction of mixture in oxidase apparatus.

Further proof of the sensitiveness of the oxidase to changes in acidity is found in columns 5 and 6 (table V), where a difference of 0 01 cc. of n/20 NaOH gave throughout the experiment a consistently greater oxidation by the less acid mixture. The figures in column 7 seem unreliable and suggest an error in determination of the acidity. Column 1 shows the great reduction in oxidation produced by increasing the acidity of the mixture in the oxidase apparatus to practically that of the unchanged extract. The figures in columns 8 and 9 are from an experiment in which extracts of healthy and diseased bark were tested at the same time; they show again the correlation between acidity and oxidase activity. This experiment was continued for 19 days, although during the last 4 all oxidation, measured by changes in mercury level, had ceased.

On the nineteenth day the negative pressure was released, all the apparatus closed again, and the test continued for three days more. The results are shown in table VI.

TABLE VI

		Healthy	Diseased
Before releasing.....	19th day of test	−2.03	−2.46
After releasing....... ..	19th " ." "	0.00	0 00
" "	20th " " "	−0.12	−0 08
" " ....'.....	21st " " "	−0.15	−0.11
		Average of 2	Average of 2

Table VI indicates that oxygen supply is a limiting factor as well as acidity. Its importance should be investigated further. On the twenty-first day, when the experiment was finally closed, it was found that the mixture from the two apparatuses containing healthy bark extract and from the other two containing diseased bark extract had increased from an original acidity of o.38 in the first case, and o.28 in the second case, to o.50 in both (acidity expressed here as before in terms of cc. of n/20 NaOH necessary to neutralize to phenolphthalein). Similar results were obtained in two other experiments in which the reaction had come to an end.

It thus appears that the gradual slowing down of the rate of oxidation in these experiments and in others conducted with similar apparatus is due to increasing acidity and not to chemical combination of the oxidase with some substance in solution, as suggested by BUNZEL (2, p. 39). If this be true, the evidence is strong that oxidases are true catalytic agents, prevented usually from bringing about indefinite catalysis by the presence or absence of something which acts as an inhibitor. It is probable that just as enzyme hydrolysis of carbohydrates and other substances reaches a condition of equilibrium because of accumulation of the products of hydrolysis, in exactly the same way, in the BUNZEL apparatus at least, oxidation ceases because of inhibition by accumulated oxidation products. These will include carbon dioxide and acetic and oxalic acids if the decomposition of pyrogallol by oxidase is like that induced by alkalies and the salts of heavy metals. In the presence of O_2 alkalies cause pyrogallol to turn brown with the formation of carbon dioxide and acetic acid; salts of mercury, silver, and gold oxidize pyrogallol to acetic and oxalic acids. It might appear that oxidation in the larger BUNZEL apparatus, when the alkali basket is used, is greater because of the removal of carbon dioxide, which would tend to increase the acidity if allowed to remain. This can hardly be true, however, since carbonic acid has a low dissociation constant and at the temperature of these experiments would be present in the solutions in comparatively low concentration. Its effect on acidity is but slight at best, and the removal of it merely shortens the time required for the reaction to come to an end. Acetic and oxalic acids are both more important, for two reasons: all of the acid formed remains in solution, and if it is acetic acid, dissociation (and the consequent inhibitory effect) is about ten times as great as for carbonic acid.

An H^+ concentration of 10^{-3} or 10^{-4} is just in the midst of H^+ concentration optima for various enzymes, as given by HÖBER (6, p. 721), varying from $10^{-1.5}$ for pepsin to $10^{-8.5}$ for esterase of the blood. No data are given for oxidase, but the work here reported on indicates that the optimum for them is much less than 10^{-3}. If the pyrogallol as used in the apparatus is just about the H^+ optimum, it is easily seen how the plant juice could make the H^+ concentration too high. Then when base is added, the strongest, most highly dissociated acids are neutralized first, and the H^+ concentration drops faster than represented by the degree of neutralization.

It may be possible to get some idea of the H^+ optimum for oxidase activity, and of the H^+ concentration of various mixtures in the apparatus by further work with indicators, but definite knowledge on these points must come finally from careful determinations by the gas-chain method.

Summary

1. Extract of apple tree bark affected with Illinois canker causes greater and more rapid oxidation of pyrogallol than does the extract of healthy bark.

2. Diseased bark extract is less acid than healthy bark extract, according to both indicator and titration figures, hence the

conclusion seems justified that oxidation is in approximately inverse ratio to the acidity of the extract in the range of concentrations here used.

3. This conclusion is borne out by the fact that addition of acid to the solution in the apparatus decreases oxidation and addition of alkali increases it.

4. Oxidases are very sensitive to small variations in acidity of the solution in the oxidase apparatus.

5. As a hypothesis in need of further proof the following is offered. The gradual slowing down of oxidation in the ·BUNZEL apparatus is brought about by accumulation of oxidation products, probably acetic and oxalic acids, and not by a using up of the oxidase through chemical combination between oxidase and oxidizable substance.

MISSOURI STATE FRUIT EXPERIMENT STATION
MOUNTAIN GROVE, MO.

LITERATURE CITED

1. BUNZEL, H. H., A simplified and inexpensive oxidase apparatus. Jour. Biol. Chem. 17:409–411. 1914.

2. ———, The measurement of the oxidase content of plant juices. U.S. Dept. Agric. Bur. Pl. Indus. Bull. 238. pp. 5–40. 1912.

3. REED, H. S., The enzyme activities involved in certain fruit diseases. Va. Exp. Sta. Rpt. 1911 and 1912. pp. 51–57.

4. APPLEMAN, C. O., Biochemical and physiological study of the rest period in tubers of *Solanum tuberosum*. Maryland Agric. Exp. Sta. Bull. 183. p. 193. 1914.

5. SCHLEY, EVA O., Chemical and physical changes in geotropic stimulation and response. BOT. GAZ. 56:483. 1913.

6. HÖBER, RUD, Physikalische Chemie der Zelle und der Gewebe. Leipzig und Berlin. 1914.

SANIO'S LAWS FOR THE VARIATION IN SIZE OF CONIFEROUS ·TRACHEIDS

IRVING W. BAILEY AND H. B. SHEPARD

In 1872 SANIO[1] published the results of an investigation upon the tracheids of Scotch pine, *Pinus sylvestris* L. From his observations and measurements he deduced five general laws which may be translated as follows:

1. In the stem and branches the tracheids everywhere increase in size from within outward, throughout a number of annual rings, until they have attained a definite size, which then remains constant for the following annual rings.

2. The constant final size changes in the stem in such a manner that it constantly increases from below upward, reaches its maximum at a definite height, and then diminishes toward the summit.

3. The final size of the tracheids in the branches is less than in the stem, but is dependent on the latter, inasmuch as those branches which arise from the stem at a level where the tracheids are larger themselves have larger tracheids than those which arise at a level where the constant size is less.

4. In the gnarled branches of the summit the constant size in the outer rings increases toward the apex, and then falls again, but here irregularities occur which may be absent in regularly grown branches.

5. In the root the width of the elements first increases, then falls, and next rises to a constant figure. An increase in length also takes place, but could not be exactly determined.

SANIO'S measurements and conclusions have been accepted by DEBARY, PFEFFER, HABERLANDT, and others and have been considered to be applicable to conifers in general.

[1] SANIO, KARL, Über die Grösse der Holzzellen bei der gemeinen der Kiefer (*Pinus silvestris*). Jahrb. Wiss. Bot. 8:401–420. 1872.

The writers recently had occasion to test the validity of SANIO's first two laws in connection with an investigation undertaken in the endeavor to secure a simple method for segregating long and short "fibers" in the manufacture of raw wood pulp. The following North American conifers were studied: *Pinus Strobus* L., *P. palustris* Mill., *Picea rubens* Sarg., *Tsuga canadensis* Carr., and *Abies concolor* Lindl. and Gord. Chips of wood were removed from the specimens and were macerated by the use of a 5 per cent solution of equal parts of chromic and nitric acid. The tracheids were separated after maceration by being shaken with water and glass beads, insuring minimum breakage. They were then kept in a mixture of 95 per cent alcohol and chloroform to prevent softening. The measurements were made with a micrometer eyepiece, 50 measurements from each chip. The results obtained are shown in the following tables:

EFFECT OF AGE ON THE LENGTH OF TRACHEIDS

TABLE I

Pinus Strobus, 120 ANNUAL RINGS, CROSS-SECTION 1 FOOT FROM GROUND

ANNUAL RING	TRACHEID LENGTHS—MILLIMETERS			ANNUAL RING	TRACHEID LENGTHS—MILLIMETERS		
	Max.	Min.	Av.		Max.	Min.	Av.
20.........	3.35	1.60	2.65	80........	5.10	2.65	3.77
30.........	3.65	2.35	3.02	90........	4.60	2.50	3.79
40.........	4.80	2.55	3.64	100........	4.65	2.50	3.85
50.........	4.20	2.35	3.30	110........	4.45	3.20	3.91
60.........	4.65	2.45	3.47	120........	5.40	3.45	4.20
70.........	4.80	2.65	3.52				

TABLE II

Abies concolor, 80 ANNUAL RINGS, ELEVATION OF CROSS-SECTION UNKNOWN

ANNUAL RING	TRACHEID LENGTHS—MILLIMETERS			ANNUAL RING	TRACHEID LENGTHS—MILLIMETERS		
	Max.	Min.	Av.		Max.	Min.	Av.
10.........	3.50	2.20	2.75	50........	5.45	3.30	4.45
20.........	3.90	2.40	3.30	60........	5.45	3.55	4.64
30.........	4.90	2.50	3.65	70........	6.30	3.50	5.02
40.........	5.65	2.15	3.58	80........	6.45	3.55	5.20

TABLE III

Pinus palustris, 230 ANNUAL RINGS, CROSS-SECTION 1 FOOT FROM GROUND

ANNUAL RING	TRACHEID LENGTHS—MILLIMETERS			ANNUAL RING	TRACHEID LENGTHS—MILLIMETERS		
	Max.	Min.	Av.		Max.	Min.	Av.
10 . .	3.00	1.80	2.42	130.... .	5.90	3.90	4.99
20 . . .	3.85	2.40	3.06	140	6.25	3.75	4.92
30...	4.50	2.80	3.70	150	6.00	4.00	4.87
40 . .	4.95	2.80	4.06	160	6.20	4.00	5.08
50 ...	5.30	2.95	4.11	170 . .	6.05	3.45	4.57
60........	6.05	3.05	4.33	180	5.75	3.40	4.10
70 . . .	5.25	2.85	3.92	190 .	5.50	3.10	4.03
80	5.70	3.05	4.41	200........	6.20	3.15	4.42
90........	6.05	3.20	4.66	210 .	6.05	3.35	4.66
100........	6.30	3.45	4.88	220 ..	5.90	3.10	4.35
110 . . .	6.05	3.60	4.57	230 .	5.20	2.85	4.00
120	5.80	3.85	4.66				

TABLE IV

Tsuga canadensis, 80 ANNUAL RINGS, CROSS-SECTION 1 FOOT FROM GROUND

ANNUAL RING	TRACHEID LENGTHS—MILLIMETERS			ANNUAL RING	TRACHEID LENGTHS—MILLIMETERS		
	Max.	Min.	Av.		Max.	Min.	Av.
10	2.30	0.95	1.63	50	4.25	1.85	3.17
20 ...	3.20	1.15	2.20	60 .	3.80	1.60	2.48
30...	3.75	2.05	2.99	70 .	3.20	1.40	2.47
40 . .	4.00	2.25	3.16	80.. . .	3.15	1.80	2.84

EFFECT OF POSITION IN THE VERTICAL AXIS ON THE LENGTH OF TRACHEIDS

TABLE V

Picea rubens, 50 ANNUAL RINGS, AVERAGE TRACHEID LENGTH IN MILLIMETERS

Annual ring	Distance from ground in feet					
	1	6	12	18	24	30
10.. .	1.68	....	....	... .	....	....
20	2.21	2.49	2.22	... .	.. ....	..... .
30 . .	2.58	2 92	2 84	2.49	1.89	. . ..
40 . .	3.17	3.00	3.48	3.31	3.23	2.71
50	3.48	3.56	3.84	3.70	3.87	3.31

EFFECT OF WIDTH OF RING UPON THE LENGTH OF TRACHEIDS

TABLE VI

Pinus Strobus, ECCENTRICITY DUE TO FASTER GROWTH ON
ONE SIDE, AVERAGE

ANNUAL RING	TRACHEID LENGTHS— MILLIMETERS	
	Narrow	Wide
10	1.63	1.66
22	2.58	2.41
27	2.51	2.53
Average	2.24	2.20

Discussion of measurements and conclusions

The results of these measurements are obviously not in accord with SANIO's first law, since no constant tracheid length was found in any of the specimens examined. Inasmuch as SANIO found a constant in all cases within the 50th ring, and in one case within the 20th, and this study found none, even within the 230th, it is evident that SANIO's first law cannot be applied to conifers, and some doubt is cast upon the accuracy of his observations upon Scotch pine. As is shown diagrammatically in the accompanying figure, the length of the tracheids increases rapidly for a period of years varying from 25 to 60. At the end of this period there is a marked falling off in the length of the tracheids, which lasts for a decade or more. Subsequently the tracheids again increase in length. In the case of the long-leaf pine, which unfortunately was the only very old material available, the tracheid length reaches a maximum at 160 years and decreases, with one marked period of recovery, during succeeding rings. The factor or factors which produce the fairly regular cycles or series of crests and depressions which occur in the long-leaf pine curve are obscure and deserve to be studied carefully by some one who has easy access to old coniferous trees. DOUGLASS'[2] and HUNTINGTON'S[3] interesting correlations

[2] DOUGLASS, A. E., Weather cycles in the growth of big trees. Monthly Weather Review. June 1909.

[3] HUNTINGTON, ELLSWORTH, The climatic factor. Carnegie Inst. Washington, D.C. 1914.

between rainfall and ring width suggest the possibility that the dimensions of tracheids may be equally sensitive to modifying climatic factors.

It is evident from Table V that SANIO's second law is applicable to *Picea rubens* as well as to *Pinus sylvestris*. However, a fact unnoted by SANIO is that the maximum average tracheid length occurs higher from the ground in rings nearer to the bark. This probably bears a relation to the fact that each successive increment is larger, that is, extends farther from the ground. The study of

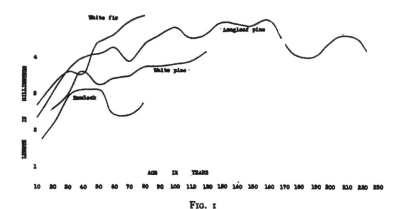

FIG. 1

the eccentric cross-section recorded in table VI seems to indicate that width of ring has no marked effect upon the length of tracheids.

A number of botanists have endeavored to make use of the dimensions of the xylem elements in the classification and identification of the secondary wood of living and fossil plants. When based upon the study of a limited amount of material taken from a given region in a tree, the measurements are significant only if compared with those secured from a homologous region in a tree which has grown under similar conditions. Average dimensions for a species must obviously be based upon a large amount of material selected from mature plants grown in different environments. However, general averages of this character are of little value for the purposes of the systematic botanist, or in the identification of woody tissues, and are only of relative value to the manufacturer of wood pulp.

BUSSEY INSTITUTION
HARVARD UNIVERSITY

BRIEFER ARTICLES

CHARLES EDWIN BESSEY

(WITH PORTRAIT)

In the history of American botany the name of CHARLES E. BESSEY will always hold a conspicuous place. When his first textbook (*Botany for High Schools and Colleges*) appeared in 1880, it introduced a new era in botanical instruction in this country. Before that date, the study of botany in the United States was bounded practically by the taxonomy of the higher plants, with such gross morphology as enabled the student to use a manual. BESSEY's *Botany* brought the atmosphere of SACH's

Lehrbuch to American colleges, and this compelled the development of botanical laboratories. This original textbook was the first of a series of texts that continued to be very influential.

Professor BESSEY was a most stimulating teacher, and perhaps no American botanist has left his mark upon so many students. Especially in government service and in agricultural colleges are these students to be found, for the teacher believed that his science should be directed toward the public service. At the same time, he was uncompromising in his views as to the difference between botany as a science and its various practical applications.

He was born in Milton, Ohio, May 21, 1845; received his Bachelor's degree at Michigan Agricultural College in 1869; and subsequently studied at Harvard University. He held only two academic positions as professor of botany: the first at Iowa State College (1870–1884), and the second at the University of Nebraska (1884–1915), where he served also as dean and at various times as acting chancellor. He died at Lincoln, Nebraska, February 25, 1915.

He was a member of many scientific societies, and was very active in promoting their interests. His colleagues recognized his great services by electing him to various important offices. Among these offices were vice-president of Section F of the American Association, a position which he filled four times, (1893, 1894, 1902, 1907); president of the Botanical Society of America (1896); and president of the American Association (1910).

Aside from his preparation of textbooks, his special interests as an investigator were the fungi, the flora of the Great Plains, plant migration, and the relationships of plant families. His later publications dealt with "plant phyla," the last upon this subject appearing in January 1914. These studies of the relationships involved a laborious assembling and comparison of material that few would care to undertake, and the conclusions as to classification were far from stereotyped. For example, instead of the conventional four great groups of plants, Professor BESSEY presented the plant kingdom in fourteen "phyla."

In addition to textbooks and papers upon the special subjects referred to, Professor BESSEY was a regular reviewer of botanical literature for *Science*, so that he was in continual contact with those interested in his subject. His reviews were characteristic of the man, for they were always kindly. Apparently he searched for the pleasant things to say, and left the unpleasant things unsaid.

The botanical fraternity of this country will miss the presence of Professor BESSEY keenly, for his frequent attendance upon scientific meetings made him well known personally to all botanists, and his very genial nature made friends of all his colleagues.—J. M. C.

CURRENT LITERATURE

BOOK REVIEWS

Ascent of sap

DIXON'S[1] volume on transpiration and ascent of sap clearly puts the evidence in favor of the cohesion theory, along with the weak points of the other theories. The chapter headings give an idea of the organization and contents of the work: (1) nature of transpiration and the ascent of sap; (2) ascent of sap in stems; criticism of physical theories; (3) ascent of sap in stems; criticism of vital theories; (4) cohesion theory of ascent of sap in stems; (5) tensile strength of the sap of trees; (6) estimate of the tension required to raise the sap; (7) osmotic pressure of leaf cells; (8) the thermo-electric method of cryoscopy; (9) method of extracting sap for cryoscopic observations; (10) osmotic pressure in plants; (11) energy available for raising the sap.

The cohesion theory of the rise of sap assumes that the water is drawn up through the stem by a pull applied at the top of the water column, and that this pull is transmitted downward through the cohering water. The pull, at least in high transpiration, is due to the evaporating power of the air, but it is limited in amount, as experiments indicate, by the osmotic pressure of leaf cells. The theory excludes living cells along the tracheary system of the stem and root from an active part in the rise of sap. At times of low transpiration, DIXON believes the living cells of the leaf blade may generate the pull by secreting water into the intercellular spaces. For the support of this theory three things require quantitative study: cohesion of sap under conditions existing in the tracheae; osmotic pressure of leaf cells; and total pull required to maintain the water column and to satisfy the transpiration loss in the highest trees at maximum transpiration. It is scarcely less important to know whether there is at all times a continuous water column in the stem of sufficient cross-section to account for the rise by cohesion, and whether actual tensions of considerable magnitude exist in this column. DIXON has apparently answered the first three points adequately, and has brought some evidence for the fourth; while RENNER has shown the existence of tensions of 10–20 atmospheres in the water of the stem tracheae of transpiring plants.

DIXON's new measurements of the tensile strength of the sap of trees as exhibited in capillary tubes are of great interest. The values range from 45 to 207 atmospheres. The latter, according to DIXON, is the highest experimental

---

[1] DIXON, H. H., Transpiration and ascent of sap in plants. 8vo. viii+216. figs. 30. London and New York: Macmillan. 1914.

CURRENT LITERATURE 75

value ever obtained for the cohesion of a liquid. The measurements were made with sap containing considerable air in solution, thus approximating the conditions in the tracheae. It is estimated that in the highest trees at maximum transpiration, the cohesion necessary to raise the sap does not exceed 20 atmospheres, a value well within the tensile strength of sap. For determining the osmotic pressure of leaf cells, DIXON has used the freezing point method. To this end, he and ATKINS have improved the current methods of extracting plant saps and have developed a thermo-electric method of determining the lowering of the freezing point. The essential new feature in the extraction of sap is the emersion of the tissue for some time in liquid air. This destroys all living semipermeable membranes and allows the solutes to be pressed out unrestrictedly with the water. The main advantage claimed for the thermo-electric method is the small amount of fluid needed, 2 cc. as against the 15–20 cc. for the ordinary Beckman apparatus. This is frequently a very important consideration for the plant physiologist. With the use of multiple juncture couples, a lowering of the freezing point by 0.0001° C. can be measured as against 0.001° C. for the Beckman. With the single juncture couples, used in the main by DIXON, a lowering of 0.01° C. can be measured. By reducing the size of the Beckman apparatus, BURIAN and DRUCKER were able to measure the lowering of the freezing point of 1.5–2 cc. of solution accurately to 0.005° C., and DRUCKER and SCHREINER by a different method were able to measure the freezing point of 0.005 cc. of a solution accurately to 0.01° C.[2] For the needs of the plant physiologist, I am inclined to think that the latter methods are more suitable.

This book is rather more a statement of the extensive and in the main extremely important work of the author on the rise of sap than a critical consideration of the whole literature of the subject. This accounts for chap. vii, which presents DIXON's cumbersome and inaccurate method of determining osmotic pressure of cells of the leaf by balancing it against gas pressure; for the slight consideration given RENNER's work, which proves a critical point for the cohesion theory, namely the actual existence under certain conditions of a stretch of 10–20 atmospheres in the water of the tracheae; for the failure to point out that by a very different method RENNER confirmed a law established by DIXON, namely the rate of flow of a water column through a stem is proportional to the push or pull applied to it; and for the absence of the significant data of HANNIG, FITTING, and others on the osmotic pressure of leaf cells.

As yet the existence of a continuous gas-free water column from root to leaf under all conditions has not been established. Until this is established, there will be a just argument against the cohesion theory of the rise of sap.—WM. CROCKER.

---

[2] HÖBER, Physikalischechemie d. Zelle, etc., p. 22. 1914; Biol. Centralbl. 33:99. 1913.

MINOR NOTICES

Plant breeding.—A new edition of BAILEY's *Plant breeding* has been prepared by GILBERT.[3] The preceding edition, the fourth, appeared in 1906. The new edition includes much of the material of the older ones, and in addition presents the results of the more recent investigations in genetics that have removed plant breeding from the simple operations of a few years ago, and have made of it a subject of rigid experimental work, with a wealth of special terminology. Professor BAILEY voices a feeling among the older botanists when he says: "The literature has now become complex and difficult, with considerable gain no doubt in a closer acquaintance with the subject, and a nearer approach to the ultimate truth; but the charm of the simple literature is largely buried, and I fear that much of our interest is now expressed in the discussion of methods and in disputing about the reasons. Yet we are accumulating knowledge, and after a time we shall come back to clarity and to a simplicity that the layman can use."

The titles of the chapters indicate the contents: The fact and philosophy of variation; The causes of individual differences; The choice and fixation of variations; The measurement of variation; Mutations; The philosophy of the crossing of plants; Heredity; How domestic varieties originate; Pollination; and The forward movement in plant breeding.

A most useful feature of the book is the very complete bibliography, which includes 59 pages of citations, extending from 1905 to 1912, inclusive. There are also 27 laboratory exercises, which will be most helpful to teachers who wish to make plant breeding a laboratory course rather than a series of lectures.—J. M. C.

The coco-nut.—Under this title COPELAND[4] has published a detailed account of one of the most valuable plants of the tropics. Naturally the book is of chief interest to those engaged in the culture of the coco-nut palm. In the introduction the origin and spread of this palm as a crop plant is considered. Assent is given to the belief that it is a native of the American tropics, and that in very early times it was carried across the Pacific and introduced into Polynesia, where it has become a crop of the first importance. From Polynesia it reached New Guinea and Malaya, but not Australia. Much more recently it has been introduced along all tropical coasts.

The chapters deal with the physiology of the plant, climate and soil, diseases and pests, selection and treatment of seed, field culture, and products. The products of the coco-nut are considered under four heads: toddy, a general name referring to the sap, which is the source of drinks of various kinds and

[3] BAILEY, L. H., Plant breeding. New edition revised by ARTHUR W. GILBERT. 8vo. pp. xviii+474. *figs. 113* New York: Macmillan. 1915. $2 00.

[4] COPELAND, E. B., The coco-nut. 8vo. pp. xiv+212. *pls. 23.* London: Macmillan. 1914. 10s.

sugar; coir, a commercial name for the fiber obtained from the husk of the
coco-nut; copra, the desiccated kernel or oil-cake; and oil.—J. M. C.

Cocoa.—The important cultural plants of the tropics are gradually becoming known from the modern standpoint, and the most recent contribution in
this field is the large volume on cocoa by VAN HALL,[5] director of the institute
for plant diseases and cultures at Buitenzorg, Java.   In the brief historical
account it is stated that the first knowledge and use of the plant came to Europe
in connection with the conquest of Mexico, where Cortez found it in extensive
cultivation under the name "cacao," as he reported it.   Some conception of
the kind of scientific data included in such a study of an important cultural
plant may be obtained from some of the titles of the chapters, such as Geographical distribution and climatic conditions, Chemistry of cocoa and cocoa
soils, Botanical characteristics of the cocoa plant, Varieties of cocoa, Cultivation of cocoa, Fermentation, Diseases and enemies, Cocoa-growing countries.—
J. M. C.

Flora of California.—Part 5 of *A flora of California* by JEPSON[6] has just
appeared.   The first parts were noticed in this journal,[7] attention being called
to the general high character of the work.   This part completes the Portula-
caceae, includes the Caryophyllaceae, and begins the Ranunculaceae.   A new
genus of Caryophyllaceae is proposed (*Eremolithia*), based upon *Achyronychia
Rixfordii* Brandegee.—J. M. C.

NOTES FOR STUDENTS

The gases in the floats of marine algae.—This subject has attracted the
attention of a number of investigators during recent years.   The fact that it
is being investigated from so many points of view gives hope that we may soon
have fairly complete knowledge of the subject.   LUCAS[8] has investigated the
subject from the point of view of the origin of the gas.   The following Austral-
ian marine algae have vesicles filled with gas: Sargassaceae (*Sargassum, Car-
pophyllum, Turbinaria, Cystophora, Cystosira, Scaberja, Phyllospora*), Fucaceae
(*Hormosira*), and Laminariaceae (*Macrocystis* and possibly *Adenocystis*).   In
young plants, and even in mature fruiting individuals in sheltered situations in
shallow water, there are no vesicles.

---

[5] VAN HALL, Dr. C. J. J., Cocoa.   8vo. pp. xvi+515 *figs. 140.*   London.   Macmillan.   1914.   14s.

[6] JEPSON, W. L., A flora of California.   Royal 8vo.   Part V, pp 465–528. *figs. 14.*
San Francisco: H. S. Crocker Co. 1914.

[7] BOT. GAZ. **49**:153.   1910.

[8] Lucas, A. H. S., The gases present in the floats of certain marine algae.   Proc.
Linn. Soc. New South Wales **36**:626–631.   1912.

There seem to be three possible sources of the gas in these vesicles: (1) atmospheric air, (2) the gases dissolved in the sea water, and (3) gases produced in the metabolism of the plants themselves. He excludes (3) because he has "never detected any gases beyond nitrogen and oxygen in the floats," and "it is hardly conceivable that any process of metabolism should yield these two gases only without any carbon compounds." He excludes (1) because his analyses (table I) do not show the nitrogen and oxygen in the vesicles to be in the proportion that they are in atmospheric air (79:21), and because the gas must be formed in the cavity *pari passu* with its growth, otherwise the vesicles would collapse.

TABLE I

ANALYSIS OF GASES IN THE FLOATS OF ALGAE

Species	Percentage of nitrogen	Percentage of oxygen	Comments
Phyllospora comosa (1).	86.0	14.0	Fresh-looking plants cast up on the shore
" " (2).....	89.4	10.6	Just cast up
" " (3).....	88.9	11.1	" " "
.. " (4).....	82.3	17.7	" " "
" " (5).....	83.2	16.8	" " "
Hormosira Banksii (1).....	88.46	11.54	Growing
" " (2)... .	88.0	12.0	Same gathering kept a day longer
Cystophora monilifera.	80.0	20.0	Floating in the sea

TABLE II

ANALYSIS OF GAS OBTAINED BY BOILING IT OFF FROM WAVE-WATER FROM THE OCEAN

	(a)	(b)
Nitrogen.................	67.12	58.33
Oxygen.................	26.03	30.55
Carbon dioxide...........	6.85	11.11

He notes that in the analyses of surface waters collected by the "Challenger" the proportion of nitrogen to oxygen varied around a mean of 2:1, and that this would be expected from the relative solubility of the two gases.

He accepts (2), although his table does not show the proportion of nitrogen and oxygen to be the proportion that he has found in sea water gas (table II), and although he finds considerable carbon dioxide in sea water gas and none in the gas from vesicles. He observes that the presence of organisms seems to alter the natural ratio of the gases dissolved from the air. He believes that the gas enters, dissolved in the water, by osmosis, and that just as marine algae show a marked preference for potassium sulphate and commonly reject much of the sodium chloride, so they may show a marked preference for oxygen, and the useless nitrogen together with the oxygen not required may be eliminated and set free in the floats "in order to serve a mechanical purpose." He suggests that since the proportion of oxygen in the vesicles is less than in atmospheric air or in air dissolved in sea water, the plant must use up some of the oxygen "for other purposes than levitation." This last suggestion evidently means that the gas is modified by the metabolism of the tissues through which it passes.

In regard to (3) we shall await with interest to see whether the absence of carbon dioxide from the floats of marine algae is confirmed by other workers on other species.

On the whole, the paper is a clear, clean-cut statement of the problem and of the results of the investigation, although in a few places its scientific value might be improved by assuming a less teleological point of view.—GEORGE B. RIGG.

Cecidiology.—Among the American papers by botanists is one by STEWART[9] on the gall caused by *Andricus punctatus* Bassett. The author has made a very careful study of the histology of the gall, which he has compared with the histology of tissues formed in the healing of wounds. In his summarization on this point he says:

The following conditions in this gall can be correlated with similar conditions in traumatic tissue: (1) a recapitulation of similar conditions of ray structure; (2) a vertical shortening of the broad rays; (3) the presence of ball-formations in the wood, which appear only in tangential sections; (4) a parting of the fibers in the vicinity of the larval chambers, similar to the condition resulting from longitudinal wounds; (5) isodiametric parenchyma cells around the base of the larval chambers with irregularly distributed fibers and other woody elements among them; (6) a great reduction in the number, or an entire lack of vessels; (7) a shortening of many of the cells of the wood; (8) absence of distinct annual rings of growth; (9) a suggestion of the return of the cambium to normal activity after a time; (10) woody inclusions in the bark.

Unfortunately, the bibliography is not satisfactory in that it includes some papers which have much less bearing on the subject than certain papers which are not included. Also, some of the statements in the first part of the paper concerning our knowledge of the anatomy of galls are misleading. The reviewer also questions the statement of the title of the paper, in that the inclusion of the generic name of the gall insect would have made the title much more suggestive to those who may wish to become informed on the literature of the subject. However, the author has brought out a number of points which are frequently overlooked by students of abnormal plant growths.

Another very excellent paper is by LUTMAN[10] on club root, in which the author discusses the relation of *Plasmodiophora Brassicae* to its hosts and the

---

[9] STEWART, ALBAN, Notes on the anatomy of the *punctatus* gall. Amer. Jour. Bot. 1:531–546. *pls. 2.* 1914.

[10] LUTMAN, B. F., Studies on club root. Univ. Vermont, Bull. 175. pp. 27. 1913.

structure and growth of its plasmodium. The parasite gains entrance to the host through the epidermis or the root hairs, and thence from cell to cell either by penetrating the cell walls or by division of the host cells, each daughter cell containing a part of the parasite. The parasite is most abundant in the cortex, but is always found in the central cylinder. The infected cells hypertrophy, but there is no serious interference in nuclear or cell divisions until the disease is well advanced, when both processes are suspended. Although the author has not made a definite statement on this point, it appears from the general discussion and the drawings that the hypertrophy is most pronounced in the primary cortex. This is to be expected, but a careful study to determine the extent to which the various groups of cells are susceptible to enlargement as a result of stimulation by the parasite would be well worth while. The discussion of the character and life history of the parasite is excellent.

A brief paper by HEALD[11] gives a description and results of inoculation experiments with a gall on *Prosopis glandulosa*. The gall is probably due to *Bacterium tumefaciens* Smith and Townsend.

Among the most important contributions to cecidology by American entomologists, we find a brief paper by COCKERELL[12] on a mite gall on the red orpine (*Clementsia rhodantha* Gray), one of the Crassulaceae growing in the high altitudes of the Rocky Mountains. The flowers "were aborted to a dense, round mass of a dark crimson-like color, consisting of excessively tuberculated floral parts inhabited by an eriophid mite." The mite is similar to or possibly the same as *E. rhodiolae* Can. of *Rhodiola rosea* of Central Europe and Italy.

There is also a paper by SEARS[13] in which the author records, figures, and gives a brief description of 63 species of galls from Cedar Point, Ohio. These galls occur on 31 species of host plants representing 22 genera.

An important taxonomic paper has been published by FELT,[14] in which he lists a number of species and also describes several new species. This paper will enable the botanist to determine a large number of species which come to their attention.

WELLS[15] describes 22 unreported galls from Connecticut and figures many of them.

FULLWAY[16] gives a very interesting taxonomic paper on gall fly parasite as a result of studies on Mrs. ROSE PATERSON BLAKEMAN's collection in Uni-

[11] HEALD, F. D., Aerial galls of the mesquite   Mycologia 4:37-38. *pl. 1.* 1914.

[12] COCKERELL, T. D. A , A mite gall on *Clementsia*. Entomol. News 25:466. 1914.

[13] SEARS, PAUL B., The insect galls of Cedar Point and vicinity. Ohio Nat. 15: 377-388. *pls. 4.* 1914.

[14] FELT, E. P , Additions to the gall midge fauna of New England. Psyche 21: 109-114. 1914.

[15] WELLS, B. W., Some unreported cecidia from Connecticut. Ohio Nat. 14: 289-296. *pls. 2.* 1914.

[16] FULLWAY, D. T., Gall fly parasite from California. Jour. N Y. Ent. Soc. 20: 274-282. 1914.

versity of California. Although this paper deals primarily with taxonomic entomology, it contains many good records of American cecidia.—MEL. T. COOK.

Cytology of the Uredineae.—FROMME[17] described the processes taking place in the formation of spores of the flax rust, *Melampsora Lini* (Pers.) Desm., thereby adding another rust to those whose life histories are fully known. Interest in these forms centers largely in the aecidium. The sequence of events here is essentially like that occurring in aecidia of the *Caeoma* type, and differs only in the formation of two sterile cells above each gamete instead of one. The gametes fuse laterally in pairs, but occasionally irregularities apparently occur, since basal cells with three or more nuclei are found, however, the mode of origin of these cells was not observed. The spermatiophores also differ from the usual form. They are divided into a number of cells, from each of which a branch bearing a spermatium arises. The spermagonia and aecidia are closely associated and mature at the same time. The formation of the uredospores and teleutospores offers nothing noteworthy.

In a preliminary paper, KUNKEL[18] gave an account of the highly interesting discovery that the widely distributed blackberry rust (*Caeoma nitens*), which as a result of infection experiments reported by TRANZSCHEL and by CLINTON has been regarded as the aecidial form of *Puccinia Peckiana*, is in reality a rust of the *Endophyllum* type. This conclusion was based upon the observation that the aecidiospores of this fungus on germination produce promycelia and sporidia instead of germ tubes. The nuclear phenomena occurring in the germinating aecidiospores are described in a later paper. The processes agree in all essentials with those in the germinating aecidiospore of *Endophyllum*. At maturity the aecidiospore contains two nuclei, these fuse just before germination begins, so that the germinating aecidiospore is uninucleate. The fusion nucleus usually passes into the germ tube, where two divisions take place. Occasionally the first division occurs in the spore, as also reported by HOFFMAN[19] for *Endophyllum Sempervivi*. Five cells are formed in the promycelium; the basal one, which is continuous with the spore cavity, contains very little protoplasm and no nucleus. Sporidia arise from the other cells in the usual way. At the time of germination the sporidia are generally binucleate as a result of a division of the original nucleus. These facts, as well as some negative results of infection experiments on blackberry leaves reported by the author, leave no doubt that the orange rust of the blackberry is an independent fungus of the type of *Endophyllum*, and not, as hitherto supposed, the aecidial generation of

[17] FROMME, F. D., Sexual fusions and spore development of the flax rust. Bull. Torr. Bot. Club **39**:113-129 *pls 2*. 1912

[18] KUNKEL, L. O., The production of promycelium by the aecidiospores of *Caeoma nitens* Burrill. Bull. Torr. Bot. Club **40**:361-366. *fig 1*. 1913.

———, Nuclear behavior in the promycelium of *Caeoma nitens* Burrill and *Puccinia Peckiana* Howe. Am Jour. Bot. **1**:37-47. *pl. 1* 1914.

[19] Rev. BOT. GAZ. **54**:437 1912.

Puccinia Peckiana. The wonder is that the true nature of this exceedingly common fungus has remained so long unknown.

Although the cell rows forming the peridial wall in the aecidia of rusts are usually considered to be homologous with the spore chains, it is generally believed that the intercalary cells characteristic of the spore chains are not produced in the peridium. This belief is shown to be erroneous by KURSSA-NOW,[20] who has investigated the development of the aecidia of a number of rusts belonging to different types, and finds that a complete homology exists between the spore chains and the cell rows of the peridium. The basal cells of the peridium give rise to a series of cells each of which divides, cutting off a small cell in the lower outer corner. These small cells, which correspond to the intercalary cells, thus lie toward the outside of the rows of cells forming the peridial wall. By the growth of the peridial cells they are crowded still farther outward. Their walls soon become gelatinous and their structure is entirely obliterated, so that they are not recognizable in any but the youngest parts of the aecidium. This fate undoubtedly accounts for their remaining long unobserved. The basal cells of the peridial rows, as well as the peridial cells themselves and the "intercalary" cells, are binucleate, like the basal cells of the spore chains, in all cases except in a form of *Aecidium punctatum* Pers. (*Puccinia Pruni-spinosae* Pers.), which had uninucleate basal cells, giving rise to uninucleate aecidiospores and peridial cells. This form resembles in this respect the aecidium with uninucleate cells which was described by MOREAU[21] on *Euphorbia silvatica*. This form, which was the first aecidium having uninucleate cells discovered, has uninucleate basal cells also.—H. HASSEL-BRING.

Resistance of mosses to drought and cold.—It has long been known that many mosses are able to survive long periods of extreme drought, although much of this knowledge has been based upon inexact and insufficient data. It is a matter of much satisfaction, therefore, to have their resistance proved by the exact and extensive results of IRMSCHER[22] based upon careful experiments with a large range of species. He finds resistance to drying related in a remarkable degree to the ordinary habitat of the species, but often surprisingly great even for aquatic and mesophytic forms. Thus, while the leaves of some species of *Fontinalis* succumbed to a week's air-drying or to five days in a desiccator, *Philonotis fontana* and three species of *Hypnum* died only after 15–20 weeks in dry air or 10–18 weeks in the desiccator. The mesophytic species from deciduous forests, like *Barbula subulata, Bryum alpinum, Mnium*

---

[20] KURSSANOW, L., Über die Peridienentwicklung im Aecidium. Ber. Deutsch. Bot. Gesells. 32:317–327. *pl. 1.* 1914.

[21] MOREAU, Mrs. F., Sur l'existence d'une forme écidienne uninucléeé. Bull. Soc. Mycol. France 27:489–492. *fig. 1.* 1911.

[22] IRMSCHER, E., Über die resistenz der Laubmoose gegen Austrocknung und Kält. Jahrb. Wiss. Bot. 50:387–449. 1912.

rostratum, and *Catherinea undulata*, withstood 8–20 weeks in dry air and 5–13 weeks in the desiccator; and in species from the drier coniferous woods, like *Bartramia ityphylla*, *Dicranum fuscescens*, and *D. scoparium*, the resistance to dry air extended to 50 weeks and in the desiccator to 40 weeks. It should be noted that after all the leaves were thus killed, many of the plants were still capable of growth by fresh shoots from the more resistant stems.

The leaves of soil-growing mosses showed drought resistance in the desiccator for periods varying from 3 weeks for *Physcomitrium pyriforme*, 7 weeks for *Bryum pallens*, and 9 weeks for *Funaria hygrometrica*, to 15 weeks for *Bryum argenteum* and 35 weeks for *Pottia lanceolata*. Rock-inhabiting species included remarkably resistant forms, as *Grimmia pulvinata*, enduring 60 weeks of exposure in the desiccator, and *Schistidium apocarpum*, with one-fourth of it leaf cells still alive after 128 weeks of continuous exposure to dry air. Species from tree trunks showed resistance of the same order as those from the rocks.

Field observations show that alternate wetting and rapid drying are more detrimental in their effects than continuous drought; also, that the same species growing under different conditions varies in its drought resistance, the more hardy being that from drier habitats. Protonemata were correspondingly variable, withstanding the action of the desiccator for periods varying from 2 weeks for the cultivated protonema of *Funaria hygrometrica*, whose natural grown protonema was twice as resistant, to 14 and 15 weeks for the natural growing protonema of *Bryum argenteum* and *Catharinea undulata*. The sporophytes were found to be almost as hardy as the protonemata.

Experimental evidence from a large number of species points to great uniformity in the power of resisting cold. At $-20°$ C. the leaves of most species were killed, whether they came from the aquatic, mesophytic, or xerophytic habitats, although in a few species the resistance extended to $-30°$ C. Plants grown at a high temperature showed decidedly less resistance to sudden cold than those from cooler situations. Attention is also directed to the relation between turgor and the freezing point of the leaf cells, and to the behavior of the cells in various solutions.

The evidences of careful methods, the large number of species under experimentation, the abundance of the quantitative data, and the logical organization of the report are to be commended.—GEO. D. FULLER.

Composition and qualities of coal.—By improved methods of making thin coal sections, JEFFREY[24] claims to have obtained new and valuable results in the study of the composition of coal. In a brief discussion of the modes of accumulation he holds the view that at present both the accumulation as the result of heaping up of the remains of successive generations of plants in peat bogs, the autocthonous hypothesis, and the accumulation by drift in lakes or lagoons, the allocthonous hypothesis, commonly take place in the temperate

24 JEFFREY, EDWARD C., On the composition and qualities of coal. Econ. Geol. 9:730–742. 1915.

climates; but that only the method of accumulation by drift takes place in the tropical climates, nothing corresponding to a peat bog of the temperate climates having been found in the tropics. As the result of microscopic examination of coal he concludes that the allocthonous hypothesis also harmonizes best with the structure of coal. This is in harmony with the view of some that the climate of the coal-forming epochs approached present tropical conditions.

The canneloid coals, such as the true cannel coals, tasmanite, and boghead coals, are composed chiefly of spores, now very much crinkled and collapsed, imbedded together with some woody material in a dark ground substance. Hitherto many of these bodies had been considered gelatinous algae, but owing to improved methods can now be identified as spores. These coals have been formed under open water, representing the muck of ancient lakes or lagoons.

The ordinary bituminous coals are composed of both woody or lignitoid material and spore or canneloid matter in varying proportions. The woody or lignitoid constituent, known in descriptive terminology as glanz coal, is found in layers and has lost completely its original organization, a condition generally observed in coals derived from vegetable débris. Carbonized wood or charcoal is the only material derived from the grosser parts of plant bodies which retain structure in coal. Between the shiny woody or lignitoid layers are lodged the duller canneloid layers, known in descriptive terminology as matt coal, consisting of a dark ground substance in which are imbedded remains of flattened spores.

Coals, therefore, may be composed of three recognizable constituents: (1) spores or canneloid, (2) modified wood or lignitoid, and (3) less commonly relatively unmodified carbonized wood or charcoal. The properties of coal, he conjectures, depend to a very large degree upon the proportions of the original constituents; coals rich in spores, such as cannels, bogheads, and oil shales, are highly bituminous, and in some form or other are the mother substance of oil and gas. The spore contents of a coal determine the fatness, and in all probability have a definite relation to its coking properties; the lignitoid constituent, on the other hand, reduces the bituminosity and coking value of coal.—REINHARDT THIESSEN.

Self-sterility in Nicotiana.—EAST[24] has studied self-sterility in hybrids between *Nicotiana forgetiana* (Hort.) Sand. and *N. alata* Lk. and Otto. var. *grandiflora* Comes. The parent plants were both self-sterile, though self-fertile plants occur in at least one of the parent species. All the hybrids tested (over 500 plants of F_1, F_2, F_3, and F_4) were self-sterile. The F_1 plants, like the parent species, had 90–100 per cent of morphologically perfect pollen, except for a single plant with only 2 per cent of good pollen. Cross-pollination between individual plants of F_2, F_3, and F_4 demonstrated a high degree of cross-fertility. There was found 1.5 per cent of apparent cross-sterility in F_2, 6 per

24 EAST, E. M., The phenomenon of self-sterility. Amer Nat. 49:77–87. 1915.

cent in F_3, and 9 per cent in F_4. Back crosses with the parents resulted in about 6 per cent of cross-sterility.

Self-sterility was found to be wholly a matter of rate of growth of the pollen tubes. The pollen germinated perfectly on stigmas of the same plants, but the pollen tubes grew at the rate of about 3 mm. per day, and in no case traversed over half the distance to the ovary in the 11 days maximum life of the flowers. Growth of pollen tubes in cross-pollinated styles, on the other hand, though starting at about the same rate, was so continuously accelerated that the ovaries were reached in 4 days or less.

The simple Mendelian explanations of self-sterility proposed by CORRENS for *Cardamine pratensis* and by COMPTON for *Reseda odorata*[25] are not applicable to self-sterility in *Nicotiana*. EAST suggests that specific "Individualstoffe" of the nature of enzymes present in the pollen grains can, except in the plant that produced the pollen and in other plants of like germinal constitution, "call forth the secretion of sugar that gives the direct stimulus" to growth of the pollen tube. The hypothesis satisfies the facts presented as regards both the total self-sterility in all generations and the slight cross-sterility, which increases from generation to generation as the percentage of plants of like germinal constitution increases. It occurs to the reviewer that, if the pollen or pollen tubes have specific abilities to call forth the growth stimulus in plants of unlike germinal constitution, while the stimulus itself (the secreted sugar perhaps) is not specific, simultaneous cross and self-pollination of the same flower should result in at least partial self-fertility. Evidence derived from such pollinations would in any case be of interest.—R. A. EMERSON.

Cytology of the Mucors.—Miss KEENE[26] has given an account of the development of the zygospores of *Sporodinia grandis*. She finds at first no essential morphological difference in the two sexual branches which give rise to the zygospore. Slight differences in size are not regarded as significant. Later the branches differ somewhat in their internal structure. The protoplasm of one branch is retracted from the cell wall, the intervening space being filled with a granular substance. Sometimes there is a slight retraction of the protoplasm of the opposite branch also. The nuclei in the sexual branches are small and have the same structure as the nuclei in the rest of the mycelium. They appear to increase in number, but divisions were not observed. When the sexual branches meet, their walls coalesce in the region of contact. At this time a portion of the protoplasm in the end of each branch is delimited either by cleavage furrows which cut in from the walls, or by vacuoles which enlarge and cut through the protoplasm to the hyphal wall. In either case a central strand remains connecting the protoplasm of the suspensors with that of the gametangium. Walls cutting off the gametangium from the suspensors grow in

[25] BOT. GAZ. 57:242–245. 1914

[26] KEENE, Miss M. L., Cytological studies of the zygospores of *Sporodinia grandis*. Ann. Botany 28:455–470. *pls. 2.* 1914.

from the hyphal wall. The opening is finally completely closed, and at the center an excess of material is deposited, giving rise to a papilla-like structure described as a "canal" by LÉGER. During this process the wall between the two gametangia is resorbed. The line of contact between the two protoplasmic bodies remains distinct for some time, owing to the presence between them of the granular material mentioned above, but finally the masses fuse. Multiple nuclear fusions appear to occur at this stage. The nuclei which fail to fuse are smaller than the fusion nuclei, and soon disintegrate. No evidence of a uninucleate stage was observed. At this time numerous oil bodies, which are regarded as being of the same nature as the elaioplasts of higher plants, appear in the protoplasm. These bodies fuse until two or three large ones are formed. The large elaioplast-like bodies the author believes to be the "sphère embryonaires" of LÉGER.—H. HASSELBRING.

Proceedings of the National Academy.—This new monthly publication began to appear with the January issue of 1915. In addition to the reports and announcements that belong to it naturally, as the official organ of publication of the National Academy, it will also serve as a medium for the prompt publication of brief original papers by members of the Academy and other American investigators. The papers will be much shorter and less detailed than those published in the special journals, and the aim of the *Proceedings* is to secure promptness of publication and wide circulation of the results of American research among foreign investigators. The editorial board includes a representative from each one of the special fields of science, the editor of the BOTANICAL GAZETTE being the botanical representative on the editorial board of the *Proceedings*.

The first two numbers contain the following botanical papers: *Phoradendron*, by WILLIAM TRELEASE (Proc. Nat. Acad. 1:30–35. 1915); The morphology and relationships of *Podomitrium malaccense*, by DOUGLAS H. CAMPBELL (*ibid.*, 36, 37); and A phylogenetic study of cycads, by CHARLES J. CHAMBERLAIN (*ibid.* 86–90). In addition to these papers that are credited to the section of botany, certain papers in genetics, physiology, and chemistry come well within the scope of present botanical interest. For example, the paper by E. M. EAST, entitled An interpretation of self-sterility (*ibid* 95–100), deals with an interesting problem of genetics among plants.—J. M. C.

Evolution of the flower.—HORNE[27] has contributed a very detailed study of the structures of the flower which he regards as indicators of phylogeny. The families specially studied are the Hamamelidaceae, Caprifoliaceae, and Cornaceae, but the principles involved have general application. He includes in his discussion also the possible applications of the various theories of evolu-

---

[27] HORNE, A. S., A contribution to the study of the evolution of the flower, with special reference to the Hamamelidaceae, Caprifoliaceae, and Cornaceae. Trans. Linn. Soc. London II Bot. 8:239–309. *pls. 28–30. figs. 13.* 1914.

tion, and especially the light shed upon his problem by the work in plant genetics. Much stress is laid upon the varying characters of the ovule and its connections, beginning with the orthotropous ovule, as relatively primitive, and advancing through "anatropal advance and specialization," which latter, by the way, is said to be accompanied by the transition from two integuments to one. It is interesting to note that in the author's judgment "no phylogenetic significance can be attached to a particular form of vascular system." Applying his criteria, HORNE concludes that both Caprifoliaceae and Cornaceae are polyphyletic, and warns the authors of "systems" that "knowledge of the phylogeny of angiosperms can only be truly advanced by the detailed morphological and experimental investigation of many more families, and then, but not till then, can ENGLER's system be replaced by a greater scheme, more nearly approximating to natural relationships."—J. M. C.

Fossil plants from Kentucky.—Six genera of fossil plants from Kentucky are the subject of an intensive study by SCOTT and JEFFREY.[28] The exact level from which these plants come is somewhat in doubt, but the evidence favors the base of the Carboniferous, although the uppermost Devonian is not excluded. The fossils belong to three groups. The first of these includes the stem of one of the Cycadofilicales known as *Calamostachys* with its petiole (*Kalymma*), another petiole of the related genus *Calamopteris*, and a petiole referred to the genus *Periastron*. All of these genera have previously been known only from the Culm of Germany, where they were found and named by UNGER, and it is of much interest to note that the same flora existed on this continent, although as far as is known the species were distinct. The second group comprises two new genera: *Stereopteris*, which is apparently the petiole of a fern, and *Archaeopitys*, which presents a new and interesting type of cordaitean stem. In a third group may be placed a cone of the usual *Lepidostrobus* type. The structure is fairly well preserved in most of these fossils, and is illustrated in the 13 quarto plates which accompany the paper.—M. A. CHRYSLER.

Phylogeny of angiosperms.—In continuing their studies of this subject, SINNOTT and BAILEY[29] have investigated the evidence to be obtained from leaves. They conclude, from paleobotanical evidence, from the correlation between the palmate leaf and the multilacunar node, and from the frequency of this type of leaf in the relatively primitive groups, that the leaf of the primitive angiosperm was palmate in type and probably lobed. They

[28] SCOTT, D. H., and JEFFREY, E. C., On fossil plants showing structure, from the base of the Waverley Shale of Kentucky. Trans. Roy. Soc. London **205**:315–373 *pls. 27–39*. 1914.

[29] SINNOTT, E. W., and BAILEY, I. W., Investigations on the phylogeny of the angiosperms. 5 Foliar evidence as to the ancestry and early climatic environment of the angiosperms. Amer Jour. Bot. **2**:1–22. *pls. 1–4*. 1915.

think that the chief factor in the evolution of the now dominant pinnate type of leaf to have been the development of the petiole. Since among woody plants the more ancient multilacunar type of node predominates in temperate regions, and the palmately lobed leaf among such plants is also almost entirely confined to temperate regions, they infer that angiosperms first appeared under climatic conditions more temperate than tropical, a climate in the Mesozoic probably found only in the uplands. Furthermore, such evidence from leaves indicates that the angiosperms have come from the palmate coniferous stock, rather than from the pinnate cycadean stock; and also that the monocotyledons were derived from some ancient palmate group of the dicotyledons.—J. M. C.

Seeds of Polygonaceae.—WOODCOCK[30] has investigated the seeds of representative genera of Polygonaceae, and has reached some interesting conclusions. He finds that the outermost layer of the nucellus becomes transformed into a nutritive jacket before fertilization, and that this layer apparently carries food material from the chalazal region to the developing endosperm. He also describes the growth of the endosperm by the activity of a "cambium-like layer," which is differentiated very soon after cell formation begins, and also calls attention to the varying position of the embryo in reference to the other structures of the seed. He concludes that in the germination of certain of the seeds which were investigated from this standpoint, the aleurone layer has a digestive function, secreting a ferment which converts the starch of the endosperm into available form for translocation. In some cases also the absorbed carbohydrate is temporarily reconverted into starch in the tissues of the embryo, the cotyledons being the principal storage region.—J. M. C.

Fossil cycads.—Miss HOLDEN[31] has studied the relationship between *Cycadites* and *Pseudocycas*. The latter genus was established by NATHORST for certain cycad-like leaves formerly referred to *Cycadites*, but differing from the latter genus in having a double instead of a single midrib, and in the fact that the pinnules are not narrowed, but if anything broaden at the point of attachment to the rachis. Miss HOLDEN reaches the interesting conclusion that *Pseudocycas* belongs to Bennettitales, as judged by the character of the stomata and the epidermal cells. She further concludes that the presence of a double or single midrib is of no diagnostic importance, and that the name *Pseudocycas* should be applied only to leaves whose cuticular structure is known.—J. M. C.

[30] WOODCOCK, E. F., Observations on the development and germination of the seed in certain Polygonaceae. Am. Jour Bot. 1:454–476. *pls. 45–48.* 1914.

[31] HOLDEN, RUTH, On the relation between *Cycadites* and *Pseudocycas.* New Phytol. 13:334–340. *pl. 3. fig. 1.* 1914.

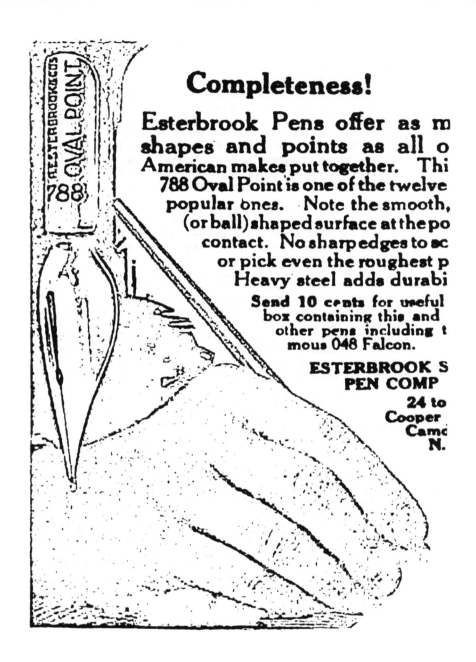
HEIMER, PH.D. and JACOB M. COHEN, A.B., LL.M.

rw Educational   *Club Director, Recreation Center 30,*
klyn   *New York*

ers and members, recreation and social
interested in social and educational work

ENTS

NICAL GA

Editor: JOHN M. COULTER

AUGUST 1915

---

and Relationships of the Araucarians.
L. Lanc

and Distribution of Eriodictyon
LeRoy Abrams and

te of Pellia epiphylla
A.

1ann Rays as an Agent for the Sterili

entous Fungi Tested for Cellulose D
Fre

VOLUME LX

NUMBER 2

THE
BOTANICAL GAZETTE

AUGUST 1915

THE ORIGIN AND RELATIONSHIPS OF THE ARAUCARIANS. II

L. LANCELOT BURLINGAME

The abietinean theory

This theory has developed more or less gradually and can best be understood by tracing the historical sequence of discovery and the ideas of relationship that have grown out of them.

1. Foremost in time and importance was the discovery that the steles of ferns, gymnosperms, and angiosperms are characterized by a leaf gap opposite the departing leaf trace. To this group was given the name Pteropsida. To the remaining groups of vascular plants the name Lycopsida was applied. This conception grew out of the investigations of the anatomy of *Equisetum* (27), the stem of angiosperms (28), and the structure and development of the stem in pteridophytes and gymnosperms (29). This distinction between these two great groups has been widely accepted by botanists and has formed one of the most fundamental objections in the minds of many (16, 53) to the lycopod theory. It has been questioned by the adherents of the latter theory (54, 61), but only in so far as to deny that a phyllosiphonic siphonostele (Araucarineae and possibly other conifers) might have arisen from a lycopod ancestry. The contention is that this type of stele is merely one of the important milestones along the evolutionary highway along which all vascular plants tend to travel. It is conceived to be in the same category as the heterosporic habit, the seed habit, and the tendency to reduce the size of the gametophytes.

It still remains true, notwithstanding that no known lycopod did actually cross the line.

2. Two years later (1904) came the statement of certain canons of evidence, some of which were well known and had already been employed by zoologists and to a certain extent by botanists. These are stated by JEFFREY (31) as follows: (1) ancestral characters that have disappeared from the vegetative axes are apt to linger in (a) reproductive axes, (b) foliar organs, (c) seedlings (ontogenetic recapitulation of the zoologists), (d) first annual ring of vigorous shoots; (2) ancestral characters may be recalled by wounding.

These canons of evidence have been consistently applied and somewhat extended in all of the subsequent work. They have been used to check conclusions derived from comparative anatomy (resemblance) and geological sequence, and in some cases practically overrule them. A really astonishing number of forms, both fossil and living, has been studied by JEFFREY and his associates in the last dozen years. Although not always without serious protest, they have been able to interpret all these forms in conformity with the general assumption that the Araucarineae have been derived from an abietineous ancestry. Much of the most important material has come from the Mesozoic of eastern North America.

A complete review of all this work is neither necessary nor profitable. Essentially the same methods have been employed in all of it. Reference will be made only to those papers in which important new facts or an advance of ideas are contained.

3. It must always be borne in mind that this school of anatomists is firmly committed to the brachyblast theory of the pine cone (19, 42, 58). From the vantage point of this conviction they extend the conception to the ovulate cones of all other conifers, and regard the spur shoot of the pines as the homologue of the assumed axillary sporangium-bearing shoot of the cones. The contention is that the spur shoot has disappeared from most modern conifers, as represented in its most primitive form in such ancient conifers as *Prepinus* and *Woodworthia*.

4. An additional canon was provided (1910) by Miss GERRY's study of the distribution of the bars of Sanio in living conifers (20). She concluded that this structure is present in the mature secondary

wood of all conifers except the Araucarineae. It has since been used by this school as the *sine qua non* in distinguishing fossil araucarian woods from those of abietinean affinities (9, 24, 25, 57).

An early application (1906) of these principles was made by JEFFREY and HOLLICK (32). Certain of the remains studied consisted of cone scales that had previously been referred to plants of such diverse relationships as *Dammara*, fossil genera belonging to Cupressineae and Taxodineae, and even to *Eucalyptus*. These scales have three basally attached and inverted seeds on their adaxial surface. There are longitudinal resinous lines on their surface. The internal structure, particularly the arrangement of the vascular supply, is very like that of *Agathis*. For these reasons they have called the plant *Protodammara*.

Closely associated with the scales were branches of *Brachyphyllum*. The authors think it probable that the branches and cone scales belong to the same plant. The branches were sectioned and referred to the Araucarineae on the ground that of the three groups (Cupressineae, Sequoiineae, and Araucarineae) which they externally resemble, only the last agrees with them in the possession of a double leaf trace, insoluble resins accompanied by mucilage, and flattened bordered pits which may rarely be alternate and biseriate. Moreover, these branches lack the alternating bands of hard bast in the phloem characteristic of all the living members of the first two groups.

The wood fragments were of two kinds. One of them agrees with the *Brachyphyllum* branches in lacking resinous tracheids and in forming traumatic resin canals. This wood is believed by the authors to be the wood of *Brachyphyllum*. The other wood has resin tracheids and does not form traumatic resin canals. The inference from these facts is that araucarians, as represented by *Brachyphyllum*, have come from ancestors with resin canals.

In the same year (1906) JEFFREY and CHRYSLER (33) described certain cretaceous Pityoxyla from the same source as the *Brachyphyllum*. These Pityoxyla appear very probably to be the wood of cretaceous pines, since they are very closely associated with typical cone scales and leaf fascicles of this genus. These pines appear to have combined the characters of hard and soft pines.

The scales and leaves resemble those of the hard pines, while the presence of abundant tangential pitting of the autumn wood is a character of soft pines. The ray cells are highly resinous, and there are no ray tracheids such as are characteristic of the Pineae. The authors point out that these are just the characters shown by the wood of the cones of hard pines. They enforce their argument from vestigial structures in the following words: "there can be little doubt that in the wood of the cones of *Pinus palustris*, for example, the general absence of marginal tracheids, the highly resinous character of the rays, and the abundant presence of tangential autumnal pits, all features of difference from the vegetative wood structure of existing hard pines, are ancestral characters, since such characters are apt to linger on in the reproductive axes. In no other way can the presence of these features in the wood of the cone be explained." They call attention to the great geological age of the pines as further support of the application of these principles. "There is good reason to believe from recent researches (33) that the genus *Pinus* in essentially its modern form, so far as the external features of the female cones go, existed as far back as the Jurassic. There is even evidence that the two great series of the hard and soft pines existed at this early period, so that the geological extension of the genus must have been much more remote."

Following up the same line of reasoning, JEFFREY (34), in a paper on wound reactions of *Brachyphyllum*, put forward the suggestion that "there is nothing inherently improbable in the derivation of the Araucarineae from an abietineous stock." He puts forward three sorts of evidence in support of this suggestion.

In the first place, he points out, the wound reactions of *Brachyphyllum* are of exactly the same character as those of *Sequoia*. In a previous investigation he was led to conclude, from the traumatic production of resin canals, taken in conjunction with their vestigial occurrence in the cone axis, first annual ring of the stem, and in the root, that resin canals were characteristic of the ancestors of *Sequoia* (30) as well as of *Abies* and certain other Abietineae (31). By combining the vestigial structures exhibited by the cones of the living araucarians with the wound reactions of *Brachyphyllum*, he is led to infer a similar ancestry for Araucarineae.

The second proof brought forward in support of this conclusion
is that the triassic *Araucarites moniliforme* (34) is reported to have
strings of flattened moniliform masses of resin in the wood. The
author thinks that such masses of resin would be produced by resin
canals larger than those of *Brachyphyllum*. Apparently there is
a reduction series in resin production in the Araucarineae. It is
abundantly secreted in the canals of the triassic *Araucarites*, less
abundantly in those of the cretaceous *Brachyphyllum* and only
when wounded, and resin canals are entirely absent in living genera.

As a third proof a still stronger claim is made again for the
antiquity of the pines. The author points to the recognized
impressions of pine leaves from the Permian onward, of hard and
soft pines after the Jurassic, and of Pityoxyla from the Carboniferous and Permian. It should be recalled that both the Permian
and Carboniferous Pityoxyla have since been rendered extremely
doubtful by the work of GOTHAN (21, 22) and of THOMSON and
ALLIN (71), though undoubted Pityoxyla are known from the late
Jurassic onward.

Araucariopitys was described (35) in 1907. The description is
based on certain leafless twigs with spirally arranged scars. They
were found in the Androvette pit (Cretaceous) in association with
"impressions of the deciduous leaf fascicles of *Czekanowskia*, a
supposed but doubtful representative of the Ginkgoales." It is
inferred, with some hesitation, that the two belong to the same
plant. It is shown that *Araucariopitys* had deciduous spur shoots
lasting, very probably, only a single year. Traumatic resin canals
were produced; the ray cells are pitted on sides and ends; the pits
are usually uniseriate, round, and remote, but may occasionally be
biseriate and alternate or opposite, in which case they are sometimes flattened. The uniseriate pits are also sometimes flattened
and in contact. It is rather difficult to credit close araucarian
affinities to this plant when one considers that it resembles a
Ginkgo externally and has the spur shoot and pitted rays of the
pines, as opposed to the slight resemblance to araucarians in the
occasional occurrence of alternate and flattened pits. The authors,
however, decide in favor of its being an araucarian on the ground of
its close association with other araucarian woods and transitional

cone scales. If this plant really is, as the authors think, an abietinean in process of transformation, it would appear to be a question whether it were headed toward *Ginkgo*, which it resembles in external features, or toward an araucarian, which it does not resemble externally and toward which it has made but a very small beginning structurally.

In 1908 *Prepinus* was described (37). The name was proposed "for this type in the belief that it is the direct ancestor of *Pinus*." "It is characterized by the possession of short shoots or brachyblasts of a generalized type, which were deciduous, but bore numerous spirally arranged instead of few verticillate fascicular leaves." "The leaves attached to the brachyblasts differed from the fascicular leaves of *Pinus* in having their paired resin canals continuous to the very base. The leaves further possessed well marked centripetal xylem. About the foliar bundles was present a complicated double sheath of transfusion tissue closely related to the centrifugal wood and resembling that found in certain of the Cordaitales." "Many of the true pines of the Cretaceous possessed the same double transfusion sheath as is found in *Prepinus*, but entirely lacked the centripetal wood which is characteristic of that genus." "The elongated pitted elements described by WORSDELL and others on the ventral side of the protoxylem in existing coniferous leaves appear rather to be the relics of the inner transfusion sheath, which is a feature of cretaceous pines, than of true centripetal xylem."

From the resemblance of the leaf structure to certain Cordaitales, the conclusion is reached that the Abietineae are "a very old, if not the oldest, family of the Coniferales." From this argument, and others already detailed, it is concluded that "the Abietineae must be considered more primitive than the Araucarineae." What at first sight appears to be a new argument in support of this contention is introduced in this paper. "The pitting of the older Araucarineae, which still survived in the Middle Cretaceous, showed a marked deviation from that found in *Agathis* and *Araucaria*, and a transition toward the type of pitting found in the Abietineae, while the oldest structurally known type of the Abietineae (*Prepinus*) shows no tendency whatever toward the araucarian type of bordered pits." There is, however, nothing new in this statement, for its entire force depends upon whether these intermediate forms are called araucarians or not. It is, of course, precisely the point at issue whether these intermediate forms are of araucarian descent on the way toward becoming Abietineae or the reverse. In fact, if this statement could be substantiated, it would completely overthrow the abietinean theory of the descent of araucarians, for this theory demands that Abietineans shall have departed in many characters, not only toward araucarians, but that this departure shall have continued until the latter were actually reached.

In 1909 SINNOTT (57) described from Second Cliff, Massachusetts, another fossil conifer, which he referred to a new genus, *Paracedroxylon*. The pits are uniseriate, remote, and round. The rays are without marginal tracheids, and the cells are thin-walled and without pits on the ends or horizontal walls. Simple pits occur on the radial walls corresponding to the half-bordered pits of the adjacent tracheids. Resin canals are normally absent, and no sure evidences were found of their traumatic production. The new genus, nevertheless, is assigned to the Araucarineae on the ground that bars of Sanio are absent. As I shall point out later, other anatomists have strongly objected to the reference of fossil woods to the Araucarineae on this ground.

In 1911 JEFFREY described the structure of the cone of *Geinitzia gracillima* (41) from the Kreischerville beds. This piece of investigation furnishes a very interesting application of the canons of evidence that have been applied in the attempt to seriate these fossil types, for it furnishes an attempt to make a comparative study of the structures of the vegetative and reproductive axes of the same fossil plant, and to apply to the results the canon of vestigial structures. The external appearance of the cones, as well as the individual scales, are very reminiscent of certain Taxodineae. The branches are thought to be *Brachyphyllum*. The structure of the cone axis is that of SINNOTT's *Paracedroxylon* (57). From these facts the conclusions follow that *Paracedroxylon* is ancestral to *Brachyoxylon;* that the evolutionary sequence must have been Abietineae (*Pityoxylon*, perhaps), *Paracedroxylon*, *Brachyoxylon*, *Araucarioxylon*, modern Araucarineae. It will be pointed out later that this scarcely accords with the strict geological sequence as known at the present writing.

Woodworthia arizonica was described in 1910 by JEFFREY as a new genus from the triassic petrified forests of Arizona (38). It agrees in all respects with *Araucarioxylon*, with the exception of possessing short shoots and the absence of persistent leaf traces. The spur shoots are thought to persist as long as the axis which bore them. The spur shoots are held to show a relationship to the pine type of conifer. The failure of the subtending leaf traces to persist indefinitely, as in living forms, is held to be an argument against this persistence being a primitive character. Notwithstanding the fact that the cretaceous *Araucariopitys* is much more abietinean in all respects and a much more modern type, the author is still disposed to cite the two as evidence of "the tendency of the Araucarineae to become more and more like the Abietineae."

In 1911 BAILEY described a cretaceous *Pityoxylon* with marginal tracheids and concluded (2) that such marginal tracheids originated in the Upper Cretaceous. In a paper published in 1913, Miss HOLDEN has extended our knowledge of the generic and geologic distribution of ray tracheids (25). She concludes from her study that (1) "ray tracheids are present normally in the Pityoxyla from the Middle Cretaceous on, and in the Abietineae"; (2) "ray tracheids are present traumatically in the Taxodineae and the Cupressineae"; (3) "on the evidence of traumatic recapitulations of ancestral characteristics, it is evident that the Taxodineae and Cupressineae are descended from the Abietineae, having sprung from that line sometime after the Middle Cretaceous"; (4) "since ray tracheids are universally absent in the Podocarpineae, Taxineae, and Araucarineae, these lines must have come off the Abietineae at some time before the Middle Cretaceous."

In 1912 JEFFREY published a very complete résumé (42) of his views and investigations. The first part of the paper deals with wood parenchyma and medullary rays. He concludes: (1) "The ancestors of *Araucaria* and *Agathis* were characterized by the possession of wood parenchyma." This conclusion rests on the facts that, though the living forms resemble the Cordaitales in the absence of wood parenchyma, it is present in the first annual rings of the root and shoot of both *Agathis* and *Araucaria*, in the early wood of seedlings, in the wood of the cones, and may be traumatically recalled in the older wood of both root and shoot. Furthermore, it is present in the more abietinean Araucarioxyla of the Cretaceous. (2) "They likewise had strongly pitted rays." This is shown by their presence in the inner portion of the cone axis of living forms and in the normal wood of those cretaceous forms (*Araucariopitys*, for example) which the author assigns to the Araucarineae. Pitted rays may also be recalled in the seedling and root by injury. (3) "The possession of these two features is quite inconsistent with their derivation from cordaitean ancestry," notwithstanding the practical identity of structure of the two groups. This argument rests partly on recapitulationary phenomena and partly on merely calling the transitional cretaceous conifers araucarians rather than abietineans, which some of them resemble far more closely.

The second part deals with "the characteristic features of the tracheids and the nature of the pitting." The conclusions are: (1) "The characteristic pitting of the wood of *Agathis* and *Araucaria*, the *Araucarioxylon* type, is not ancestral but more recently acquired." This conclusion is based on the fact that the multiseriate, flattened, and appressed pits of the mature wood of living araucarians and of Cordaitales is replaced in the inner wood of the cone and seedling axis of living genera and in the innermost wood of the stem of mesozoic forms by a type of pitting with the pits less frequently multiseriate, flattened, or appressed, but often uniseriate, remote, and round. (2) Since bars of Sanio are absent from the mature wood of living genera (see Thomson 70 for a contrary opinion) and from the wood of mesozoic Araucarioxyla, but are present in the wood of the cones, it follows that they are a feature of the ancestors of the Araucarineae. In anticipation of objections to be urged later, it may be mentioned here that the author admits their absence in the stem of the mesozoic forms, in the seedling, and probably in the leaf trace, in all of which they should be found in accordance with theoretical expectations. (3) "On the basis of comparative studies of the tracheids of the Araucarineae they cannot be regarded as primitive representatives of the coniferous order."

The third part deals with resin canals. Two reasons are alleged for thinking them to be features of the ancestors of araucarians: (1) Though they are practically absent from living araucarians, "interesting vestigial resin canals appear in the vascular supply of the lowermost abortive cone scales, attached to the peduncle of the cone, and die out before the cone scale supply leaves the wood of the peduncular axis." (2) Traumatic resin canals occur in the wood of some mesozoic woods which the author assigns to the Araucarineae because of the lack of the bars of Sanio and the possession of a modified type of pitting. The pitting in the *Araucariopitys* type, as has already been pointed out, is very little like that of araucarians and very much like that of abietineans, as are, in fact, its other characters. The *Brachyoxylon* type is rather more reminiscent of araucarian affinities but still not beyond challenge.

The fourth part treats of the foliar trace and the pith, and presents a final summing up of conclusions. In regard to the leaf traces and pith the conclusions are: (1) "This persistence of the leaf trace [that is, in mesozoic forms] seems to be a characteristic of all woods of the true *Araucarioxylon* type, and, as has been particularly indicated by THISTLETON-DYER (66) and SEWARD (54), is likewise a feature of the trunks of the living genera *Agathis* and *Araucaria*." (2) In the *Brachyoxylon* type from the Cretaceous, which is more abietinean in the rays, pitting, and in the formation of traumatic resin canals, the traces persist for a short time only. (3) In the seedling axis of *Agathis australis* the leaf trace is less persistent. (4) The leaf trace is more persistent in *Araucaria Bidwillii* than in *Agathis australis*, the former of which is assumed to be the more primitive type. (5) It follows from the preceding that persistent leaf traces are not an ancestral feature of the Araucarineae. (6) In regard to the pith I am not at all sure that I apprehend clearly JEFFREY'S position. He records the usual presence of sclerotic diaphragms in the pith of mesozoic forms, and finds them absent in the pith of the seedlings and cones of living forms. Sclerotic nests are said to occur in the latter, perhaps as a vestige of the diaphragms of the earlier forms. He says further that "it is moreover obvious that medullary diaphragms are equally characteristic of both the older Araucarineae and of the Abietineae living and fossil.   Their presence in the older araucarian types, consequently, is one more piece of evidence in favor of the derivation of the araucarian tribe from abietineous ancestors." (7) The author reiterates his belief in the compound nature of the ovulate cone and in its essential unity throughout the group. He says, "it is perfectly clear that not only in the more primitive species of the living genus *Araucaria*, but also in the cones of the mesozoic representatives, the araucarian female cone, like that of the other tribes of conifers, was originally composed of cone scales with a double system of bundles, independently emanating from the cone axis and of inverse orientation." (8) In regard to the male gametophyte he says, "certainly we should not expect to find the primitive type of pollen tube formation in a group in which the pollen no longer reaches the apex of the ovule," and "the peculiar method of germination of the pollen is an unmistakable stigma of aberration." "The contents of the pollen tube likewise vouch for the highly specialized condition of the Araucarineae.   Here the two prothallial cells common to the Abietineae and the equally ancient Ginkgoales become proliferated into a large number, doubtless in correlation with the extreme length and meandering course of the pollen tube. Moreover, the absence of a stalk cell in connection with the setting off of the body cell, which gives rise to the two sperm cells, is a clear and outstanding feature of aberrancy." These views are in a measure a modified restatement of those stated by JEFFREY and CHRYSLER in 1907 (36). JEFFREY is equally convinced that the female gametophyte is not primitive but aberrant.

In a very complete study of *Agathis* (19) EAMES has reached conclusions closely paralleling those of JEFFREY in regard to the specialization of the gametophytes and the interpretation of the structure of the ovulate cone of the araucarians. An interesting and important feature of this investigation is the very complete reduction series which he has worked out in the supposed development of the apparently simple scales of *Agathis*. Beginning with *Arthrotaxis cupressoides* with a completely double vascular supply, he traces the gradual fusion of the independent bundles through other species of the same genus, *Cunninghamia sinensis*, and finally reaches the condition found in *Agathis*. He also points out

that a complete series is exhibited in the genus *Araucaria* with
A. Bidwillii standing as the most primitive and *A. brasiliensis* and
A. imbricata as the most specialized. In preceding papers of this
series the reviewer has discussed EAMES's views in relation to the
gametophytes (6) and the embryo (7).

Although SINNOTT (59) concludes from his study of the podocarps that they have been derived from the Abietineae directly,
and not through the Araucarineae, his view is incidentally interesting in that it points out that on this assumption the podocarps
become the primitive members of the group instead of the *Saxego-
thaea-Microcachrys* forms. In his diagram it would appear that he
thinks this whole assemblage derived from the ancient Araucarineae,
and that they had arisen by an approximately equal split of some
ancient Abietineae, the other arm being the modern abietineans.
From the text, however, it appears that "the close series of forms
from *Podocarpus* to *Saxegothaea* is very suggestive as offering a
key to the evolutionary development of the modern Araucarineae."
The argument turns on the interpretation of the vascular supply in
these forms as a reduction series and the epimatium as the equivalent of a reduced ovuliferous scale. He calls attention to the
already well known gametophytic resemblances, which his own
studies have rendered more apparent, as evidence of a relationship
between podocarps and Abietineae. In like manner he minimizes
the points of difference. He recognizes that his series can be read
in the other direction, and calls attention to the necessity in that
case of recognizing and accounting for what would be numerous
parallel developments in the two lines.

In the concluding section of his paper on the *Araucarioxylon*
type (42) JEFFREY sums up the conclusions for the whole theory as
follows: (1) "The Araucarineae cannot have been derived from the
Cordaitales since they possessed primitively a number of features
which, so far as our knowledge goes, never existed in the cordaitean
stock." (2) "The *Araucarioxylon* type is derived from ancestral
forms which possessed opposite pitting, bars of Sanio, strongly
pitted rays, and horizontal and vertical resin canals." (3) "The
primitive existence of these features in the ancestral type from which
Araucarioxylon has been derived shows clearly that it has taken its

origin from the abietineous *Pityoxylon* type." (4) "This conclusion is entirely confirmed by a consideration of the reproductive
structures, both sporophytic and gametophytic." (5) "Any
hypothesis as to the origin of the Coniferales in general must start
with the Abietineae as the most primitive tribe."

This theory has received from time to time certain incidental
criticism in connection with the work of investigators who have
considered that their results justify other interpretations. A
number of these have already been mentioned in the presentation
of the lycopod and cordaitean theory. It is in the nature of things
that the facts which form the support of one theory are usually the
facts that refuse to adjust themselves easily to others. Objections
of this sort have found their proper place in the preceding paragraphs. I shall now mention some of the more specific objections
that have been made.

It has already been pointed out that the protagonists of the
lycopod theory hold the ovulate cone of the araucarians to be simple.
The abietinean theory is circumscribed by the necessity of proving
it compound. The weight of opinion, at least so far as numbers go,
among those who have investigated the subject appears very
decidedly to favor the idea that the ovulate structures of podocarps
and araucarians are homologous in structure and simple. It
appears from the work of EAMES and SINNOTT, already quoted
above, that if the abietinean theory prevails they can be explained
as a reduction series. On the other hand, if this theory were not in
question, it appears that most investigators would decide in favor
of simplicity of structure. Aside from the authors already mentioned, TISON (76) and NOREN (44) have expressed themselves
in favor of a simple explanation.

The writer has in earlier papers called attention to the inade-
quacy of the explanation offered by JEFFREY and CHRYSLER (36) of
the more numerous prothallial cells in the gametophytes of podocarps and araucarians. These authors suggested that the greater
number of these cells might be a coenogenetic adaptation to the
extensive pollen tube. Aside from the reasons for thinking that the
tube itself has not undergone any such coenogenetic development
as this theory suggests (by implication) for the araucarians, it is

entirely inapplicable to the podocarps and *Abies* (26), which has recently been shown to form regularly a considerable proportion of pollen grains with 3 or 4 prothallial cells. Either these widely separated cases are to be explained as a heritage from more or less remote ancestors, or as remarkable examples of the revival of abandoned structures, or as the still more remarkable origination of apparently useless structures. In JEFFREY's paper on the *Araucarioxylon* type (42) he speaks of the pollen grain and male gametophyte as clearly aberrant in its germination, prothallial cells, and absence of a stalk cell. Whether it is aberrant or not is doubtless somewhat a matter of opinion. That a stalk cell is not formed is an error so far as the statement concerns *Araucaria brasiliensis*. I have figured in a previous paper the division which results in stalk and body cells (5). *Araucaria* resembles *Podocarpus* (4) exactly in respect to the manner of this division. The axis of the spindle is transverse in both cases and the resulting cells lie side by side above the prothallial cells. Because this division occurs late in the development of the male gametophyte, the cell wall and cell identity of the stalk cell are soon lost in both genera. At the time of shedding, only the body cell retains its cell identity, the other nuclei being free in the common cytoplasm and often indistinguishable from one another.

The reference of some of the mesozoic fossils to Araucarineae has met with rather severe criticism. JEFFREY's reference (40) of *Yezonia* and *Cryptomeriopsis* (62) to *Brachyphyllum* and *Geinitzia* respectively has met with opposition from their authors (63). Dr. STOPES, notwithstanding the dissimilarity of the cones, is inclined to agree that there is a considerable structural resemblance between *Brachyphyllum* and *Yezonia*. She dissents entirely from the opinion that *Cryptomeriopsis* is an araucarian. She has not stated an opinion as to whether it is or is not identical with *Geinitzia* as described by JEFFREY (41). She is emphatic, however, in thinking that it differs very little from the modern *Cryptomeria*.

STILES (61) has also criticized the reference of *Geinitzia* (41) and *Paracedroxylon* (57) to the Araucarineae. He is particularly severe on the use of the bars of Sanio as a final criterion of relationship. The soundness of this criticism has since been emphasized by

THOMSON's discovery of these bars of Sanio in the ordinary wood of araucarians (70). They have also recently been reported in the cycads (56). In this group they are said to occur when the pits are scattered, but not when alternate and crowded.

THOMSON (70) has suggested that the araucarian affinities of some of the mesozoic transitional forms with traumatic resin canals would be equally well explained as having descended from the cordaiteans as by the assumption that they are acquiring araucarian characters. If I understand this suggestion correctly, it assumes that these forms are not directly related to araucarians at all, but are really abietineans that still retain some cordaitean characters and have acquired or are in the process of acquiring the characters of modern abietineans.

On the principles of evidence

After having set forth the evidence that has been adduced by various writers in support of the several theories, it is now pertinent to return to the problem originally proposed by the quotation from JEFFREY. Are there any general principles of evidence or are there not? Are all sorts of evidence of equal value? Shall any class of evidence be excluded, as is done in our law courts? Such questions as these must be answered by every botanist before he can properly proceed to sound inferences from the facts uncovered by his investigation. The arguments set forth in the preceding sections may be conveniently grouped under the following heads: (1) resemblance or likeness, (2) geological sequence, (3) vestigial structures, (4) ontogenetic recapitulations, (5) traumatic reversions, (6) abnormalities or monstrosities.

1. *Resemblance or likeness means relationship.*—This appears to be the most fundamental principle in the minds of the great majority of writers on this and other phylogenetic problems. That this is a sound principle is unquestioned. No fact in our biological experience is better grounded than that "like begets like." That the parent and child may differ in minor points is an everyday experience, but that they ever differ by large differences is not believed. The theory of evolution itself is founded firmly on these two well known facts of general likeness with slight variations.

From this it follows that affinity is roughly proportional to resemblance. Resemblance in one or a few points may mean a slight degree of affinity, or it may merely show a case of parallel development. Heterospory and the seed habit are excellent illustrations of the latter. If, on the contrary, two plants are very similar in all of their organs, the resemblance is usually considered unimpeachable evidence of close affinity. The classification of all our living plants is almost exclusively based on this principle.

A corollary of this principle is that neither of two supposedly related plants must possess any organ or structure which cannot be reasonably derived from the homologous organ or structure of the assumed ancestor.

2. *The geological sequence* should roughly conform to the proposed evolutionary sequence in the development of a modern group from an ancient one. It is obvious that if all the intermediate forms have been fossilized and all discovered, this agreement would be exact and complete. Such conditions doubtless never occur. Since evolution of related forms cannot be supposed to run exactly parallel in different lines, it follows that the discovered fossils from any given horizon might be expected to show one structure or organ ahead in one and another in another. An important corollary of this has frequently been insisted on by COULTER (14, 15, 16). One line of plants may run ahead along a certain line and remain practically stationary for ages in some other. For these reasons and because the fossil record is always very incomplete, inferences from geological sequence must always be subject to considerable doubt.

3. *Vestigial structures.*—By this is meant that anatomical characters that once were general through the entire plant are likely to be retained in certain supposedly primitive regions of the plant, such as the root, cotyledons, cone axes, and leaves. The use of this principle is attended with very considerable difficulties and may frequently lead to very erroneous conclusions. The difficulties lie in two assumptions that must be made in its application. First, we must assume that the stem structure of a paleozoic (let us say) plant was also present in the cone axis. Then when we find this same structure in the cone axis of a modern plant, we must again assume

that this structure is a retained one and not a newly acquired one. If we could avoid the first assumption (that is, if we knew the structure of the ancient cone), we should not need this principle, but could apply the principle of resemblance. Neither is it always easy to decide the correctness of the second assumption. Its limitations, therefore, are clear. It is useful in enabling us to infer a likeness which we do not know actually to have existed. It is, consequently, of much lower value than a direct comparison of known structures. Its highest possible value would equal that of a direct comparison between the homologous parts of the two plants, while its lowest value is actually zero.

4. *Ontogenetic recapitulations.*—This principle assumes that there will be formed in the juvenile stages of a plant or animal organs or structures that were characteristic of the adult ancestral forms. In the form I have stated it, this principle is almost certainly invalid. This is the form in which it is commonly applied. What is probably true is that related animals and plants resemble one another and their common ancestor at all stages from the egg to the adult, inclusive, in all those organs and structures which have neither been lost since the separation 'from the parent stock nor added to either of the descendants. This principle is not infrequently applied in such a manner as to deprive the conclusions of any real validity whatever. In so far as it possesses validity at all, it owes it to a direct comparison of homologous structures in the same stage of development.

5. *Traumatic reactions.*—When a plant or animal is wounded it not infrequently reacts by forming organs or structures that differ from those usually formed. In some cases these structures are such as are thought to be identical with those of its ancestors. There is no a priori reason why they should be reversions, so far as I can see, unless the original structures were introduced into the sum of the hereditary qualities through wounding in the first place. Pruning a grape vine or a fruit tree usually induces a yield of larger fruit or even a greater total quantity. Increased physiological activity is a very common result of wounding, but it does not therefore follow that this is an ancestral quality of the stock. Some of these responses may represent ancestral conditions, but it seems to

me to be very unsafe to infer that any particular one is so, without independent proof thereof.

6. *Monstrosities or abnormalities.*—So far as concerns inherited abnormalities, it must be true that in the long run the progressive changes must have much exceeded the reversions, else evolution from the simpler ancestors to the more complex descendants of today could not have occurred. This argument cannot be applied in the same way to non-heritable abnormalities, though there is no obvious reason why they should follow any different law of probability. By themselves, abnormalities afford evidence of little weight, since it is impossible to say whether they represent reversions or other chance variations.

Conclusions

We may now attempt to apply these criteria of evidence to the arguments that have been offered by the various supporters of each of the theories. So far as concerns the lycopod theory and the cordaitean theory, it is readily seen that each of them is founded on certain more or less striking resemblances. The evidence, then, is valid so far as the principles are concerned. The weakness of each theory lies in the necessity of certain more or less plausible explanations that must be accepted before the resemblances are evident. The lycopod theory must explain away the very apparent difference in the stelar structures of the two groups and must show how the pine cone has been evolved from the simple cone of a club moss. Without repeating what has already been set forth in other parts of this paper, let it suffice to recall the very large number of points of difference that must be explained away and the comparatively few points of likeness relied on to establish a relationship.

On the contrary, the points of resemblance between araucarians and Cordaitales are numerous and striking. The points of difference are few, and for the most part more easily explained away than those that confront the preceding theory. The geological record appears also to favor the cordaitean theory, for none of the fossil forms are known to approach lycopods in any character more closely than do the modern forms. In fact, it would appear from our present knowledge that the fossil forms were less like them than the

modern ones.   This would strongly indicate parallel development
of similar structures in the two groups.   On the evidence, then, as
it stands, we must decide that the cordaitean theory is much the
more probable.

In respect to the abietinean theory we meet a different state of
affairs.   It is not claimed that Abietineae are more like cordaiteans
than araucarians, but that it can be shown that the likenesses of the
latter have been secondarily acquired.   This is not parallel development.   The theory assumes that a considerable number of characters underwent extensive modification during the evolution of
the primitive conifers (that is, abietineans).   Multiseriate bordered
pits of the cordaiteans became uniseriate, remote, and rounded.
The thin-walled unpitted ray cells became thick-walled and pitted.
Resin canals were evolved.   Then this primitive stock is conceived
to have split into two lines, one of which continued its evolution
along the same lines as the parent stock.   The other line (araucarians) faced about and began the reacquisition of the characters
that had been lost.   It almost completely regained the original type
of pitting, lost all trace of its pitted rays, and almost totally lost
the ability to produce resin canals in the wood.   The history of
other characters is much the same.   The adherents of this theory
do not seek to deny these resemblances nor the necessity of showing
that they have been secondarily acquired.   I have already set forth
the evidence through which they believe that they have proved
this astounding evolutionary sequence.   So far as resemblance
or likeness goes, the cordaitean theory is far and away the more
probable.

On the basis of fossil history there is not much to choose in
respect to antiquity of the two families, though the Araucarineae
have, perhaps, at present the more certain record in the older
rocks.   Still, specialized abietineans are known so far back that
we must assume their origin to have been very much farther back.
I shall speak of the bearing of the transitional fossils of the Mesozoic
after discussing the remaining canons of evidence, because it is
only in the light of the inferences made in accord with them that
these fossils can be made to support this theory.   On the basis
of geological sequence they favor the cordaitean theory, inasmuch

as the older forms are more araucarian and the more recent ones more like modern Pinaceae, particularly certain Taxodineae and Sequoiineae.

Though the argument from resemblance and geological sequence is unfavorable to this theory, there are, nevertheless, a great many known facts that can be best explained in accordance with it. The evidence derived from a study of vestigial structures, recapitulations, traumatic reversions, and monstrosities largely favors this theory, though many facts are known which appear to be inconsistent with it.

The pitting of the tracheids in the ovulate cones has been interpreted by THOMSON and JEFFREY in exactly opposite ways. The former calls attention to the multiseriate (3–5 rows) cordaitean pits of the older wood, and the latter to the uniseriate wood of the first few tracheids. To make matters worse, the former calls attention to araucarian pitting in abietinean cones and roots. This could, of course, be explained as a heritage from the cordaiteans.

The argument from recapitulation is hardly more fortunate. The seedling pine lacks spur shoots, just as does *Araucaria*, and hence spur shoots are not ancestral; but on the other hand the seedling *Araucaria* has abietinean pitting in the earlier annual rings.

Similarly, wounding an araucarian produces no resin canals, though it should if this theory be true; though wounding *Brachyphyllum* did. In other cases it does recall or induce abietinean characters.

The use that has been made of the bars of Sanio appears to the reviewer to fall in a class by itself. From the time of LINNAEUS' classification of flowering plants down to the present many such artificial distinctions have been proposed. They have usually been short-lived. Inasmuch as a bar of Sanio must have been at some time acquired, one would suppose that if you traced the ancestral line backward it would gradually fade out. In that case there would certainly be Abietineae somewhere along the line that lacked this structure. THOMPSON's discovery of its presence in mature secondary wood of modern araucarians renders its use of very doubtful value. It seems very unlikely that its mere absence from a mesozoic form otherwise unlike an araucarian is a character of sufficient importance to justify its inclusion in the Araucarineae.

The argument from abnormalities as applied to this theory is subject to attack on the historical side as well as on the ground of inherent probability. That a given abnormality represents always a reversion is an assumption that no one seriously maintains. For example, six toes are not uncommon in mankind. No one believes that this is an ancestral character any more than brachydactyly, where there are fewer parts than usual. An extra digit in hoofed animals is, on the contrary, usually looked on as a reappearance of an ancestral condition, because this condition is believed on other and trustworthy grounds actually to have occurred in this evolutionary line. In respect to the known conditions of the cones that may be supposed to be ancestral to modern pines, there is not a scintilla of evidence that they were any nearer the brachyblastic condition than their modern representatives. As I have already pointed out, the case is still more difficult in regard to the Cordaitales, the supposed remote ancestors, where the condition should, theoretically, be well developed. The strength of this theory is that it explains the pine cone, and its weakness is that it makes it necessary to apply the same explanation to other cones and shoots where it is very much less satisfying.

To sum up, the argument for the abietinean theory, therefore, is seen to be of a less convincing kind. Moreover, there are other arguments of the same kind that favor the cordaitean theory.

It seems clear to the writer, also, that there are well defined rules of evidence in the investigation of phylogenetic problems, but that the conclusions attained through their application have far less certainty than a chemical analysis or a mathematical prediction of a comet's course. With care the latter attains a high degree of certainty. The degree of probability in phylogenetic inquiries is more nearly that pertaining to the verdict of juries in our law courts. They are both always subject to attack by the introduction of new evidence.

Summary .

1. The science of phylogeny possesses fairly adequate and reasonably trustworthy rules of evidence.

2. The degree of relationship is most clearly indicated by a detailed and accurate comparison of all the structures of the plant in all the stages of development, and is roughly proportional to the number and exactness of the resemblances.

3. Conclusions derived from direct comparisons should be checked carefully by the geological record.

4. Direct comparisons may be supplemented by indirect comparisons instituted through the use of more or less valid conclusions derived from the presence of supposed vestigial structures in primitive regions and from recapitulationary phenomena. Such indirect comparisons afford much less certain conclusions.

5. Reversions to ancestral conditions may sometimes occur under normal conditions or be experimentally produced by wounding or unusual conditions of growth. Conclusions based on evidence of this sort have little weight unless supported by other more reliable sorts of evidence.

6. Gymnosperms as a group resemble one another much more closely in very many ways than any one of them resembles any other group. They are, therefore, monophyletic. Since the cycadophytes are almost certainly derived from a filicinean ancestry, it follows that all are ultimately traceable to the same source.

7. The conifers closely resemble Cordaitales and are probably derived from them.

8. Araucarineae resemble the Cordaitales far more closely than do any other conifers, and are probably derived directly from them. This conclusion is consistent with the geological record.

9. The transitional conifers of the Mesozoic are either araucarians or cordaiteans well on their way toward Pinaceae. Some of them may be actually ancestral to such Taxodineae as *Cryptomeria* and *Sequoia*.

10. The Abietineae are very old and are derived either directly from the Cordaitales or from the very ancient members of the Araucarineae.

STANFORD UNIVERSITY
CALIFORNIA

LITERATURE CITED

1. BAILEY, IRVING W., The structure of the wood in the Pineae. BOT. GAZ. 48:47–55. *pl. 5.* 1909.

2. ———, A cretaceous *Pityoxylon* with marginal tracheids. Ann. Botany 25:315–325. 1911.

3. BENSON, MARGARET, *Miadesmia membranacea* Bertrand; a new paleozoic lycopod with seedlike structure. Phil. Trans. Roy. Soc. London B **199**: 409–425. *pls. 33–37.* 1908.

4. BURLINGAME, L. L., The staminate cone and male gametophyte of *Podocarpus*. BOT. GAZ. **46**:161–178. *pls. 8, 9.* 1908.

5. ———, The morphology of *Araucaria brasiliensis.* I. The staminate cone and male gametophyte. BOT. GAZ. **55**:97–114. *pls. 4, 5.* 1913.

6. ———, *idem.* II. The ovulate cone and female gametophyte. BOT. GAZ. **57**:490–508. *pls. 25–27.* 1914.

7. ———, *idem.* III. Fertilization, the embryo, and the seed. BOT. GAZ. **59**:1–39. *pls. 1–3.* 1915.

8. CAMPBELL, D. H., Mosses and ferns. New York. 1905.

9. ———, Plant life and evolution. New York. 1911.

10. CHRYSLER, M. A. (with JEFFREY), The microgametophyte of the Podocarpineae. Amer. Nat. **41**:355–364. 1907.

11. ———, (with JEFFREY), On Cretaceous Pityoxyla. BOT. GAZ. **42**:1–15. *pls. 1, 2.* 1906.

12. ———, Tyloses in tracheids of conifers. New Phytol. **7**:198–294. *pl. 5.* 1908.

13. ———, The origin of the erect cells in the phloem of the Abietineae. BOT. GAZ. **56**:36–50. 1913.

14. COULTER, JOHN M., Recent advances in the study of vascular anatomy. Amer. Nat. **43**:219–230. 1909.

15. ———, Evolutionary tendencies among gymnosperms. BOT. GAZ. **48**: 81–97. 1909.

16. ——— (with CHAMBERLAIN), Morphology of gymnosperms. Chicago. 1910.

17. ———, History of gymnosperms. Pop. Sci. Monthly. February 1912.

18. ———, Phylogeny and taxonomy. Amer. Nat. **46**:215–225. 1912.

19. EAMES, ARTHUR J., The morphology of *Agathis australis*. Ann. Botany **27**:1–38. *pls. 1–4.* 1913.

20. GERRY, ELOISE, The distribution of the bars of Sanio in the Coniferales. Ann. Botany **24**:119–124. *pl. 13.* 1910.

21. GOTHAN, W., Über die Wandlungen der Hoftüpfelung bei den Gymnospermen, im Laufe der geologischen Epochen und ihre physiologische Bedeutung. Sitzungs. Gesells. Naturf. Freunde Berlin. no. 2. 1907.

22. ———, Die fossiler Holzresten von Spitsburgen. Kungl. Svensk. Vetensk. Handl. **45**: no. 8. 1910.

23. HOLDEN, RUTH, Ray tracheids in the Coniferales. BOT. GAZ. 55:56–65. *pls. 1, 2.* 1913.

24. ———, Permian and triassic fossils from New Brunswick and Prince Edward Island. Ann. Botany 27:243–255. 1913.

25. ———, Coniferous jurassic woods from Yorkshire. Ann. Botany 27:535–545. 1913.

26. HUTCHINSON, A. H., The male gametophyte of *Abies*. BOT. GAZ. 57:148–153. 1914.

27. JEFFREY, EDWARD C., The development, structure, and affinities of the genus *Equisetum*. Mem. Boston Soc. Nat. Hist. 5:155–190. *pls. 26–30.* 1899.

28. ———, The morphology of the central cylinder in the angiosperms. Trans. Canad. Inst. 6:1–40. *pls. 7–11.* 1900.

29. ———, The structure and development of the stem in Pteridophyta and Gymnosperms. Phil. Trans. Roy. Soc. London B 195:119–146. *pls. 1–6.* 1902.

30. ———, The comparative anatomy and phylogeny of the Coniferales. I. The genus *Sequoia*. Mem. Boston Soc. Nat. Hist. 5:441–459. *pls. 68–71.* 1903.

31. ———, *idem.* II. The Abietineae. Mem. Boston Soc. Nat. Hist. 6:1–37. *pls. 1–7.* 1904.

32. ——— (with HOLLICK), Affinities of certain cretaceous plant remains commonly referred to the genera *Dammara* and *Brachyphyllum*. Amer. Nat. 40:189–216. *pls. 1–5.* 1906.

33. ——— (with CHRYSLER), On cretaceous Pityoxyla. BOT. GAZ. 42:1–15. *pls. 1, 2.* 1906.

34. ———, The wound reactions of *Brachyphyllum*. Ann. Botany 20:383–394. *pls. 27, 28.* 1906.

35. ———, *Araucariopitys*, a new genus of araucarians. BOT. GAZ. 44:435–444. *pls. 27–30.* 1907.

36. ——— (with CHRYSLER), The microgametophyte of the Podocarpineae. Amer. Nat. 41:355–364. 1907.

37. ———, On the structure of the leaf in cretaceous pines. Ann. Botany 22:207–220. *pls. 13, 14.* 1908.

38. ———, A new araucarian genus from the Triassic. Proc. Boston Soc. Nat. Hist. 34:325–332. *pls. 31, 32.* 1910.

39. ———, A new *Prepinus* from Martha's Vineyard. Proc. Boston Soc. Nat. Hist. 34:333–338. *pl. 33.* 1910.

40. ———, On the affinities of the genus *Yezonia*. Ann. Botany 32:767–773. *pl. 65.* 1910.

41. ———, The affinities of *Geinitzia gracillima*. BOT. GAZ. 51:21–27. *pl. 8.* 1911.

42. ———, The history, comparative anatomy, and evolution of the *Araucarioxylon* type. Proc. Amer. Acad. 48:531–571. *pls. 1–8.* 1912.

43. LLOYD, FRANCIS E., Morphological instability, especially in *Pinus radiata*. BOT. GAZ. 57:314–319. *pl. 14.* 1914.

44. NOREN, C. O., Zur Kenntniss der Entwicklung von *Saxegothaea conspicua*. Svensk. Bot. Tidsk. 2:101–122. *pls. 7–9.* 1908.

45. OLIVER, F. W., On the origin of gymnosperms (at the Linnaean Society). New Phytol. 5:68–70, 147. 1906.

46. PENHALLOW, D. P., North American species of *Dadoxylon*. Proc. Roy. Soc. Canada 6:51–79. 1900.

47. POTONIÉ, H., Lehrbuch der Pflanzenpalaeontologie. Berlin. 1899.

48. RENAULT, B., Structure comparée de quelques tiges de la flore carbonifère. Nouv. Arch. Mus. II. 2:213–348. *pl. 8.* 1879.

49. SCOTT, D. H., On *Spencerites*, a new genus of lycopodiaceous cone. Phil. Trans. Roy. Soc. B 189:83–106. *pls. 12–15.* 1897.

50. ———, On *Cheirostrobus*, a new type of fossil cone from the Lower Carboniferous strata. Phil. Trans. Roy. Soc. B 189:1–34. *pls. 1–6.* 1897.

51. ———, The seedlike fructification of *Lepidocarpon*. Phil. Trans. Roy. Soc. B 194:291–333. *pls. 38–43.* 1901.

52. ———, On the origin of gymnosperms (at the Linnaean Society). New Phytol. 5:141–145. 1906.

53. ———, Studies in fossil botany. London. 1909.

54. SEWARD, A. C., and FORD, SYBILLE, The Araucariae recent and extinct. Phil. Trans. Roy. Soc. B 198:305–411. *pls. 23, 24.* 1905.

55. SEWARD, A. C., On the origin of gymnosperms (at the Linnaean Society). New Phytol. 5:73–76. 1906.

56. SIFTON, H. B., Cycad pitting. Science N.S. 39:261. 1914.

57. SINNOTT, EDMUND W., *Paracedroxylon*, a new type of araucarian wood. Rhodora 11:165–173. *pls. 80, 81.* 1909.

58. ———, Some features of the anatomy of the foliar bundle. BOT. GAZ. 51:258–271. *pl. 17.* 1911.

59. ———, The morphology of the reproductive structures in the Podocarpineae. Ann. Botany 27:39–82. *pls. 5–8.* 1913.

60. STILES, WALTER, The anatomy of *Saxegothaea conspicua*. New Phytol. 7:209–222. 1908.

61. ———, The Podocarpeae. Ann. Botany 26:443–514. *pls. 46–48.* 1912.

62. STOPES, MARIE C. (with FUJII), Studies on the structure and affinities of cretaceous plants. Phil. Trans. Roy. Soc. B 201:1–90. *pls. 1–9.* 1910.

63. ———, A reply to Professor JEFFREY's article on *Yezonia* and *Cryptomeriopsis*. Ann. Botany 25:269. 1911.

64. ———, A new *Araucarioxylon* from New Zealand. Ann. Botany 28:341–350. *pl. 20.* 1914.

65. STRASBURGER, ED., Histologische Beiträge III. 1891.

66. THISTLETON-DYER, SIR W. T., Persistence of leaf traces in Araucarineae. Ann. Botany 15:547–548. 1901.

67. THOMSON, ROBERT BOYD, On the origin of gymnosperms (at the Linnaean Society). New Phytol. **5**:144-145. 1906.

68. ———, On the pollen of *Microcachrys tetragona*. BOT. GAZ. **47**:27-29. *pls. 1-2.* 1909.

69. ———, The megasporophyll of *Saxegothaea* and *Microcachrys*. BOT. GAZ. **47**:345-354. *pls. 22-25.* 1909.

70. ———, On the comparative anatomy and affinities of the Araucarineae. Phil. Trans. Roy. Soc. B **204**:1-50. *pls. 1-7.* 1912.

71. ——— (with ALLIN), Do the Abietineae extend to the Carboniferous? BOT. GAZ. **53**:339-344. *pl. 26.* 1912.

72. ———, On the origin of the resin tissue of the conifers. Science N.S. **39**:261. 1914.

73. ———, The spur shoot of the pines. BOT. GAZ. **57**:362-384. *pls. 20-23.* 1914.

74. ———, Reviews of Miss HOLDEN's papers. BOT. GAZ. **57**:80-83. 1914.

75. THOMPSON, W. P., The origin of ray tracheids in the Coniferae. BOT. GAZ. **50**:101-116. 1910.

76. TISON, A., Sur le *Saxegothaea conspicua*. Caen. 1909.

77. WORSDELL, W. C., On the origin of gymnosperms (at the Linnaean Society). New Phytol. **5**:146-147. 1906.

78. ———, The structure of the female flower in Coniferae. Ann. Botany **14**:39-82. 1900.

TAXONOMY AND DISTRIBUTION OF ERIODICTYON

LeRoy Abrams and Frank J. Smiley

(WITH THREE FIGURES)

In 1905, Chancellor Jordan (7), in reviving the isolation theory, presented the following general law: "Given any species in any region, the nearest related species is not likely to be found in the same region nor in a remote region, but in a neighboring district separated from the first by a barrier of some sort."

Botanists were not agreed as to the applicability of the law to plants, and Lloyd (8) even asserted that it "would be more in harmony with the facts in the case as understood by the botanists if stated in the converse form." At the time we were inclined to accept the law and offered (1) a number of illustrations in support. But it was obvious that very few data on the distribution of plants in this country were available. Few attempts had been made to map accurately the distribution of closely related species. Distributional notes in the manuals were of such a general nature that they were largely useless for such a study, and many of the herbarium specimens were but little better, the average collector's data being far too meager and often confusing. The need for careful distributional studies of closely related species seemed imperative, and with this in mind we have been collecting data on a number of Pacific Coast genera.

Eriodictyon was selected as one of the genera for these distributional studies for various reasons. The species constitute a single clearly defined natural group instead of an aggregate of groups, as is often the case with the larger genera; they are all evergreen shrubs readily detected in the field at all times of the year; and, finally, their center of distribution is in southern California, the region with which we are most familiar.

As originally recognized by Bentham (2) and Gray (4) and as recently delimited by Brand (3), the genus *Eriodictyon* comprises a small natural group of sclerophyllous shrubs peculiar to California and the Southwest, where they are commonly called

"Yerba Santa." All the species are essentially Upper Sonoran, but range in their zonal distribution from the upper edge of the Lower Sonoran to the Transition.

Propagation may be by seed or, at least in *E. californicum*, it may be by suckers; whole colonies or thickets are often connected by lateral roots that lie only three or four inches below the surface of the ground. In the xerophytic habitat where these plants usually grow, ability to propagate in this manner is an effective factor, for seedlings have difficulty withstanding the long dry seasons.

The leaves present striking examples of adaptations to xerophytic conditions, which fall into two categories. One type, represented by *E. californicum*, has the upper surface covered by sessile glands that give a smooth varnished appearance, and the lower surface conspicuously reticulate-veined with the stomatic surface in the sunken meshes clothed more or less densely with a short grayish tomentum. The palisade tissue is also prominent, comprising several tiers of cells. Other xerophytic adaptations in this type are mainly in the reduction of leaf surface; in *E. californicum*, an inhabitant of central and northern California, the normal leaf is flat and 10–15 mm. wide, while in *E. angustifolium*, of the Arizona mountains, where xerophytic conditions are more severe, the leaves are revolute and only 2–5 mm. wide.

In the other foliage type, represented by *E. crassifolium*, both surfaces of the leaves are clothed with thick-walled unicellular trichomes that form a dense mat over the entire surfaces. In *E. tomentosum* these trichomes stand out at right angles to the surface for a short distance, then turn sharply and interlace, forming a dense felt just above the surface; by this arrangement a sort of "ramada" is produced which shuts off the sun's rays and the hot winds, but allows slow diffusion. The palisade layer is much less developed than in the glutinous type, comprising a single tier of cells instead of several tiers. The leaves are also broader and never revolute.

It is evident that the members of the genus are adapted to xerophytic conditions, and that two diverse methods of meeting these conditions have been evolved. That the two types have sprung from a common stock is also evident, for, as we shall show, there are forms still existing that comprise an almost unbroken series from the glutinous to the tomentose type.

FIG. 1.—*Eriodictyon crassifolium* Benth.: type specimen in the Bentham Herbarium at Kew Gardens.

Since *E. crassifolium*, a typical tomentose form, inhabits the San Diego region, and *E. californicum*, a glutinous form, the more humid central and northern California, it would indicate that a direct environmental influence was the causal factor in evolving the two types. That is to say, if specimens of the two types were grown under the same environment they would eventually become alike, but that is a point that remains as an enticing future problem. From the present studies, however, we learn that other members of the glutinous series occur in the San Bernardino Valley and even in the vicinity of Ensenada, Lower California, where they are subjected to more severe xerophytic conditions than are members of the tomentose type in the Los Angeles region. These distributional facts demonstrate that the glutinous forms may be as well fitted to a xerophytic environment as those of the tomentose group. We may with equal propriety, therefore, suggest that the two groups represent two strains of a common stock that have developed independently. This theory is borne out further by the fact that forms of the two groups growing in contiguous territories more or less intergrade where they meet; while in isolated regions, where only one strain is represented, no marked variation occurs even in strikingly different environmental conditions. As an example, *E. californicum* is the only species represented on the Santa Cruz Mountain peninsula, while no related form extends north of the Santa Barbara region. On this peninsula various climatic conditions are at hand, and if the glutinous and tomentose types simply represent more or less xerophytic conditions, we should find some indications of the two strains; but such is not the case. Plants have been observed associated with the madroño, California black oak, thimbleberry, and similar plants in mesophytic conditions, where the average annual rainfall is approximately 40 inches and where the rainless summers are ameliorated by frequent coast fogs. Other plants have been observed growing in typical xerophytic chaparral, where the annual rainfall is less than half as much and summer fogs very infrequent. The only difference detected between the plants from these two stations was found in the stature of the plants and in the size of the leaves.

It is not possible in this preliminary work to prove or disprove the "Jordan law," but it is evident that the species of *Eriodictyon* conform remarkably. We wish to reiterate, however, that mere

FIG. 2.—*Eriodictyon tomentosum* Benth.: type specimen in the Bentham Herbarium at Kew Gardens.

isolation has not been the primary factor in evolving these forms.
That would hardly be argued by anyone.   Isolation is no more a
causal factor in variation or mutation than is artificial segregation,
but like artificial segregation, as practiced by all breeders and experimental evolutionists, it is of prime importance in preserving new
variations or mutations.   We saw at the outset that our method
of investigation was not such as to divulge the real underlying
factors that have brought about the different forms of *Eriodictyon*.
It must be supplemented with experimental studies, but whether
it is possible to conduct these experiments successfully remains
a future problem.   For the present it has been our aim to ascertain,
first of all, what forms exist in the group selected, and so far as
possible where and under what conditions they grow.   With data
of this nature available, the experimental work can be undertaken
by anyone with a clearer conception of the problem, and with more
likelihood of interpreting rightly the results of his experiments.

ERIODICTYON Benth.   Bot. Voy. Sulphur 35. 1844.

Low shrubs with thin shreddy bark and persistent, alternate,
toothed or rarely entire leaves, tapering at the base to a more or
less evident petiole, or sessile in one species, of firm coriaceous
texture.   Flowers in a terminal, usually naked, panicle of scorpioid
cymes.   Sepals linear, not enlarged above.   Corolla funnelform
to nearly campanulate, pale or dark violet or white.   Filaments
more or less adnate to the corolla and included, usually hirsute
and of irregular length.   Ovary 2-celled by the meeting of the
dilated placentae in the axis.   Capsule first loculicidal then septi-
cidal, thus 4-valved, each valve with a short beak and closed on
one side by the adherent dissepiment or half-partition.   Seeds
brown or black, finely reticulate with ridges running lengthwise, and
connecting them many cross-bars.

A genus of eight species restricted to southwestern United States
and adjacent Mexico, where they extend from southern Oregon
and southern Nevada and Utah through Arizona and California
to northern Lower California, but they belong essentially to the
California element, and six of the species occur within and are
practically confined to that state.

In addition to the species included in these studies, there are two other plants which were transferred by GREENE (5, 6) from

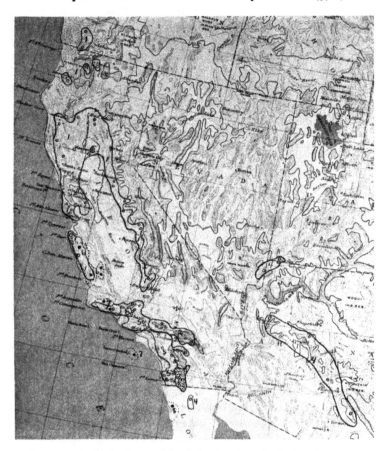

FIG. 3.—Distributional map of *Eriodictyon:* each species is given the same number as in the text; species 1–5 are outlined by an unbroken line and the stations cited in the text are marked by a circle; species 6–8 are outlined by a dotted line and the stations by a black dot.

Nama to *Eriodictyon. Nama Parryi* Gray agrees with typical *Eriodictyon* in seed character and essentially in fruit, and perhaps should be placed here rather than in *Nama*, but it is an ill-scented herbaceous perennial of a totally different habit from true *Eriodictyon* and at least worthy of subgeneric distinction. The other species, *Nama Lobbii* Gray, has seeds merely muriculately papillose without any traces of longitudinal ridges, a seed character of the *Nama* type, and it is with that genus that it is most closely related.

Leaves petioled; herbage glutinous or tomentose, not hirsute.
  Herbage not grayish or hoary tomentose throughout; upper surface of the leaves glabrous and glutinous.
    Branches glabrous and glutinous or those of the panicle sparsely pubescent, reticulations evident on the lower surface of the leaves.
      Corolla tube about 12 mm. long, lavender-purple, much exceeding the nearly glabrous sepals   1. *E. californicum.*
      Corolla tube 5–7 mm. long, little or not at all exceeding the sepals, white.
        Leaves mostly broadly lanceolate, not revolute; sepals densely pubescent; filaments adnate to the corolla tube half their length   2. *E. trichocalyx*
        Leaves mostly narrowly linear and revolute; sepals usually sparsely pubescent; filaments adnate one-third their length   3. *E. angustifolium*
    Branches tomentose, becoming more or less denuded with age in *lanatum.*
      Leaves lanceolate, firm-coriaceous and more or less revolute, densely white-tomentose beneath, obscuring the reticulations.
          4. *E. lanatum.*
      Leaves broadly oblanceolate, not revolute; sparsely soft-tomentose beneath; reticulations evident.
          5b. *E. c. denudatum.*
  Herbage more or less densely tomentose throughout.
    Calyx and corolla not glandular-pubescent; corolla not constricted at the throat.
      Leaves densely covered with a silvery tomentum.
          5. *E. crassifolium.*
      Leaves clothed with a dull gray tomentum or sometimes nearly glabrous on the upper surface. 5a. *E. crassifolium nigrescens.*
    Calyx and corolla with stalked glands; corolla more or less constricted at the throat.
      Stamens equaling the corolla tube, this slightly constricted at the throat   6. *E. tomentosum.*
      Stamens half as long as the corolla tube, this strongly constricted at the throat   7. *E. Traskiae.*
Leaves sessile; herbage more or less hirsute......... 8. *E. sessilifolium.*

1. ERIODICTYON CALIFORNICUM (Hook. & Arn.).   Torr. Bot. Mex. Bound. 148. 1859.

Wigandia californica Hook. & Arn.   Bot. Beechy 364. 1840.
Eriodictyon glutinosum Benth. Bot. Voy. Sulphur 36. 1844.
Eriodictyon glutinosum var. *serratum* Choisy, DC. Prod. 10:483. 1846.
Eriodictyon californicum forma *linearis* Brand, Univ. Calif. Pub. Bot. 4:224. 1912.
Eriodictyon californicum subsp. *glutinosum* Brand, in ENGLER, Pflanzenreich 59:141. 1913.
Eriodictyon californicum forma *latifolia* Brand, *loc. cit.*

An erect shrub 1–3 m. high, with mostly erect branches, the older branches and trunk clothed with light brown shreddy bark, the branchlets glabrate and glutinous, terete or sometimes angled: leaves lanceolate or oval, sometimes linear-lanceolate, 4–10 cm. long, 1–5 cm. wide, dentate or undulate, glabrous and glutinous above, whitened beneath with a fine indument between the reticulations, the midvein prominent, its main branches usually anastomosing at the margin; petioles gradually narrowed to the base, sometimes winged: branches of the inflorescence glabrous: sepals with a few scattered hairs or glabrous, linear, $\frac{1}{2}$ as long as the corolla: corolla funnelform, about 12 mm. long: stamens unequal, in sets of three long and two short or sometimes the reverse: style half as long as the corolla tube: capsule globular, usually covered with a white gum: seeds brown, about 6–12 maturing in a capsule.

BRAND'S two forms, *latifolia* and *linearis*, may occur in almost any locality where plants are subjected to different exposures. They are adjustments of the individual to local environmental conditions, and are not worthy of a taxonomic designation.

TYPE LOCALITY.—Described from specimens collected by DOUGLAS in the coastal region of central California, between Sonoma and Monterey.

DISTRIBUTION.—Inhabits the Upper Sonoran and Lower Transition zones, growing on clay, sandy, or rocky soils. Ranges from the southern end of the Sierra Nevada in Kern County and the Coast Ranges of San Luis Obispo County northward to the Siskiyou Mountains, where it reaches its northern limit, so far as known, in the vicinity of Wimer, Jackson County, Oregon. In the central Sierra Nevada it has been collected well within the Transition zone in Yosemite Valley (*Abrams* 4563) between Mirror Lake and Kenneyville, but associated with a number of other Upper Sonoran intruders. In the coastal region it is seldom met with in the fog belt or typical redwood region, or again on the dry hot eastern slopes of the Inner South Coast Ranges.

SPECIMENS EXAMINED.—Oregon, Jackson County: Wimer, *Hammond* 291; Dunn's Butte, near Ashland, *Walpole* 250. Josephine County: Waldo, *Piper* 6215.

California.—Siskiyou County: hills west of Yreka, *Heller* 7996; dry thickets, near Yreka, *Butler* 1405. Humboldt County: Hupa Indian Reservation, alt. 500 ft., *Chandler* 1352; between Three Rivers and the mouth of Willow Creek, *Tracy* 3360. Mendocino County: near Cummings, *Davy* and *Blasdale* 5336; Round Valley, alt. 400 m., *Chestnut* 165; near Ukiah, *Bolander* 3912; *Purdy*, 1897. Tehama County: Cooper's South Fork of Elder Creek, *Ward* 108; Fort Reading, *Newberry*. Colusa County: College City, *Miss King* 1905. Butte County: Butte Creek, *Mrs. Austin* 1811; Little Chico Creek, *Mrs. Austin* 296. Plumas County: Shoo Fly Bridge, *Mrs. Austin*, August 1893. Lake County: Eel River, one mile below Hullville, *Heller* 6032; Clear Lake, *Torrey*, 1865. Sonoma County: Little Sulphur Creek, *M. S. Baker*, June 11, 1898. Marin County: Mount Tamalpais, *Michener* and *Bioletti*, May, 1892; *Eastwood*, May 30, 1898; Bolinas Ridge, *Palmer* 2339. Solano County: Putah Bluffs, *Jepson*, May 1891; Gates Cañon, near Vacaville, *Heller* and *Brown* 5380. Contra Costa County: northeast of Mount Diablo, *Brewer* 1133; southeast of Mount Diablo, *Brewer* 1158; Antioch, *Brandegee*, June 1908. Alameda County: Oakland, *Chestnut*. San Mateo County: Belmont, *Davy*, June 17, 1893; Lake San Andreas, *Elmer* 4801; King's Mountain road, *Miss Randall* 164; Searsville Ridge, *Dudley*, June 4, 1895. Santa Clara County: hillsides near Stanford University, *Dudley*, April 1893; between Saratoga and Los Gatos, *Dudley*, April 16, 1893; Black Mountain, *Rutter* 8; foothills west of Los Gatos, *Heller* 7384; near Stanford University, *Abrams*, May 1898, May 1901. Santa Cruz County: upper part of San Lorenzo road near the river, *Dudley*, May 7, 1893; Glennwood, *Horace Davis*, September 1907; Boulder Creek, *Bailey*, October 13, 1891; Soquel Gulch, *Mrs. B. H. Thompson*, June 7, 1902; Lorenzo Cañon, *Mrs. B. H. Thompson*, July 6, 1902. Monterey County: Monterey, *Parry*, 1850; Pescadero ranch, *Brewer* 670; near Mission Soledad, *Brewer* 583; Santa Lucia Mountains, *Plaskett*, May 1898; Tassajara, *Elmer* 3204. Calaveras County: Gwin Mine, *Jepson*, October 6, 1902. Amador County: Agricultural Station, alt. 2000 ft., *Hansen* 187; New York Falls, alt. 1500 ft., *Hansen* 1647; vicinity of Ione, alt. 200–500 ft., *Braunton* 102. Mariposa County: Mariposa, *Congdon*, June 1883; Yosemite Valley between Mirror Lake and Kenneyville, *Abrams* 4563; Kingsley, *Miss Hook*, May–June 1905. Madera County, Chiquito Creek, alt. 6000 ft., *Blethen*, June 30, 1912; North Fork and vicinity, *Griffiths* 4538. Tulare County: Giant Forest, *Brandegee*, August 1905. Kern County: North Fork of Kern River, alt. 750 m., *Coville* and *Funston* 1034; Kern River Cliffs, *Dudley* 767; vicinity of Havilah, *F. Grinnell* 275; Johnson's Cañon, Walker Basin, *F. Grinnell* 30, 116.

2. Eriodictyon trichocalyx Heller, Muhlenbergia 1:108. 1904.

Eriodictyon angustifolium pubens Gray, Proc. Am. Acad. 17:224. 1882.

Eriodictyon californicum subsp. *australe* Brand, Engler; Pflanzenreich 59:141. 1913, in part.

Eriodictyon californicum var. *pubens* Brand, *loc. cit.*

Eriodictyon glutinosum var. *intermedium* Parish, Brand, *loc. cit.*, published as a synonym.

Low shrub 0.5–1.5 m. high, with the erect branches and branchlets glabrous or nearly so and glutinous: leaves broadly lanceolate to linear-lanceolate, 5–10 cm. long, 1–3 cm. wide, firm coriaceous, flat and dentate or sometimes slightly revolute, glabrous and glutinous on the upper surfaces, paler beneath with a close fine tomentum within the conspicuous reticulations: branches of the inflorescence pubescent: calyx densely pubescent, nearly equaling the corolla tube: corolla 5–6 mm. long, its tube funnel-form, white, densely pubescent without.

This species was first described by Gray as a variety of *angustifolium*, but afterward (4) considered by him as one of the intermediate forms between that species and *californicum*. Brand, basing the distinction mainly on the relative length of the free and adherent parts of the filaments, considered it as a variety of *californicum*, and restored *angustifolium* to specific rank. That the species is closely related to both *angustifolium* and *californicum* is evident, but the tendency to approach *californicum* is not through intergradation with typical *californicum*, but with *crassifolium nigrescens*. It is through this variety that all three of the species, *californicum*, *crassifolium*, and *trichocalyx*, form almost a complete series of intergradations. With this as the status of these forms, as demonstrated now by a large series of specimens, there are two ways of expressing their relationship taxonomically, either *crassifolium*, *trichocalyx*, and *californicum* should be considered as strains of a single species, or each should be given specific rank, and their more extreme geographical forms varietal rank, admitting that they do have a tendency to intergrade in contiguous territory. We have chosen the latter course because relationship can be shown sufficiently accurate by the simple binomial and trinomial method without the introduction of such burdensome combinations as *Eriodictyon californicum australe pubens coarctatum* Brand. Such combinations are as cumbersome as the pre-Linnaean system and less significant, for they need not be in any way descriptive.

Type locality.—"Seven Oaks Camp, San Bernardino Mountains."

Distribution.—This species grows in stony or sandy soil in the chaparral of the San Gabriel and San Bernardino mountains. It extends from elevations of 1000 ft. on the sandy plains of the San Gabriel and San Bernardino valleys to 8000 ft. on the south slope of San Antonio Mountain, and to nearly the same altitude on the south side of the divide between Seven Oaks and Bear Valley. On the desert slopes of the mountains it has been collected as low down as the juniper belt along Rock Creek.

SPECIMENS EXAMINED.—Los Angeles County: Mount Wilson, alt. 5600 ft., *Abrams* 2607; Claremont, *Chandler*, May 11, 1897; Rock Creek, desert slopes of the San Gabriel Mountains, *Abrams* and *McGregor* 526; Mount San Antonio, alt. 8000 ft., *Abrams* 1939, 2607. San Bernardino County: Lytle Creek, San Gabriel Mountains, alt. 800 m., *Leiberg* 3335; near San Bernardino, *Vasey* 438; *Coville* and *Funston* 107; sandy plains, near Colton, *Pringle*, May 27, 1882; Hogback, San Bernardino Mountains, *Parish* 2977 (approaching *crassifolium nigrescens*); Holcomb Valley, San Bernardino Mountains, *Shaw* and *Illingsworth* 85; Victorville, *Hall* 6188. Riverside County: Banning, *Brandegee*, May 14, 1895; Whitewater, alt. 1126 ft., *Parish* 2976. Lower California: no locality, *F. E. Fish*, May 7, 1883 (this is herbarium sheet no. 45671 of the National Herbarium. The locality is not given and two labels are attached, one purporting to be ORCUTT'S and the other FISH'S, but neither is apparently original. The specimens are not in flower, but seem to belong here); near Ensenada, *Jones* 3739 (this plant together with a specimen collected by *Brandegee*, May 20, 1893, on San Pedro Martir Mountain, BRAND described as *californicum* subsp. *australe* var. *pubens* subvar. *coaractatum*. Jones's specimen may well be placed with *trichocalyx*, although the specimen at hand is too scrappy to warrant more than a guess, but *Brandegee's* specimen is strikingly different. It is much like typical *angustifolium* in foliage characters and size of flowers, but the filaments are a little more united to the corolla tube. In this respect it seems intermediate between *angustifolium* and *trichocalyx*).

3. Eriodictyon lanatum (Brand) Abrams, sp. nov.

Eriodictyon californicum subsp. *australe* var. *lanatum* Brand, in ENGLER, Pflanzenreich 59:142. 1913.

An erect branching shrub, 0.6–2 m. high, with the branches more or less permanently tomentose: leaves thick coriaceous, slightly revolute on the entire, undulate or dentate margins, 2–7, mostly about 5, cm. long, 8–20 mm. wide, tapering from near the middle to both ends, glabrous and glutinous above, at least in age, densely covered beneath with a white tomentum obscuring the reticulations: branches of the inflorescence pubescent; flowers crowded in the scorpioid cymes: sepals densely white-pubescent, about 2.5 mm. long: corolla funnelform, 7–8 mm. long, much exceeding the calyx, pale purplish-blue or nearly white, the tube pubescent without.

TYPE.—The type of the variety is *Abrams* 3632, collected in chaparral between Campo and Jacumba. The description of the species is drawn from a sheet of the same collection deposited in the Dudley Herbarium.

DISTRIBUTION.—Southern California from the vicinity of Toro Mountain, Riverside County, southward through the chaparral region, especially on the desert slopes of the Cuiamaca Mountains, to the northern boundary of Lower California.

SPECIMENS EXAMINED.—Riverside County: near Toro Mountain, *Leiberg* 3199. San Diego County: Jacumba Hot Springs, *Mearns* and *Schoemfeldt* 3261, 3288; Mountain Springs, *Mearns* and *Schoemfeldt* 3207; Cameron's ranch, Laguna, *Mearns* and *Schoemfeldt* 3702; San Felipe, *Brandegee*, April 16, 1895; Colorado Desert, in the foothills, *Brandegee*, April 13, 1896; Potrero, *Alderson*, May 1893; Palm Creek, Colorado Desert, *Brandegee*, April 18, 1895; Laguna Mountains, *Brandegee*, June 20, 1904; in chaparral between Jacumbo and Campo, *Abrams* 3632; Campo, *Hall* 9424. Lower California: Nachoguero Valley, *Mearns* and *Schoemfeldt* 3463.

4. ERIODICTYON ANGUSTIFOLIUM Nutt. Jour. Acad. Nat. Sci. Phil. II. 2: 181. 1848.

Eriodictyon glutinosum angustifolium Torr., GRAY, Syn. Fl. 2:176. 1878.

Low erect shrub, o.6 to 2 m. high, the branches often crowded, glabrous and glutinous: leaves narrowly linear to narrowly linear-lanceolate, 5–10 cm. long, 3–10 mm. wide, entire or inconspicuously dentate, revolute, glabrous and glutinous above, canescent and reticulated beneath: branches of the inflorescence glabrous and glutinous or sparsely pubescent: cymes racemosely or corymbosely arranged: sepals linear, nearly glabrous or somewhat hirsute: corolla nearly campanulate, its tube only about 5 mm. long, not exceeding the calyx: filaments united only ⅓ their length: seeds black, slightly longer than those of *californica*.

TYPE LOCALITY.—"On the Sierra of Upper California [Arizona]."

DISTRIBUTION.—In the Upper Sonoran chaparral of southern Nevada and southern Utah southward through Arizona, and also in Lower California in the San Pedro Martir region.

SPECIMENS EXAMINED.—Utah: Silver Reef, alt. 3500 ft , *Jones* 5149, 5176; nearly at the head of the grade 5 miles above Bellevue, *Jones* 5001; Sandy, 4 miles east of Leeds, alt. 3400 ft., *Jones* 5214. Nevada: Charleston Mountains, alt. 4000–5000 ft., *Purpus* 6074; Bunkerville, Virgen River, *Goodding* 746. Arizona: Diamond Valley, *Purpus*, May–October 1898; "hills near Cactus Pass in the western part of New Mexico" (probably in western Arizona on Bill Williams Fork), *Bigelow*, 1853; Pinal Mountains, *Jones*, May 24, 1890;

Oak Creek, *Rusby* 239; no locality, *Palmer*, 1869; Santa Catalina Mountains, *Lemmon*, May 1881; Copper Basin, *Toumey* 199-*b;* Cheno Valley, *Toumey* 199-*a;* canyon 2 miles below Pagumpa, alt. 4000 ft., *Jones* 5089. Lower California: Trinidad Valley, *Belding*, May 1885; Vallederos Creek, *Brandegee*, May 29, 1893; San Pedro Martir Mountain, *Brandegee*, May 4 and 20, 1893.

5. ERIODICTYON CRASSIFOLIUM Benth. Bot. Voy. Sulph. 35. 1844.

Eriodictyon tomentosum of various authors, not Benth.
Eriodictyon crassifolium subsp. *Grayanum* Brand, in ENGLER, Pflanzenreich 59:139. 1913.
Eriodictyon crassifolium var. *typica* Brand, *loc. cit.*

Shrub 1–4 m. high, with spreading branches, the twigs, both surfaces of the leaves, and calyx densely hoary or silvery tomentose: leaves 7–15 cm. long, 2–5 cm. wide, reticulate beneath, not revolute, entire or crenate, dentate or sometimes shallowly lobed: sepals about 5 mm. long, narrowly linear: corolla 10–15 mm. long, rather broadly funnelform, pale bluish purple, pubescent without: seeds smaller than in *californicum* and indented.

TYPE LOCALITY.—"San Diego."
DISTRIBUTION.—Typical *crassifolium* inhabits the dry gravelly or sandy mesas and foothills in the vicinity of San Diego and extends northward to Santiago Canyon, Santa Ana Mountains, in the coastal region, also on the mesas and foothills between the Santa Ana and the San Jacinto mountains, where it extends to the desert slope in the vicinity of Palm Springs. In the Los Angeles region, extending along the coastal slopes of the San Gabriel Mountains to the Santa Monica and San Fernando mountains, the leaves are covered with a somewhat less dense and shorter tomentum, giving a dull gray instead of a hoary tone. These plants often have the flowers reduced in size and grade fairly gradually into the variety *nigrescens*. This more or less intermediate type extends on northward to the Tehachapi region.
SPECIMENS EXAMINED.—Kern County: San Emidio Canyon, *Davy* 2026; San Emidio Potreros, alt. 5000 ft., *Hall* 6391; vicinity of Fort Tejon, *Xantus de Vesey* 94; Canada de las Uvas, alt. 1050 m., *Coville* and *Funston* 1142. Los Angeles County: Mount Lowe, *Grant* 417; Little Santa Anita Canyon, *Abrams* 2627; San Fernando Mountains, near Chatsworth, *Abrams* 1363; Santa Monica Mountains, *Hasse*, June 1892; Los Angeles, *Brewer* 39; Glendora, *Braunton* 293. Orange County: Santiago Canyon, *Miss Bowman*, June 1899; *Hall* 9402. Riverside County: western base of the San Jacinto Mountains, *Hall* 2006; Hemet, *Hall* 561; Valle Vista, San Jacinto Mountains, *Hall* 1107; Elsinore, *Abrams* 5052; Tahquitz Canyon, near Palm Springs,

Dudley, December 25, 1903; near San Jacinto, *Leiberg* 3214; *Berg*, April 3, 1904; Menifee, *Miss King*, 1893. San Diego County: San Diego, *Barclay* (type); *Cooper* 498; *Palmer*, 1875; *Pringle*, April 26, 1882; *Dunn*, April 21; *Jones* 3143; Mission hills, near San Diego, *Abrams* 3432; Point Loma, *Chandler* 5067 (this specimen resembles the less tomentose form of the Los Angeles region, a condition undoubtedly due to the greater humidity on the promontory than on the adjacent mainland); Encinatas, *Brandegee*, March 28, 1894; Witch Creek, *Alderson*, April 1894; San Ysabel, *Henshaw* 148; near Bennington, *Dudley*, January 1908.

5*a*. ERIODICTYON CRASSIFOLIUM NIGRESCENS Brand, in ENGLER, Pflanzenreich **59**: 140. 1913.

Leaves smaller and comparatively narrower, dull gray green with a shorter and much less dense tomentum, usually crenate-dentate: corolla narrowly funnelform, about 6 mm. long, white and densely hairy without.

This variety is intermediate between *crassifolium* and *trichocalyx*. The intergradation with typical *crassifolium* is complete but less marked with *trichocalyx*, although the form originally described by GRAY as *angustifolium* var. *pubens* partakes of both and might be placed in either category.

TYPE LOCALITY.—Acton, Los Angeles County.

DISTRIBUTION.—Chaparral-covered slopes of the Liebre and San Gabriel mountains, especially in the Soledad Pass region.

SPECIMENS EXAMINED.—Acton, *Elmer* 3596, 3598; Kings Canyon, Liebre Mountains, *Dudley* and *Lamb* 4343; Oakgrove Canyon, Liebre Mountains, *Abrams* and *McGregor* 322.

5*b*. ERIODICTYON CRASSIFOLIUM denudatum Abrams, var. nov.

Leaves 10–15 cm. long, 2–3.5 cm. wide, flat and more or less dentate on the margins, green and glabrate but not evidently glutinous on the upper surface, soft tomentose beneath, as also the branches and inflorescence: calyx densely clothed with silky hairs, about one-third the length of the corolla tube: corolla lavender, densely pubescent without, 8–10 mm. long.

This variety is the extreme type of the *crassifolium* group; it merges into the variety *nigrescens* and through the Los Angeles form into typical *crassifolium*, while the denuded upper surface of the leaves suggest close affinity with *californicum*. It occurs in the cross ranges of Santa Barbara and Ventura counties, the region from whence *californicum*, *crassifolium*, and *trichocalyx* diverge. It is natural, therefore, according to the Jordan law, that we should find here the intermediate forms.

TYPE.—Red Reef Canyon, Topatopa Mountains, alt. 2800–3500 ft., Ventura County, California, *Abrams* and *McGregor* 159, June 8, 1908. The type sheet is deposited in the Dudley Herbarium of Stanford University.

DISTRIBUTION.—Chaparral-covered hills and mountains of Santa Barbara and Ventura counties.

SPECIMENS EXAMINED.—Santa Barbara County: Santa Inez Mountains, near Santa Barbara, *Brandegee*, 1888. Ventura County: Ojai and vicinity, *Peckham*, April 13, 1866; Sisar Canyon, Topatopa Mountains, *Abram* and *McGregor* 65; Red Reef Canyon, Topatopa Mountains, *Abrams* and *McGregor* 142, 159. Kern County: Vicinity of Fort Tejon, *Abrams* and *McGregor* 300 (approaches the Los Angeles form of *crassifolium*).

6. ERIODICTYON TOMENTOSUM Benth. Bot. Voy. Sulph. 36. 1844.

Eriodictyon niveum Eastw. Proc. Cal. Acad. III. 1:130. 1898.

Eriodictyon crassifolium subsp. *Benthamianum* var. *niveum* Brand, in ENGLER, Pflanzenreich 59:140. 1913.

An erect, branching shrub with the herbage hoary throughout with a dense feltlike tomentum or rarely becoming more or less denuded and green: leaves thick, elliptic-ovate or obovate, 4–6 cm. long, 1–3 cm. wide, cuneate at base, acute or obtuse at apex, entire, crenate, or even coarsely dentate, veins scarcely evident on the upper surface, beneath reticulate-rugose; panicle terminating a usually elongate naked peduncle widely branched or simple; flowers crowded, nearly sessile: calyx lobes linear-subulate, equaling the corolla tube, clothed with white silky hairs with a few stalked glands interspersed: corolla white or pale violet, 4 mm. long, urceolate, glandular-hirsute without, the tube slightly contracted below the very short spreading lobes: stamens with the free portion of the filaments short, inserted below the throat; anthers oval 1 mm. long.

Until now the identity of *E. tomentosum* has been misunderstood. TORREY (9) first considered it and *crassifolium* as conspecific, with the remark, "We have specimens that are intermediate and Dr. PARRY informs me that he has seen them in California passing into each other." We have failed to find in the field or in any of the herbaria any intermediate forms, nor is there any evidence that PARRY ever collected true *tomentosum*. Geographically and structurally, aside from the superficial character of hoariness, these two species are more distinct than *crassifolium* and *californicum*. TORREY's conclusions were followed by GRAY and other botanists until GREENE (6) discovered a plant in Monterey County which he believed to be the true *tomentosum*.

Miss EASTWOOD, red scovering the same plant and wishing to solve the problem, sent specimens of her Monterey plant and also *crassifolium* material from San Diego to Kew to be compared with the types.   The person at Kew who investigated the matter for her replied that "*Eriodictyon crassifolium* and *E. tomentosum* are conspecific, and your plant is apparently an undescribed species." This seemed final, but from our distributional studies we found that the northern limit of true *crassifolium* is far south of the region visited by DOUGLAS, who first collected *tomentosum*.   We know that DOUGLAS collected in the vicinity of Mission San Antonio where *niveum* is found, and about Santa Barbara where *Traskiae* and *crassifolium denudatum* grow.   Which of these was the true *tomentosum* was not evident without access to the type, but that one of them and not *crassifolium* would prove to be it seemed probable.   We wrote, therefore, to Dr. OTTO STAPF, of the Kew Herbarium, who furnished us with a photograph of the type of both *crassifolium* and *tomentosum*, and also fragments for study.   The evidence was clear; *tomentosum* and *niveum* were found to be one and the same.

TYPE LOCALITY.—Not given in the original publication, but collected by DOUGLAS probably in the vicinity of Mission San Antonio, Monterey County.

DISTRIBUTION.—On the chaparral-covered eastern slopes of the Santa Lucia Mountains, extending southward through the middle foothill region of Monterey and San Luis Obispo counties.

SPECIMENS EXAMINED.—Monterey County: without locality but probably from the Jolon region, *Douglas* 1833; Jolon, *Vasey* 439, *Mrs. K. Brandegee*, June 8, 1909; Tassajara Hot Springs, *Elmer* 3210; San Antonio Creek, above the Mission, *Dudley*, May 11, 1895; Cholome, *Lemmon;* Arroyo Seco, Santa Lucia Mountains, *Dudley*, January 1, 1896; China Camp, on road to Tassajara Hot Springs, alt. 3500 ft., *Cox*, July 1908; near Soledad, *Congdon*, June 1881. San Luis Obispo County: between Pozo and La Panza, *Miss Eastwood*, June 10, 1902; Santa Margarita Mountains, east of pass on road to San Luis Obispo, *Dudley*, April 3, 1903; central coast ranges, *Palmer*, 389.

7. ERIODICTYON TRASKIAE Eastw.   Proc. Cal. Acad. III. 1:131. 1898.

Eriodictyon crassifolium subsp. *Benthamianum* var. *Traskiae* Brand, in ENGLER, Pflanzenreich 59:140. 1913.

An erect branching shrub 1–2 m. high, clothed with a hoary feltlike tomentum except on the calyx: leaves oblanceolate to elliptic-ovate, 5–12 cm. long, 1.5–4 cm. wide, acute at apex, narrowed at base to a petiole 1 cm. long or more, dentate, veins obscure on the upper surface, reticulate-rugose beneath: panicle usually much branched, bearing short congested terminal cymes, or these elongated; flowers short-pediceled, crowded: sepals narrowly linear, 4–5 mm. long, dark colored, and glandular-hirsute: corolla purple, the tube equaling the calyx, contracted at the throat and base: the lobes irregularly orbicular, these and the upper part of the tube glandular-hirsute without: stamens inserted in the middle of the tube, nearly sessile: ovary glandular-hirsute.

This species is closely related to *tomentosum* and possibly should be considered a variety. The two are best differentiated by the position of the stamens.

TYPE LOCALITY.—"On one volcanic upland on Santa Catalina Island, California, at an elevation of about 1500 feet."

DISTRIBUTION.—Santa Catalina Island and on the mainland in the Santa Inez Mountains, Santa Barbara County. Why it should occur in practically unmodified form on the mainland and the islands is not easily explained, although the fact that it is found on the islands in only one station and that composed of only a few bushes would indicate that it had been transported, possibly by birds of passage, from the mainland. It is significant in this connection that a number of insular species are also found on the mainland in the Santa Barbara region.

SPECIMENS EXAMINED.—Santa Barbara County: Painted Cave ranch, *Eastwood*, May 9, 1908; Sisquoc, *M. S. Baker*, July 1895; La Cumbre trail, Santa Inez Mountains, *Abrams* 4311; Santa Inez Mountains, *Eastwood*, May 17, 1904; *Hall* 7845; bushy hills, head of Santa Inez River, *Hall* 7830; near Santa Barbara, *Brewer* 296; *Elmer* 4017; *Brandegee*, 1888. Los Angeles County: Santa Catalina Island, *Blanche Trask*, May 1897 and September 1910.

8. ERIODICTYON SESSILIFOLIUM Greene, Bull. Cal. Acad. Sci. 1:201. 1885.

A shrub 1–2 m. high with very leafy glandular-pubescent and hirsute branches: leaves lanceolate-oblong, 6–12 cm. long, 2–4 cm. wide, acute at apex, sessile and truncate or cordate-clasping at base, coarsely serrate, glutinous and nearly glabrous to densely hirsute on the upper surface, sparsely to densely hirsute beneath and short-tomentose between the veins: inflorescence an open or somewhat congested cymose panicle: sepals villous-hirsute and glandular, about 5 mm. long: corolla funnelform, 12 mm. long, lilac-purple, pubescent without but not glandular: stamens equal, their filaments densely clothed with bristly hairs: capsule globular: seeds dark brown or nearly black.

Considerable variation in the amount of hirsute pubescence suggests a plastic condition or possibly two varietal strains, but too few collections have been made and too little is known of its geographical range to warrant any new propositions to designate these forms.

TYPE LOCALITY.—"All Saints Bay, Lower California."

DISTRIBUTION.—On canyon slopes in the coastal region of northern Lower California, especially in the neighborhood of Ensenada.

SPECIMENS EXAMINED.—Lower California: Todas Santas, *Fish*, June 2, 1883; La Gruella Canyon, *Orcutt*, July 14, 1885; Guadaloupe Canyon, *Orcutt*, January 20, 1882, and June 2, 1883; Burro Canyon, *Brandegee*, April 22, 1893.

STANFORD UNIVERSITY
CALIFORNIA

LITERATURE CITED

Only the references mentioned in the general discussion are included; those of taxonomic importance are placed after each species.

1. ABRAMS, L. R., The theory of isolation as applied to plants. Science 22: 836–838. 1905.
2. BENTHAM, G., Botany of the voyage of the Sulphur. pp. 35, 36. 1844.
3. BRAND, A., Monograph of the family Hydrophyllaceae. ENGLER, Pflanzen-reich. 59:138–143. 1913.
4. GRAY, A., Synoptical flora of North America. 2^1:175. 1878; also ed. 2, 2^1:419. 1886.
5. GREENE, E. L., New or noteworthy species. Pittonia 2:22. 1889.
6. ——, Studies in the botany of California and parts adjacent. Bull. Cal. Acad. 1:201–202. 1885.
7. JORDAN, D. S., The origin of species through isolation. Science 22:545–562. 1905.
8. LLOYD, F. E., Isolation and the origin of species. Science 22:710–712. 1905.
9. TORREY, J., Botany of the Mexican boundary. p. 148. 1859.

GAMETOPHYTE OF PELLIA EPIPHYLLA

CONTRIBUTIONS FROM THE HULL BOTANICAL LABORATORY 206

, A. H. HUTCHINSON

(WITH PLATES I–IV AND ONE FIGURE)

Three of the species of *Pellia*—*P. epiphylla*, *P. calycina*, and *P. endivaefolia*—show morphological differences, especially with respect to the apical cell, which would suggest a generalized or possibly an unstable ancestral form. No detailed study of any of these species has been reported. During the investigation upon which this account is based, it has been found that in *P. epiphylla* not only are there transitions in the method of growth, but also that the development of the antheridium may follow any one of several divergent lines.

Antheridium

It has been generally accepted that "the antheridium of *Pellia* is larger than that of *Aneura*, but its development is very similar, except that the stalk is multicellular, as it is in other *Anacrogyneae*."[1] In addition to this method of development, *P. epiphylla* shows young antheridia having characters of Marchantiales; moreover, the spermatogenous initials may be cut out in the same way as the primary axial cell of the archegonium.

The division of a dorsal cell, the third or fourth from the apical cell, by a horizontal cross-wall is the first evidence of an antheridial initial. The outer of the two cells formed divides again, giving the three cells of the antheridial row—the basal cell, the stalk initial, and the outer cell; the latter by successive divisions gives rise to the wall cells and spermatogenous cells. Meanwhile, the dorsal cells, immediately surrounding the antheridial group, divide and become papillate, thereby producing a ring-shaped involucre (figs. 2, 3). The outer cell of the antheridial group divides next. The position of the wall is significant; if it is vertical and median,

[1] CAMPBELL, D. H., Mosses and ferns   New York 1905 (p 92).

the successive divisions follow an antheridial sequence; if, however, the wall is inclined and somewhat removed from the central
position, the cell divisions which follow are similar to those of a
developing archegonium.   In the former case the vertical wall is
followed by a curved wall on either side, which cuts the vertical
wall as shown in figs. 7 and 8.   Two similar walls, rotated about
the central axis through an angle of 90° (figs. 8, 9), complete the
separation of the peripheral region from the central spermatogenous
region.   The first two of these walls may be nearly parallel at the
base (fig. 7), or they may be at right angles; similarly the second
pair.   Such an antheridium is characteristic of the Jungermanniales.

Occasionally the outer cell, mentioned above, is divided into
quadrants by walls at right angles to the vertical wall (figs. 10, 11),
in which case four wall cells are cut off by periclinal divisions,
giving also four spermatogenous cells.   The process is similar to
that characteristic of *Sphaerocarpus* or Marchantiales.

Fig. 14 illustrates the result of a combination of these two
methods of development.   In fig. 15 is shown a double antheridium;
the two halves have become completely separated by the vertical
division and each has developed independently.   The process may
be compared to the characteristic development of the double
antheridia of Anthocerotales.

When the first wall formed in the outer cell is inclined and
somewhat removed from the median position, it is followed by a
second and a third wall, each of which is similarly placed, but
revolved with respect to each other through an angle of 120° about
the vertical axis (figs. 17, 18, 26, 27, 28).   A transverse wall divides
the central cell into the cap cell and the spermatogenous initial
(fig. 19).   The characteristic archegonial development is followed
until the massive spermatogenous group begins to be formed (figs.
22, 28), instead of the axial row.

Occasionally this critical third wall of the antheridium is inclined
inward instead of outward, as described above (fig. 23).   Two
walls similarly inclined complete the separation of the peripheral
region from the central spermatogenous initial (figs. 24, 25).   This
form is similar to that last described with the exception of the inclination of the walls and the resulting lack of the cap cell.

The differences in these antheridia are emphasized by the fact that in the first form (fig. 8) there are two spermatogenous initials, paired as in Jungermanniales; in the second form there are four, arranged in the form of a quadrant (fig. 10), as in Marchantiales; and in the third and fourth there is but one spermatogenous initial (figs. 10, 24, 25).

That the structures described above are developing antheridia, and not archegonia, is evidenced by their position and by the presence of an individual involucre. The antheridia are single and dorsal, while the archegonia are grouped in the terminal pocket.

Archegonium

The position of the archegonial group is of considerable morphological importance. CAMPBELL, with reference to the work of JANCZEWSKI, states: "The archegonia are formed in groups just back of the apex but he [JANCZEWSKI] does not seem to have been able to detect any relation between them and the apical cell such as obtains in *Aneura*, but it is possible that such a relation does exist." As mentioned above, the archegonia are terminal and inclosed in a "pocket," which is formed by a cup-shaped involucral growth. As will be described more fully, the archegonia arise from an apical group of cells, any of which may become an archegonium initial. There is no regular succession in the formation of archegonia; apparently old and young organs are indiscriminately intermingled. Since the apical group ceases to function as such after the production of archegonia, *P. epiphylla* may be regarded as truly acrogynous. The involucre is produced by cells which are cut off laterally by the apical group (fig. 40), and pushed out very much as the wings during the previous period of growth; in this event, however, the lateral cells are forced out on all sides to form a complete inclosure.

The structure and development of the archegonium conforms, in general, to the characteristic liverwort form. Some specific characters may be noted.[2] "After the archegonial mother cell is cut off it does not divide at once by vertical walls, but a pedicel is first cut off [fig. 31]; after which the upper cell undergoes the

[2] CAMPBELL, D. H , Mosses and ferns. New York. 1905 (p. 90).

usual divisions." This character serves to emphasize the similarity which exists between the archegonium as shown (figs. 31–33), and the antheridium (figs. 16–21). Except for the position and presence or absence of the involucre, these organs would be difficult to differentiate until the spermatogenous group or the archegonial axial row, as the case may be, begins to develop. The cap cell gives rise to a group consisting of more than the usual number of cells. The first division of the cap is often simultaneous with the division which gives rise to the primary neck canal cell and the primary ventral cell (figs. 32, 35), and four cap cells may appear in cross-section when there are but three neck canal cells and a ventral cell (fig. 36). Fig. 38 shows a group of cap cells, 6 in cross-section, which may be compared with those which form the neck of the archegonium of Filicineae. JANCZEWSKI reports that the number of neck canal cells may be as high as 16 or even 18; 9 is the greatest number seen by the writer (fig. 38). The venter becomes massive before fertilization; it may be 2 or 3 cells in thickness; a many-celled stalk is also formed (fig. 37). In the young archegonium the neck has usually only 5 vertical rows of cells. The cells originating from the third wall cell do not divide until the archegonium approaches maturity (fig. 39).

Methods of growth

A certain form of apical cell may usually be given as characteristic of a genus or even of a larger group. In *Pellia*, however, there is no such conformity; the apical cell of *P. calycina* has four cutting faces, two lateral, a dorsal, and a ventral—the cuneate apical cell. The dolabrate apical cell of *P. endivaefolia* has but two cylindro-convex cutting faces; while that reported as characteristic of *P. epiphylla*, the lenticular cylindric apical cell, has a posterior convex and two lateral cutting faces. In the last named species, however, there are several methods of growth; these cannot be sharply delimited, but for clearness five rather distinct forms may be taken as characteristic of successive periods of growth.

During the time of intra-capsular gametophytic division a massive body is formed. There is no regional growth, but all cells have an equal power of division. This period of growth is of short duration. The gametophyte body retains this massive form until after the resting period.

The second period of growth begins by the formation of a cuneate apical cell. A terminal cell is cleft by a wall inclined about 30° with reference to the longitudinal axis of the spore; a second inclined wall bisects the first at an angle of approximately 90°; lateral walls complete the cuneate apical cell (fig. 41). The latter cuts off dorsal and ventral segments (fig. 42) as well as lateral segments (fig. 43). Such a method of growth is characteristic of the young gametophyte until about the time of antheridium formation.

The growth of the third period is by means of a lenticular cylindric apical cell (fig. 47). The transition is somewhat irregular. Sometimes the apical cell is more or less equally divided; then the two halves simultaneously cut off cells (fig. 45) which correspond to the dorsal and ventral segments of the preceding form. The thallus in cross-section has the appearance of being medianally divided by a wall. Another transition form is shown in fig. 44. As the thallus becomes thicker the posterior angle becomes greater, the two faces being finally replaced by one which is curved. The cells cut off from the posterior face divide rapidly; as many as 6 segments may be formed before the next division occurs. The rapid division of the laterally placed cells causes the wings to be protruded outward and forward (figs. 46, 48, 49). There has been much discussion regarding the method of branching. HOFMEISTER[3] believed that the central papilla ("Mittellappen") shown in fig. 14 was the seat of the chief apex; hence that there is no true dichotomy. LEITGEB[4] states that the origin of the central papilla is from a marginal cell. When branching takes place, the apical cell, instead of cutting off a lateral segment, divides equally or almost so; each of these daughter cells assumes apical characters. The central papilla is produced by the crowding together of lateral segments from the two apical cells. It later develops into a central lobe corresponding to fused wings. Branching is essentially

[3] HOFMEISTER, W., Higher Cryptogamia. Ray Society. 1862.

[4] LEITGEB, H., Untersuchungen über die Lebermoose. 1882. Vols. II and III. Jungermannieen.

dichotomous; although the thallus may appear to have a central axis along the main line of growth, it is a matter of comparative rapidity in growth rather than origin. This form of apical cell is continued throughout the antheridial period.

The fourth period of growth is terminal and regional; it is con-cerned with the production of the archegonial pocket. The cells surrounding the original apical cell assume the power of cutting off lateral and posterior segments. The region of growth is in the form of a terminal disk. Lateral segments are crowded out on all sides to produce the continuous, cup-shaped involucre. A cross-section in any plane is similar to a horizontal section through the growing region at the time of branching (fig. 49). This growth is checked by the production of archegonia; any of the surface cells of the pocket may produce an archegonium (fig. 40). About the time of fertilization the last period of growth begins.

Any further growth is in connection with the developing sporo-phyte. Starch accumulates in the cells surrounding the foot and a massive growth takes place. Usually only one sporophyte develops in each pocket; a rather thick calyptra is formed about it, and sterile archegonia are carried along; these are to be seen attached to the surface of the calyptra. The first growth of the gametophyte is massive, similarly the last.

Relation of antheridium and archegonium

The relationship of the various forms of antheridia and their relation to the archegonium is demonstrated by the occurrence of antheridia in a single species, *Péllia epiphylla*, which are similar in development to each of these organs. The Marchantiales condi-tion is generally to be found among the first antheridia; the Junger-manniales form is most dominant at the middle period, occurring, however, throughout the complete antheridial period. The arche-gonial form is usually found among the last antheridia to be pro-duced. There is an evident time relation between these forms. Moreover, one may be regarded as derived from another through a series of progressive sterilizations (text fig. 1).[5] In the form

s Cf. HOFMEISTER, The higher Cryptogamia. Untersuchungen über die Leber-moose. Vol. II (*Pellia*).

characteristic of Marchantiales (figs. *A*, *B*, *C*), the vertical median wall (1) is followed by two walls (2 and 3) at right angles to the former, thereby forming quadrants (*A*, *B*, *C*, *D*). Four periclinal walls (4, 5, 6, 7) form a sterile wall cell and a spermatogenous cell

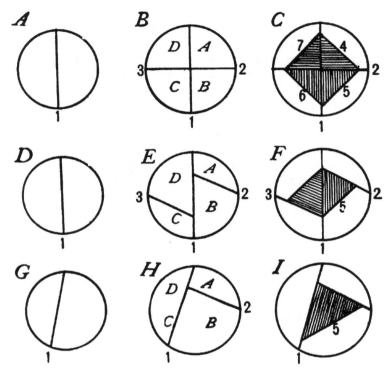

FIG. 1.—Explanation in text

from each quadrant. In the Jungermanniales form (figs. *D*, *E*, *F*) the median vertical wall is followed by two walls corresponding to the second and third above, but somewhat inclined (2, 3, fig. *E*). Periclinal walls (5, 7) form two centrally placed spermatogenous cells. The quadrants *A* and *C* have become sterile wall cells; only the quadrants *B* and *D* give rise to spermatogenous cells. In the

archegonial form (G, H, I) the first wall (1) is not median, but somewhat laterally placed and inclined; there is no wall no. 3; the
section corresponding to the quadrants C and D does not divide,
but remains as a sterile wall cell. Wall 2 again cuts off a section
corresponding to quadrant A, which also persists as a sterile wall
cell. A periclinal wall (5) divides the section corresponding to
quadrant B into a wall cell and a spermatogenous cell. Three
of the quadrants have become sterile. Starting with the Marchantiales form of antheridia, by the sterilization of alternate
quadrants the Jungermanniales form is derived; and by the
sterilization of three quadrants the archegonial form results. The
sterilization sequence corresponds to the time sequence as described above.

Only one step is lacking in this series to complete the transition
from the antheridium of Marchantiales to the archegonium, namely
the reduction of the spermatogenous mass into a single row of cells,
only one of which, the egg, shall be functional. Such transition
forms have been described by DAVIS[6] in Marchantiales. The
writer has seen such "archegonia" with gamete masses in sections
prepared by Dr. W. J. G. LAND. *The antheridial forms of P. epiphylla furnish evidence that the various types of antheridia and the
archegonium have had a common origin*, possibly gametangia resembling in structure the antheridium of Marchantiales.

Relationships

The genetic relationships of *P. epiphylla*, because of its diversity,
present a rather complex problem. With respect to its relation
to other species of the genus *Pellia*, if the other species retain
throughout life the form of apical cell ascribed to them, it would
follow that *P. epiphylla* has branched sooner from the general
line of progress and has retained and developed generalized characters. As a member of the Jungermanniales, this species is
acrogynous in the sense that growth is checked, apical growth is
stopped, by the production of archegonia. It differs from the more
characteristic Acrogynae in having several regions of growth, each
of which is checked in the same way. It is possible that acrogyny

[6] DAVIS, B. M., The origin of the archegonium. Ann. Botany 17: 477-492. 1903.

has been reached by several lines of development; hence close relationship with "Acrogynae" is not necessarily implied. If the development of the antheridium may be taken as a basis of classification, it would seem that *Pellia* arose from the main line of advance before the two branches, Jungermanniales and Marchantiales, became separate, even before the archegonium and antheridium became definitely differentiated in their methods of development.

From *Pellia* we have evidence regarding the relation of Jungermanniales and Marchantiales. There is in the life history a transition from the cuneate apical cell of Marchantiales form to the lenticular cylindrical apical cell found in certain Jungermanniales; similarly, the antheridium, already discussed, affords strong evidence for the existence of such a relation. The evidence tends to indicate that Marchantiales are primitive and that Jungermanniales are derived.

Summary

The antheridium.—The development varies. The dominant method is that characteristic of Jungermanniales; forms occur, not infrequently, which are like the antheridium of Marchantiales, while others are like the archegonium in their early development.

The archegonium.—The archegonia are produced from cells of the apical group and occur in an archegonial pocket. The diversities from the regular form are few; the large number of neck canal cells, the extreme development of the cap, the frequent reduction of the number of tiers of neck wall cells to five, and the somewhat massive venter may be noted. The outer cell of the two resulting from the division of the archegonial initial divides horizontally before the vertical wall is formed.

Methods of growth.—Several periods of growth may be recognized, each having a specific method of growth: the massive; the period of the cuneate apical cell extending until antheridium formation; the period of the lenticular cylindric apical cell, or the antheridial period; the period of regional apical growth, or the period of archegonium production; and the second period of massive growth, or the period of sporophyte dependence.

HUTCHINSON on PELLIA

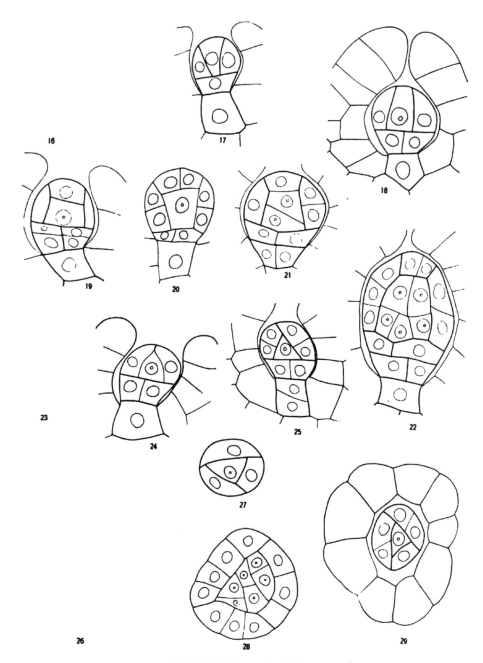

HUTCHINSON on PELLIA

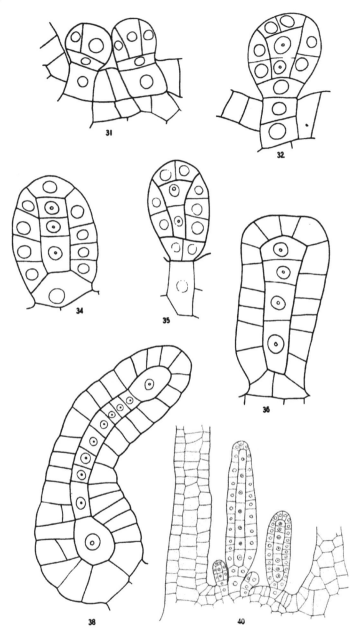

HUTCHINSON on PELLIA

15

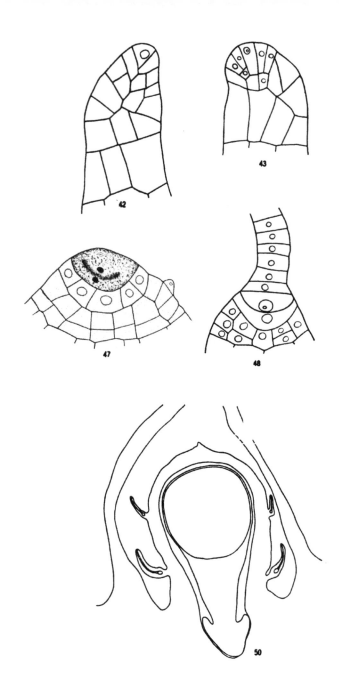

42

43

47

48

50

HUTCHINSON on PELLIA

I am greatly indebted to Professor JOHN M. COULTER for suggestions and criticisms received, and to Dr. W. J. G. LAND, under whose direction the work was done.

UNIVERSITY OF CHICAGO

EXPLANATION OF PLATES I–IV

All drawings were made with the aid of the Abbé camera lucida. Original magnification of figs. 15, 40, and 50 was 500; of all other figures 1730. The reduction in reproduction is one-third.

FIGS. 1–29.—The antheridium of *Pellia epiphylla*.

FIGS. 1–6.—Longitudinal sections of a young antheridium, such as is characteristic of Jungermanniales.

FIGS. 7–9.—Cross-sections of the same.

FIGS. 10–12.—Cross-sections of antheridia of *Pellia* similar to those of *Sphaerocarpus* or Marchantiales.

FIGS. 13, 14.—Sections through the stalk (fig. 13) and spermatogenous region (fig. 14), the latter illustrating two methods of growth.

FIG. 15.—A double antheridium.

FIGS. 16–22.—Longitudinal sections of antheridia showing a development similar to that of an archegonium.

FIGS. 23–25.—Similar to last, with the elimination of the cap cell.

FIGS. 26–29.—Cross-sections of the same.

FIGS. 30–40.—The archegonium.

FIGS. 30–38.—Longitudinal sections.

FIG. 39.—Cross-section.

FIG. 40.—Cross-section of archegonial pocket.

FIGS. 41–50.—Growth structures.

FIG. 41.—A cuneate apical cell in young gametophyte which has just escaped from the spore coat.

FIGS. 42, 43.—Sections of sporeling showing the segmentation of the cuneate apical cell; fig. 42, vertical; fig. 43, horizontal section.

FIGS. 44, 45.—Transition forms of apical cells.

FIGS. 46–48.—Sections of the lenticular cylindric apical cell, also showing the region surrounding; fig. 46, a transverse vertical section; fig. 47, a longitudinal vertical section; fig. 48, a marginal longitudinal vertical section, showing how the lateral segments produce the wing or lobe.

FIG. 49.—A section through a region of apical growth.

FIG. 50.—A young sporophyte inclosed by a somewhat massive calyptra and the involucre; sterile archegonia on calyptra.

THE SCHUMANN RAYS AS AN AGENT FOR THE STERILIZATION OF LIQUIDS

W. T. BOVIE

(WITH ONE FIGURE)

The early investigators were unsuccessful in their attempts to use light as a sterilizing agent, partly because of a lack of knowledge concerning the nature of the action of ultra-violet light upon living organisms, and partly because of the inefficiency of the sources of light at their command.

FINSEN and his co-workers greatly extended our knowledge of the action of ultra-violet light upon living organisms, and KUCH (6) in 1905 (working in the laboratories of the firm of W. C. HERAEUS) invented the quartz mercury vapor arc, a source of light very rich in the ultra-violet rays.

NOGIER and THEVENOT (8) in 1908 were the first to make use of the quartz mercury vapor arc in the study of the biological action of light, and COURMONT and NOGIER (3) in 1909 advocated the use of the quartz mercury vapor arc for the commercial sterilization of potable waters. They found that water which was contaminated with *Bacillus coli* and *B. Eberth* was sterilized in 1–2 minutes by a quartz mercury vapor lamp 3 dm. in length, using 9 amperes of current. The water passed through an iron tube which inclosed the quartz lamp.

In the same year HENRI and STODEL (5) claimed to have sterilized milk with a quartz mercury vapor lamp. They performed many experiments and were certain that the milk was sterilized and that there were no bad effects on the milk such as are produced in the sterilization of milk by heat.

COURMONT and NOGIER (2) pointed out the fact that ultra-violet light does not penetrate colloidal solutions, and DORNIC and DAIRE (4), in a paper on the use of ultra-violet rays for sterilization in the brewing industries, pointed out that the light could not sufficiently penetrate milk or cream to sterilize them, and that, even if it did, the method would not be practical, because the ozone formed would be harmful to the milk and cream. They were not able to sterilize water completely by ultra-violet light, but they found that the number of bacteria was greatly reduced by the treatment.

BILLON-DAGUERRE (1), nephew of the inventor of the daguerreo-type, described in 1909 a new quartz lamp for the sterilization of liquids. The source of light was a quartz discharge tube filled with rarefied hydrogen gas. Other gases were also used. Such a lamp, according to BILLON-DAGUERRE, emits Schumann rays. In a paper published in 1910 he figured an improved lamp and asserted that with this lamp water containing 29,000 bacteria (*Bacillus coli*) per cc. could be sterilized at the rate of 5 liters per minute. The discharge tube of the improved lamp was 25 cm. long and 20 mm. in internal diameter. It was excited by the current from the secondary of an induction coil which gave a spark 15 mm. in length; the primary of the induction coil was operated by a current of 2 amperes at 6 volts. He said that the lamp was more than 20 times as efficient as the mercury vapor arc.

URBAM, SCAL, and FEIGE (9) questioned the efficiency of Schumann rays for sterilizing, because of their small penetrating power. They recommended a lamp with terminals of carbon and of aluminum for sterilizing large quantities of water.

LYMAN (7) studied the absorption by water of the Schumann rays and showed that light of wave-length 1750 Ångström units was completely absorbed by a layer of water 0.5 mm. thick. He pointed out the improbability of the results claimed by BILLON-DAGUERRE.

The Schumann rays are undoubtedly very active chemically, but they have such a small penetrating power that it seemed improbable to Professor LYMAN (at whose suggestion the writer undertook this work) that in a sterilizer such as described by BILLON-DAGUERRE the bacteria would be exposed to sufficient light to kill them. BILLON-DAGUERRE passed the water, which he was attempting to sterilize, through his apparatus at so rapid a rate that sufficient exposure for sterilization seems impossible, since the organisms in the water could be affected by the rays only during the times when they were in very close proximity to the discharge tube.

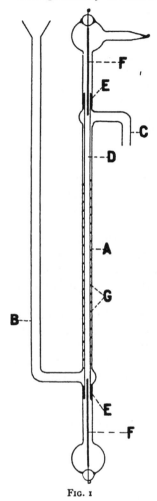

FIG. 1

In order to test the efficiency of BILLON-DAGUERRE's method, a sterilizer similar to the one which he described was constructed by the writer, and distilled water, contaminated with various bacteria, was passed through it. Although the water passed through at a much slower rate than the one given by BILLON-DAGUERRE, the water was not sterilized. Another sterilizer was then constructed in which the organisms could receive a much longer exposure; the discharge tube was longer, the volume of liquid exposed at any one time was much less, and the chance of the bacteria coming in close contact with the discharge tube was much greater, as the liquid was caused to circulate around the discharge tube. The construction of the improved sterilizer is shown in fig. 1. A glass tube A with bulbs on each end, into which electrodes F were sealed, had side tubes as shown at B and C. The electrodes were of aluminum. The tube B served as the inlet and C as the outlet for the liquid under treatment. Within the glass tube there was a very thin-walled tube of quartz (D). The distance between the outside of the quartz tube and the inside of the glass tube was 0.3 mm. The quartz tube was sealed to the glass tube at E, E with DeKotinski cement, so that the space between the two tubes was not in communication with the remainder of the interior of the glass tube. A gold wire was drawn down exactly to fit this space. The wire was coiled spirally around the quartz tube as shown at G. The distance between the inlet tube B and the outlet tube C was 16 cm. The quartz tube was 5 mm. in outside diameter, and the gold wire made 15 turns about the quartz tube. The solution under treatment was thus spread into a thin layer, 0.3 mm. in thickness, and took a spiral course through the sterilizer. The discharge tube was filled with rarefied unwashed hydrogen, containing carbon dioxide as an impurity, since a discharge tube filled with pure hydrogen would not emit light of wave-lengths between 2000 and 1675 Ångström units. The carbon dioxide which the tube contained filled in this gap. This is important, for it is highly improbable that the strong 1600 lines of the hydrogen spectrum, because of their small penetrating power, could be effective in sterilizing the water. Any virtue which the sterilizer might have must be due to its emitting light between wave-lengths in 1850 and 1675 Ångström units.

A discharge tube filled with carbon dioxide is short-lived. After the tube has been excited a few hours, the carbon dioxide disappears and is replaced by hydrogen. To insure a carbon dioxide spectrum, the discharge tube was filled and pumped before each experiment.

The discharge tube was excited by a current of 14 milliamperes, and Cambridge city water was passed through the sterilizer at the rate of 10 cc. per minute. By this treatment the bacterial count of the water was reduced from 50 to 4 per cc., and the fungal count from 3 to 1 per cc. In none of the experiments was the water completely sterilized.

It is apparent that a hydrogen discharge tube such as is described by BILLON-DAGUERRE is not an efficient source of light for photo-sterilization.

LABORATORY OF PLANT PHYSIOLOGY
HARVARD UNIVERSITY

LITERATURE CITED

1. BILLON-DAGUERRE, Stérilisation des liquids par les radiations de très courte longuer d'onde. Compt. Rend. 149 and 150: 1909.

2. COURMONT et NOGIER, Sur la faible pénétration des rayons ultraviolets à travers des liquides contenant des substances colloides. Compt. Rend. 149:364. 1909.

3. ———, Sur la stérilisation de l'eau potable au moyen de la lampe en quartz à vapeurs de mercure. Compt. Rend. 148:523. 1909.

4. DORNIC et DAIRE, Contribution à l'étude de la stérilisation par les rayons ultraviolets. Application à l'industrie beurrière. Compt. Rend. 149:354. 1909.

5. HENRI et STODEL, Stérilisation du lait par des rayons ultraviolets. Compt. Rend. 148:582. 1909.

6. KUCH et RETCHINSKY, Photometrische und spektralphotometrische Messungen am Quicksilberlichtbogen bei hohem Dampfdruck. Ann. Phys. und Chem. 20:563. 1906.

7. LYMAN, T., The sterilization of liquids by light of very short wave-lengths. Nature 84:71. 1910.

8. NOGIER et THEVENOT, Congrès pour l'avancement sci. Clermont. 1908.

9. URBAM, SCAL, et FEIGE, Sur la stérilisation de l'eau par l'ultraviolet. Compt. Rend. 158:770. 1910.

SOME FILAMENTOUS FUNGI TESTED FOR CELLULOSE DESTROYING POWER[1]

FREEMAN M. SCALES

Every year large quantities of cellulose in the various combinations occurring in plant tissue are returned to the soil. In this tissue there is also a great deal of carbohydrate material, such as sugar, starch, pectose, and hemicellulose. These substances are more easily available and will under ordinary circumstances be all used up before the cellulose is attacked. In spite of this abundance of easily available food, at the end of several months practically all the plant tissue will be disintegrated and split into soluble substances and humus. The destruction of this large quantity of cellulose in such a comparatively short time means that the cellulose destroying organisms must work very vigorously. By special methods of culture, cellulose dissolving bacteria have been obtained which hydrolyze this complex material very rapidly, but when plating on cellulose agar directly from a soil for the isolation of cellulose destroying organisms the filamentous fungi usually grow more abundantly and destroy more cellulose than do the bacteria. In order to determine whether the addition of cellulose to soil makes any difference in the number of molds in it, 2 per cent of cellulose was added to 200 grams of soil which was moistened to the optimum with distilled water and then incubated for 30 days at 30° C. The initial count of this soil on cellulose agar was 20,000 mold colonies. A check sample which received no cellulose, but otherwise had the same treatment as the one described, gave on the same kind of media a count of 100,000 mold colonies, while the sample which received cellulose showed a mold content of 200,000,000. · It is evident from these data that the filamentous fungi are an important factor in the dissolving of cellulose in the soil.[6]

It is a well known fact that filamentous fungi are very numerous in woodland soils and also that they are abundant in acid arable

[1] Published by permission of the Secretary of Agriculture.

soils where the reaction is naturally favorable for them. MARCHAL[2] found numerous mycelia of molds in acid humus soils which were naturally rich in organic material, and he believed that they took an active part in the mineralization of organic nitrogen under these conditions, but that in an arable soil under active cultivation the molds were relatively few in number owing to the alkaline reaction and the absence of large quantities of organic matter. While the filamentous fungi are naturally thought of as occurring in great numbers in acid soils, still the recent work of DALE[3] in England indicates that in some cultivated alkaline soils they are just as numerous. JENSEN[4] at Ithaca found a great variety of filamentous fungi in alkaline cultivated soils. Lately TRAAEN[5] has reported the results of his investigation of the species and physiological activity of some of the soil fungi of Norway. In our own investigations on the destruction of cellulose by bacteria and filamentous fungi, we worked almost entirely with alkaline soil that had been under active cultivation, and found among the fungi occurring in both great variety and enormous numbers many species active cellulose destroyers. The physiological activities of these organisms on organic material, whether nitrogenous or not, are well recognized. All these data indicate that the filamentous fungi are important factors in carrying on the biological processes in cultivated soils.

In further study[6] of the destruction of cellulose by microorganisms, 19 species of cellulose destroying filamentous fungi were identified and two new ones found. One of these species was found[7] to produce a very active cytase. The present work was

[2] MARCHAL, EMILE, Sur la production de l'ammoniaque dans le sol par les microbes. Bull. Acad. Roy. Belgique III. **25**:741. 1893.

[3] DALE, ELIZABETH, On the fungi of the soil. Ann. Mycol. **10**:452–477. *pls. 5.* 1912; **12**:33–62. *pls. 5.* 1914.

[4] JENSEN, C. N., Fungous flora of the soil. N.Y. (Cornell) Agric. Exp. Sta. Bull. 315. pp. 415–501. 1912.

[5] TRAAEN, A. E., Untersuchungen über Bodenpilze aus Norwegen. Nyt. Mag. Naturv. B **52**:20–121. *pl. 1.* 1914.

[6] McBETH, I. G., and SCALES, F. M., The destruction of cellulose by bacteria and filamentous fungi. U.S. Dept. Agric., Bur. Pl. Ind. Bull. 266. 1913.

[7] KELLERMAN, K. F., Formation of cytase by *Penicillium pinophilum.* Bur. Pl. Ind Circ. no. 113. pp. 29–31. 1912.

undertaken for the purpose of determining more species of the filamentous fungi that are capable of exercising this function.

Dr. CHARLES THOM, mycologist of the Bureau of Chemistry, kindly supplied us with the 30 species of *Penicillium* and 10 species of *Aspergillus* that were used in this work. The cellulose destroying power of these organisms was determined with two different sources of nitrogen. An ammonium sulphate cellulose agar and a peptone cellulose agar were used for this purpose. The ammonium sulphate medium was prepared by the method described in a previous publication,[6] and the peptone medium by substituting 1 gram per liter of peptone for the ammonium sulphate.

The stock cultures were kept on Czapek's agar. The spores from fresh growths on this medium were added to freshly poured duplicate cellulose agar plates of both media. The plate cultures were inoculated in four places. Test tube cultures on both media were also made in duplicate at the same time. The Petri dishes were placed in moist chambers and then put in the incubator along with the tube cultures, which were kept at 28–30° C. for two months. They were then examined for an enzymic zone either around the colony on the plates or underneath it in the tubes. The size of this enzymic area varied considerably in the different cultures, ranging from a very thin clear zone less than 1 mm. to one 24 mm. deep in some of the tubes. Although the depth of the clear zone was fairly constant in the duplicate tubes, this phase of the work will not be emphasized until more data are obtained.

As the surface growth on cellulose agar is frequently scanty, with few definite microscopic differences, it seemed best to transplant from the plates and tubes onto Czapek's agar, in order to check each organism that destroyed cellulose against a growth on Czapek's agar made from the stock material of this species. A microscopic examination was made in each case.

The results are given in the accompanying table. The apparent failure of a number of the organisms fermenting cellulose with ammonium sulphate as the source of nitrogen to do the same with peptone as the source of nitrogen may be accounted for in two ways. One is that the organisms produced such an abundant sterile growth in the medium that any slight clearing was obscured;

the other is that the carbon in the peptone was utilized, being more easily available, and the cellulose was not attacked.

TABLE I

RESULTS OF THE TEST TO DETERMINE THE CELLULOSE DESTROYING POWER OF SOME FUNGI WITH DIFFERENT SOURCES OF NITROGEN

+ indicates cellulose destroyed; − indicates cellulose not destroyed.

Species	Ammonium sulphate	Peptone
1a. Penicillium luteum Zukal	+	+
2a. " pinophilum Hedg...................	+	−
3a. " rugulosum Thom....................	+	+
4a. " sp. no. 2670 Thom... .	+	−
5a. " purpurogenum O. Stoll.... . . .	+	+
6a. " duclauxi Dela.	+	....
7b. " commune Thom..	−	−
8b. " biforme Thom.....................	−	−
9b. " sp. no. 66 Thom.	−	−
10b. " sp. no. 13 Thom...	+	+
11c. " expansum Link.	+	+
12c. " sp. no. 2694 Thom . .	+	−
13d. " chrysogenum Thom..	+	+
14d. " notatum West	+	....
15d. " sp. no. 12 Thom...	+	....
16e. " camemberti Thom. .	−	−
17e. " camemberti var. rogeri Thom	−	−
18f. " intricatum Thom...	+	+
19f. Spicaria simplisissima Jensen....	+	−
20. Penicillium lanosum West.....	+	+
21. " claviforme Bainier.	+	+
22. " granulatum Bainier... . ..	+	+
23. " roqueforti Thom.	+	+
24. " spinulosum Thom........... ...	+	−
25. " funiculosum Thom........ ..	+	....
26. " . lilacinum Thom.. .	−	−
27. " divaricatum Thom....... .	−	−
28. " sp. no. 64 Thom....................	+	+
29. " sp. no. 3505 Thom....	+	−
30. Scopulariopsis repens Bainier....	+	−
31. Aspergillus candidus Link	−	−
32. " clavatus Desm..... .	+	−
33. " flavus Link	+	+
34. " fumigatus Fresen...................	+	+
35. " nidulans Eidam	+	−
36. " niger VanTiegh........... . .	+	+
37. " oryzae Ahlb....	+	−
38. " Wentii Wehmer.	+	−
39. " sp. no. 144 Thom*...	+	+

*Another culture of this species sent in from the Riverside Experiment Station, California, by Mr I. G McBeth, gave the same reactions

Some cultures of *Actinomyces* that have been accumulated for a classification of this group were also tested, and of 31 cultures, 8 dissolved the cellulose agar containing ammonium sulphate as a source of nitrogen. These cultures all show marked microscopic differences and are without doubt different species. These 8 organisms may be of the same species as some of the 12 cellulose destroying species which KRAINSKY[8] has recently described.

SOIL BACTERIOLOGY LABORATORY
U.S. DEPARTMENT OF AGRICULTURE

[8] KRAINSKY, A., Die Aktinomyceten und ihre Bedeutung in der Natur. Centralbl. Bakt. 41:649–688. 1914.

BRIEFER ARTICLES

THE RELATIVE IMPORTANCE OF DIFFERENT SPECIES IN A MOUNTAIN GRASSLAND

A careful study has been made by the writer through four seasons upon an area of dry grassland in a mountain park at Tolland, Colorado. The park (altitude 8889 feet) is a widened area in the valley of South Boulder Creek. The dry grassland covers most of those parts of the park that lie from 10 to 50 feet above the stream level.

As a part of the investigation 16 quadrats, each a meter square, were staked off and examined from time to time through the four seasons of study. Most of these quadrats are on morainic material, but a few are on the upper creek terraces. In the quadrats 64 species of plants were found, while the entire dry grassland area of the park showed 62 species in addition.

Estimates were made, as explained in a former paper,[1] of the amount of bare ground in each quadrat and also the area covered by each species of plant. By combining the figures for the different quadrats, the relative abundance of the various plants was determined. While the figures for the less frequent species are of little value, those for species occurring in any considerable number of quadrats show well the relation of these plants to the composition of the association as a whole. It is certain that every plant of frequent occurrence in the dry grassland of the park is represented in a number of quadrats.

The data for July 1913 are gathered together in table I. These midsummer records have been selected for presentation as probably of greater interest than would be the figures for spring or autumn. In certain genera two or more species of similar ecological nature have been put together as one item, the names arranged in order of importance.

According to the records, the two species present in the quadrats in greatest abundance are *Artemisia frigida* and *Aragallus Lambertii*, but *Muhlenbergia gracilis* and *Carex stenophylla* are almost equally important. Plants of the different species of *Carex* taken together cover a larger part of the area than do those of any single grass genus, but the grasses as a whole are of much greater importance than the sedges. The most abundant grasses are the species of *Muhlenbergia*,

[1] Bot. Gaz. **57**:526–528. 1914.

TABLE I

PERCENTAGE COMPOSITION OF THE DIFFERENT QUADRATS, JULY 1913

	I	II	III	IV	V	VI	VII	VIII	XI	X	XII	XIIa	XIII	XIV	XV	XVI	Total	Percentage of ground covered	Percentage of total vegetation
Bare ground	30	33	50	13	20	28	20	45	20	10	15	15	20	25	25	30	399	24.94	......
Small gray lichen							1								2	1	10	0.63	0.84
Selaginella densa				8	10	3	20	10	2	5	5				5		69	4.31	5.75
Agropyron violaceum	2	2				2			2	3	2				10		41	2.56	3.41
Agrostis hiemalis	2																2	0.13	0.17
Avena americana			1						5	5			10				16	1.00	1.33
Blepharineuron tricholepis											2						2	0.13	0.17
Bromus Pumpellianus						1				2	2						7	0.44	0.59
Danthonia Parryi and intermedia									5	2	5		10			4	21	1.31	1.75
Festuca ingrata, saximontana, rubra	5			5	12	19			10		5	26				3	92	5.75	7.67
Koeleria cristata	2				10	5			15		5	5			16	1	38	2.38	3.17
Muhlenbergia gracilis and subalpina			15		12				15		10	25					97	6.06	8.08
Poa interior, crocata, rupicola, sub-purpurea	3		3	5	5	3			2	10	15				2	5	53	3.31	4.41
Sitanion elymoides			3						1		4	1					5	0.31	0.41
Stipa comata, minor, Nelsoni					2				29		20				16		32	2.00	2.66
Carex filifolia								21	2	5			20		2	20	41	2.56	3.41
Carex stenophylla and pennsylvanica	2	25	4		3	1			2		2	2	20		2	1	100	6.25	8.33
Juncus balticus				2		5				5	2						22	1.38	1.84
Allium recurvatum							1						1				1	0.06	0.08
Comandra pallida					3				2	2	2						12	0.75	1.00
Eriogonum umbellatum					5				10	5	5				2		20	1.25	1.67
Arenaria Fendleri				1		1		1				1		1	1	2	8	0.50	0.66
Cerastium occidentale					1			1	2	5	3	1		1	1	5	31	1.94	2.58
Silene Hallii		2										9				1	4	0.25	0.33
Erysimum Wheeleri			3												1		1	0.06	0.08
Sedum stenopetalum	2	2		1	3	1	1	3	2	2	2	2	1		5		25	1.56	2.08
Saxifraga rhomboidea												1					1	0.06	0.08

TABLE I—*Continued*

	I	II	III	IV	V	VI	VII	VIIᵃ	VIII	XI	X	XII	XIIᵃ	XIII	XIV	XV	XVI	Total	Percentage of ground covered	Percentage of total vegetation
Potentilla ... strigosa, ...ina	1	2		1			1		1	1		5				11		17	1.06	1.41
Potentilla Hippiana, gracilis, Nuttallii	2	2		2			2				5							21	1.31	1.74
...us deflexa	17			12	5	10			10									19	1.19	1.58
...us Lambertii		5	5	4		2				1	5	5	5	5	20		2	79	4.94	6.58
Aragallus Richardsonii	20									1								37	2.31	3.08
Astragalus Parryi										1								1	0.06	0.08
Pseudocymopterus tenuifolius				1	1		1	1					1					3	0.19	0.25
...iam ...am				2			20				5				10	1		20	1.25	1.66
...le ...inta																		1	0.06	0.08
...la Parryi	1		5	2	1	2	3	5		1	2		2		2	1		4	0.25	0.33
...Mia Bal eri				2	5		2	1				2				2	10	26	1.62	2.16
...ris ...us							3	2			5							18	1.12	1.49
...im ...us						5	2	2				3						5	0.31	0.41
...ea ...lata						8	2	5			5		5				4	12	0.75	1.00
...la ...al ea																	5	16	1.00	1.33
Antennaria ...na				1	1		1	2		1					2		1	5	0.31	0.41
Artemisia aromatica	10	5	5	2	5	5	4	1	8	5	10	5	2	5		3	4	16	1.00	1.33
Artemisia frigida	1	25	3	5	5	8					5	1	2	2	2	3	5	90	5.62	7.49
Chrysopsis foliosa	1					5		2			5	1	1				1	28	1.75	2.33
Erigeron formosissimus																		1	0.06	0.08
Erigeron trifidus	2	1	2	2		5		1				2		2			8	13	0.81	1.08
Gaillardia aristata								1										10	0.63	0.84
Solidago decumbens																		5	0.31	0.41
Troximon glaucum		2						1										3	0.18	0.24
Total	100	100	100	100	100	100	100	100	100	100	100	100	100	100	100	100	100	1600	99.97	99.94

followed in order by the species of *Festuca, Poa, Agropyron, Koeleria,* and *Stipa.* If the dry grassland were to be named by its chief generic constituents, it would be called a *Carex-Artemisia-Aragallus-Muhlenbergia* association.

From table II it will be seen that in the midsummer of 1913 two-

TABLE II

THE CHIEF CONSTITUENTS OF THE VEGETATION OF THE QUADRATS

Names of plants	Percentage of ground covered	Percentage of vegetation
Selaginella densa..	4.31	5.75
Agropyron violaceum.	2.56	3.41
Festuca ingrata, saximontana, rubra...	5.75	7.67
Koeleria cristata.	2.38	3.17
Muhlenbergia gracilis and subalpina...	6.06	8 08
Poa crocata, interior, rupicola, subpurpurea	3.31	4.41
Stipa comata, minor, Nelsoni.	2.00	2.66
Carex filifolia	2.56	3.41
Carex stenophylla and pennsylvanica. .	6.25	8.33
Cerastium occidentale.	1.94	2.58
Aragallus Lambertii	4.94	6.58
Aragallus Richardsonii............	2.31	3.08
Artemisia frigida.....................	5.62	7.49
Total...........................	49 99	66 62

TABLE III

THE MOST IMPORTANT PLANT FAMILIES AND THEIR PART IN THE VEGETATION OF THE QUADRATS

Names of families	Percentage of ground covered	Percentage of vegetation
Poaceae (19 species) .	25.38	33.82
Carduaceae (10 species)	11.49	15.30
Cyperaceae (3 species).	8.81	11.74
Fabaceae (4 species)	8.50	11.32
Selaginellaceae (1 species).	4.31	5.75
Total...........................	58.49	77 93

thirds of the vegetation of the quadrats was made up of plants listed in 13 items in the first table. A certain few plants that are widely distributed do not make up a large percentage of the ground cover and are therefore not listed in table II. The most important of these are *Juncus balticus, Arenaria Fendleri, Sedum stenopetalum, Mertensia Bakeri,* and *Chrysopsis foliosa,* each of which was found in 9 or more of the quadrats.

The five plant families best represented in the quadrats are shown in table III.—FRANCIS RAMALEY, *University of Colorado, Boulder, Colo.*

CURRENT LITERATURE

Plant poisons and stimulants

The first volume of the Cambridge *Agricultural monographs*[1] covers a very much more restricted field than the title would lead one to expect, since it deals only with the effects, mainly upon the higher plants, of compounds of the five elements, copper, zinc, arsenic, boron, and manganese. Miss BRENCHLEY has previously published some of the results of work with these compounds which she has been carrying on since 1907 at the Rothamstead Experimental Station, [2, 3] these results being here brought together in connection with a résumé of certain portions of the related literature.

An introductory chapter of six pages points out that the classification of the elements into the three groups, nutritive, toxic, or indifferent as respects their action upon plants, no longer holds, and expresses the belief that no such simple grouping is possible. The second chapter describes the water culture methods employed by the author in her work, discusses the comparative advantages and disadvantages of water, sand, pot, and field experiments, emphasizes the necessity for caution in comparing results obtained by these different methods, and asserts that "all crucial experiments have always been and must always be done in water cultures." Then follow five chapters, each devoted to discussion of the physiological effects of compounds of one of the five elements employed. The subdivisions of these chapters deal with such topics as the occurrence of the element in higher plants, its effects upon growth when present alone in water cultures, when present along with nutrient or non-nutrient salts or with insoluble substances, its effects upon growth in soils, its action upon algae and fungi, and its effects upon germination of seeds and spores. A four-page chapter entitled "Conclusions" and a bibliography of 182 titles complete the work.

The general conclusions reached are that compounds of copper, arsenic, and in all probability those of zinc also, do not exert stimulatory effects in any concentration when added to water cultures of higher plants, but are toxic at all concentrations having a discoverable effect. A stimulatory effect of each of

[1] BRENCHLEY, WINIFRED M., Inorganic plant poisons and stimulants. 8vo. pp. 110. *figs. 19* Cambridge: University Press. 1914.

[2] BRENCHLEY, WINIFRED M., The influence of copper sulphate and manganese sulphate upon the growth of barley. Ann. Botany 24:571–583. *pl. 47.* 1910.

[3] ———, On the action of certain compounds of zinc, arsenic, and boron on the growth of plants. Ann. Botany 28:283–301. 1914.

these salts for such fungi as *Aspergillus* and *Penicillium*, and in the case of zinc, for the lower algae also, is definitely shown by the literature.   Boron and manganese compounds in low concentration are stimulatory for higher plants; peas show stimulation by boric acid to a greater degree, by manganese to a less degree, than does barley.   *Aspergillus, Saccharomyces*, and green algae are indifferent to high concentrations of boric acid, while there is no evidence in the literature that it is stimulatory at any concentration.   The contradictory results obtained by RICHARDS and LOEW and SAWA as respects the effects of manganese upon fungi are reviewed, but no opinion is expressed.

The introductory chapter states that "a voluminous literature has arisen around the subject, and in the present discussion some selection has been made with a view to presenting ascertained facts as succinctly as possible.   No attempt has been made to notice all the papers; many have been omitted perforce; it would have been impossible to deal with the matter within reasonable limits otherwise."   A successful attempt of this sort requires that one have at least as great familiarity with what is left out as with what is included.   But it must be said that Miss BRENCHLEY has not this knowledge, and that the character and importance of the omitted literature is such as to make the book very far from "a succinct presentation of the ascertained facts."   For while the various chapters include discussions of the effects upon the fungi and the algae of the five elements under discussion, the citations made betray unfamiliarity with the literature; not only do they fail to give any idea of the enormous extent of the work done in this field, but they are by no means those which add most, either to the body of observed facts or to our present conceptions of the nature and effects of salt action upon lower forms.   To cite important omissions would be to fill pages of this journal, and an example or two must suffice to illustrate a general situation.   Thus, there is nowhere in the several sections dealing with change of form in algae and fungi when grown in salt solutions any mention of osmotic pressure as a significant factor; indeed the words do not occur in the book, and there is no reference to the work of KLEBS, LIVINGSTON, and the host of others who have furnished our present knowledge of this phase of the subject.   Again, while the effects of copper sprays upon foliage, a subject which has long received particular attention at the hands of American physiologists and pathologists, is discussed to the extent of three pages, no American worker finds a place among the seven whose work is cited, and the most recent paper mentioned appeared in 1908.   Such omissions would be surprising in any case; they become inexcusable when we recall that the very recent paper of PICKERING and the DUKE OF BEDFORD, certainly readily accessible to any English worker, not only furnished important results of work done at the Woburn Fruit Farms, but also admirably summarized the results of others up to the time of its publication.   That the author reaches conclusions as to the effects of Bordeaux mixture upon assimilation and transpiration which are diametrically opposed to those of most recent workers, as for example REED and DUGGAR, is therefore not surprising.   All the sections dealing with the action of salts upon the lower plants are of this character, not only failing wholly to present the literature adequately, but making such bizarre selections therefrom as to force the conclusion that the papers discussed were brought together, like Cain's rejected sacrifice, "unculled, of such as came to hand." It is sincerely to be regretted that these sections of the book were not submitted to some physiologist or pathologist whose acquaintance with this field might have made of them a reliable summary of our present knowledge.

If those portions of the book which are concerned with higher plants had been intended simply as a compilation of results of authors who have recorded visible effects of salts on plants, they would have fallen short of the mark by reason of the many and important omissions both of American and of German work. But the work professes to be critical, and this it emphatically is not. The author nowhere gives expression to her own beliefs or convictions in clear, unmistakable terms, nor does she evaluate for us the ideas of others. Conclusions of the earlier workers, filled as they are with mistaken interpretations of results, are repeated without comment, and unproven assumptions and exploded theories stand side by side with established fact. Consequently, the reader not already thoroughly familiar with the literature will at once lose his bearings and grope his way blindly through a maze out of which he will carry at least as much of fundamental error as of "ascertained facts."

But it is in the sections dealing with the mode of action of toxic compounds upon protoplasm that the book is most vitally defective. The last ten years have been years of extraordinary advance in the study of protoplasmic permeability and of its modifications under the action of external agencies, and the facts gained in this field have been utilized by a host of workers in formulating theories of toxic and antitoxic action as phenomena arising primarily from modification of permeability. The literature dealing with these subjects is readily accessible and has, moreover, recently been summarized in new editions of CZAPEK's *Biochemie* and HOEBER's *Physikalische Chemie;* consequently there are few American physiological laboratories in which the new knowledge has failed to find its way into undergraduate instruction. Consequently it is amazing to find that the author of the book under review has nowhere mentioned any portion of this literature; that LEPESCHKIN, TRÖNDLE, CZAPEK, HOEBER, and RUHLAND are not mentioned; that the conceptions of salt action which we owe to WOLFGANG OSTWALD, MOROWITZ, FLURI, SZÜCS, PAULI, DeRUFZ DE LAVISON, and HANSTEEN CRANNER, with a host of others, nowhere appear, and that the author's ideas of the whole subject are of the character generally held before 1900, when the ion-proteid theory of PAULI and LOEB appeared. A few examples will suffice to indicate this. The discussion of the nature of stimulation confounds increased permeability to water and consequent increase in green weight with true stimulation of protoplasmic activity, as on pp. 2, 3, 75, and this confusion exists throughout. One reads (pp. 40–41) that "it is very striking to see the desperate efforts that badly poisoned pea plants make to reproduce themselves . . . . In the greater strengths of such poisons as zinc and copper sulphate [sic] root growth is checked from the outset, but usually a very little shoot growth is made, and one frequently obtains ridiculous little plants about an inch high bearing unhappy and diminutive flowers, which are occasionally replaced [succeeded is meant] by equally unhappy and miniature fruits." The author fails to apprehend that the source and fount of all this woe is a lowered permeability for water, with a resultant development in which lack of water is the limiting factor. This is also the case with the plants of *Pisum*, *Phaseolus*, and *Zea* described on p. 18; the high concentration of copper sulphate here employed totally inhibited root development, but the resultant strong growth of tops, instead of being due to "stimulation of the shoots by some physiological process or other," as the author thinks, is exactly what we find wherever inhibition of root development, however caused, permits the utilization of the foodstuffs present in the seed solely in the development of aerial parts. The most remarkable passage in the book, however, is undoubtedly that on p. 27, in which we are told that "so long as the solution of copper salts is dilute enough, the absorption layer of the root, acting as a semipermeable membrane and upheld by the resistant protoplasm, is able to keep the copper out of the plant and to check its toxicity. As soon as a certain limit is reached the copper exercises a corrosive influence upon the outer layer of the root whereby its functions are impaired, so that it is no longer able efficiently to resist the entry of the poison. As the concentration increases it is easy to conceive that the harmful action should extend to the protoplasm itself." Just what this "absorption layer" may be the author does not tell us; it seems not to be protoplasmic, and inasmuch as it seems to combine the active rôle of Horatius at the bridge with the more passive function of the Holland dikes, physiologists will regret that we are told no more of its origin, functions, and relation to cell wall and protoplasm than has been quoted.

That the book is exceedingly disappointing will be obvious from these quotations. Physiologists have been awaiting the appearance of a résumé of the whole subject of toxic action which would bring together the extensive observations of the older literature and unify and explain them in the light of the new knowledge of the nature and behavior of the protoplasmic membrane. The author who successfully undertakes the task must have kept abreast of the literature in many fields; with the tremendous advances made in physical chemistry and in our knowledge of the colloids, no less than with the work done directly with the subject in hand, for most help will come from the literature of these related fields. It is just here that Miss BRENCHLEY has failed; her book merely collects a mass of observations which it will be the task of some future physiologist possessing wide training and a modern point of view to organize and explain.

There are various minor omissions and slips of the pen. Thus while formulae for two nutrient solutions are given on p. 13, we are nowhere told, in the various sections dealing with experiments in which plants in nutrient solutions were employed, which of these was used in any particular case The description of HASELHOFF's experiments (p. 25) leaves the reader to wonder whether each sample of soil was extracted with 375 liters of water, divided into 15 equal quantities, or with 25 liters used 15 successive times. The substitution of "below" for "above" in line 8, p. 30, would lead the unwary reader to conclude that the toxicity of copper is decreased as the concentration of the solution increases. The citations of literature are not numbered, and when a given author has a number of papers listed in the bibliography, the reader has no means of knowing what particular citation contains the results quoted in the text. While physiologists have universally adopted the terminology of the normal solution, concentrations are here written as parts per thousand or million, and when quotations from authors who follow the modern custom are made, normal solutions are sometimes converted to parts per million, sometimes have equivalents parenthetically introduced, and are sometimes carried over unchanged. Consequently, comparison of the figures is impossible without resorting to calculation. In the graphs showing dry weight of plants grown in various concentrations of salts, some one concentration, as 1/100,000, is chosen as a "unit," and other concentrations are written in the graph as multiples or fractions of this unit, making easy reference impossible. Those hyphenated and immortal acquaintances of our early youth, "carbo-hydrate" and "photo-synthesis," greet a surprised public in these pages once more, after a generation of absence from the ken of physiologists.

The concluding paragraph gives naïve expression to a point of view which physiologists had supposed to be happily confined to a certain rapidly decreasing class of agricultural workers in this country. After discussing the results obtained by physiological experimentation with manganese and boric acid and those obtained in field trials of stimulatory fertilizers, the author says, "the possibility now exists that in some respects the two lines of work are converging and that the more purely scientific line will have a big contribution to make to the strictly practical line." Those of us who regard plant physiology as the science of economic plant production had thought that the artificial lines of demarcation between "scientific" and "practical" work had long ago disappeared, and that future progress was to be made, not independently or along convergent lines, but by the common utilization of scientific facts and methods in the cooperative attack upon common problems. In view of the fact that the long series of publications from the Rothamstead Experimental Station have set a standard to which comparatively little of the agricultural literature of this country has attained, it is surprising that this book, with its lamentable deficiencies in grasp of subject-matter and in point of view, should have been issued as the initial number of a series of monographs whose general editors are Professor WOOD, of Cambridge, and Director RUSSELL, of Rothamstead. Closer editorial supervision would have withheld the book from publication in its present form, and it is to be hoped that immediate and radical revision may make of it an acceptable and trustworthy guide to the literature of this interesting field.—JOSEPH S. CALDWELL.

NOTES FOR STUDENTS

Experimental embryology.—Morphologists and experimental workers have been aware for some time of a need of greater cooperation between their respective lines of research. Too often morphological or cytological studies are pursued without reference to important physiological conditions, while conclusions are drawn from experimental work which would not be warranted by morphology and cytology. It is becoming increasingly evident that results can be more properly interpreted in the additional light afforded by supplementary researches in a related field. The value of this cooperative method is emphasized in a recent paper by KUSANO[4] on angiosperm embryology. A favorable form for such study was found in the orchid *Gastrodia elata*. Since the inflorescence develops at the expense of material stored in a tuber, it is a simple matter to maintain the plant under normal nutritive conditions. Some of the results of KUSANO's research, which is still in progress, are cited below.

The normal development in *Gastrodia* is as follows: A subepidermal archesporial cell becomes the megaspore mother cell and undergoes the two maturation divisions. In some cases reduction, which is said to consist in a simple pairing and separation of the chromosomes on the equator of the spindle, fails to occur, so that the functioning megaspore and gametophyte are sometimes haploid and sometimes diploid. The embryo sac contains only 4 nuclei, 3 of which are organized as an egg apparatus. This reduced condition is regarded as an economical specialization correlated with the peculiar vegetative habit. Many irregularities which occur are related to poor nutrition. At fertilization, which occurs only in haploid sacs, one of the male nuclei fuses with the egg nucleus, while the other fuses with the single polar and a synergid nucleus. The fertilized egg forms the usual undifferentiated proembryo, which is nourished through the suspensor and large nucellar cells. The endosperm nucleus does not divide. The following time schedule was determined: two days before the flower opens the ovule is yet in a rudimentary stage; the embryo sac is completed 3 days after bloom; fertilization occurs the 4th day after pollination; the fertilized egg divides the 5th day; the seed is completed the 14th or 15th day; the capsule dehisces about the 23d day. This exceptionally rapid development (for Orchidaceae) is correlated with the fact that it occurs at the expense of stored food. Occasionally two archesporial cells arise in a single ovule and undergo complete development.

An extended series of experiments led to the following conclusions: After a few days the fertilizing power of the pollen is lost and the ovules become

---

[4] KUSANO, S., Experimental studies on the embryonal development in an angiosperm. Jour. Coll. Agric. Tokyo 6:7-120. *pls. 5-9. figs. 28.* 1915.

incapable of fertilization. Although the development of ovules and embryo
sacs is promoted by the presence of pollen tubes, they are completed in unpol-
linated flowers unless nutritive conditions are too severe   In such flowers
the egg apparatus may remain in healthy condition 8 or 9 days.   In some
cases it then degenerates; in other cases the ovule forms an embryoless seed,
the number of these seeds in the fruit depending upon nutrition.   Such embryoless seeds may develop from ovules with either the haploid or the diploid egg.

If the floral axis is separated from the tuber, scarcity of food material
causes an imperfect development of ovarial and ovular tissue in pollinated
flowers, but it does not interfere with the number of ovules going into seed
formation nor with the development of the embryo.   If pollinated flowers are
cut from the floral axis and kept moist, fertilization occurs but the ovary does
not grow.   Some of the ovules develop into seeds with imperfect coats, while
others degenerate after one or two embryonal divisions have occurred.   The
effects of reduced nutrition are here manifest.   Of all the parts the embryo is
the least liable to be retarded in development by poor nutritive conditions.   It
is evident that the embryo can develop without the accompanying development
of the ovule or ovary tissue, and that the seed coat may likewise form without
the development of the embryo or ovary wall.   The growth of the latter,
however, seems to be dependent upon the development of the ovules.

Self-pollination appears to be as effective as cross-pollination in *Gastrodia*.
The pollinium will germinate within the cavity of the ovary and effect fertilization.   Pollination with foreign (*Bletia*) pollen resulted in fruits and seeds of
normal form and size, but the seeds were without embryos; no fertilization
occurred.   If both *Gastrodia* and *Bletia* pollen are placed on the same stigma
the fertilizing activity of the *Gastrodia* pollen is greatly hindered.

With regard to parthenocarpy, the author recalls the distinction between
vegetative (WINKLER) or autonomic (FITTING) parthenocarpy, in which
embryoless fruits are formed without the agency of any external factor, and
stimulative (WINKLER) or aitionomic (FITTING) parthenocarpy, in which the
formation of embryoless fruit is induced by pollen or some other agent.   Both
types occur in *Gastrodia;* the aitionomic fruits are of normal size, while the
autonomic ones are much smaller.   The former appear to be well developed
because the pollen or other agent stimulates nutritive activity; it seems that
the size of the fruit may depend upon the intensity of the stimulus and also
upon its duration.   Parthenocarpic development of the ovary is dependent
upon ovular development, the amount of seed apparently governing the size of
the capsule.

Polyembryony occurs frequently in *Gastrodia*, and is correlated with
delayed pollination.   One embryo arises from the fertilized egg, and the other
probably from a fertilized synergid.

Although diploid eggs occur frequently in *Gastrodia* their apogamous
development was not observed in any case.   It is interesting to note, however,
that KUSANO saw several cases in which the nucleus of a haploid egg under-

went division. This is the first step in true parthenogenesis, but it is never accompanied by cell division and never leads to embryo formation. Other works on apogamy are cited, but the author believes the cytological facts regarding such matters are yet too few to warrant the formulation of a hypothesis on the evolution of parthenogenesis from amphimixis.

The value of contributions of this sort is obvious. The correlation of physiological conditions and morphological phenomena is clearly shown. This should lessen the morphologist's frequent neglect of physiology, and, on the other hand, should lead to a more careful checking up of experimental results, especially those in plant breeding, by morphological study.—L. W. SHARP.

Chromosomes and Mendelian inheritance.—STURTEVANT[s] presents a recapitulation with much new data bearing upon the "coupling" and "repulsion" of Mendelian genes in the fruit fly (*Drosophila ampelophila*), and ably discusses the bearing of these breeding results upon the chromosome interpretation of Mendelian phenomena. The large number of "cross-overs" (that is, changes from coupling to repulsion and vice versa between two given genes) and the relatively small number of chromosomes in the fruit fly, makes this organism very favorable material for such a study. Over 40 Mendelian characters of the fruit fly have been studied by MORGAN and his students, and these characters form four groups, so related to one another that all of the characters within one of these groups show "linkage" with one another; while those which have been sufficiently studied are independent of genes included in any one of the other groups. Each of these groups of characters is believed to be carried by a single pair of homologous chromosomes. On the basis of the relative number of cross-overs between different genes, considered two by two, the number of cross-overs which may be expected in any untried combination among the same series of "linked" genes may be readily calculated. Each gene is assumed to occupy a definite position or "locus" in the chromosome, and these loci are represented as forming a linear series whose distances from one another is measured by the relative frequency of cross-over. When cross-overs between two genes are rare, the two loci involved are assumed to be very near each other, and when cross-overs are frequent it is assumed that the two loci in question are correspondingly removed, though still lying in the same chromosome. No less than six of these loci have been established in a single chromosome, by a fairly adequate amount of data, and the correspondence between calculated distances and the observed numbers of cross-overs is convincing as to the fundamental value of this method of representation Furthermore, the discrepancies between the observed and calculated results are so consistently in the same direction that they make possible another important

---

s STURTEVANT, A H., The behavior of the chromosomes as studied through linkage. Zeitschr. Ind. Abstamm. Vererb. 13:234–266. 1915.

generalization; namely that the occurrence of one cross-over in a chromosome lessens the likelihood that a second cross-over will take place in the same chromosome. This phenomenon the author describes as "interference." Following MORGAN, the author explains the phenomenon of crossing over as due to the twisting together of homologous chromosomes, and the failure to completely untwist when the chromosomes are separated—the "chiasmatype" of JANSSENS. Two cross-overs, or even three, may take place in the same chromosome when a series of loci sufficiently removed from one another are involved, but the frequency of such plural cross-overs is correspondingly low. While the percentage of cross-overs between any two "linked" characters was fairly constant in most of the material which has been studied by MORGAN and his students, the author points out that in certain strains there was a great deal of variation in the intensity of linkage. A part of this variability seems to be hereditary, but it is also suggested that some of the variation is probably due to conditions of food, etc. Most of the data regarding this variability are withheld for presentation and discussion in subsequent publications.

The author discusses the relation of chromosomes to Mendelian inheritance, and gives a list of 17 species of plants and animals in which clear cases of linkage have been described, and also a list of chromosome numbers which have been found in 25 species of plants and animals used in genetic experiments. He points out, as has been done by a number of geneticists, that each unit character is directly or indirectly due to the action of numerous Mendelian genes, and that each gene may and usually does affect a number of characters. The terminology used by the author, following that of MORGAN, is in one respect essentially the reverse of the one now most widely used, in that the symbols chosen to represent any Mendelian pair are based on the recessive instead of the dominant character; thus, instead of Cc for the factor for color, Ww is used, intending to suggest that in the absence of W a white individual results. This is just as usable a method of formulation as that now in general use, and has only the disadvantage that would be due to any such reversal of terminology. It has the strong pedagogical advantage that any dominant character is less likely to be misconstrued by the non-specialized reader, as the sole result of the single gene represented by the symbol.—G. H. SHULL.

A case of obligate symbiosis.—RAYNER[6] has discovered a very interesting case of obligate symbiosis between *Calluna vulgaris* and one of the mycorhiza. He has carried his investigations into careful experimental work, so that the details of the symbiosis in connection with the life history of *Calluna* have been discovered. It seems that infection by the fungus takes place shortly after germination, the source of the infection being the testa. This infection does not cease with the development of the mycorhiza in the roots, but affects

---

[6] RAYNER, M. C., Obligate symbiosis in *Calluna vulgaris*. Ann. Botany 29:97–153. *pl. 6. figs. 4.* 1915.

all parts of the seedling, from whence it extends to all parts of the mature plant; that is, into the tissues of the stem, leaf, flower, and fruit. The ovary becoming infected, the mycelium enters the seed coats of the developing seeds, but the embryo and endosperm are free from infection. The fungus was isolated in pure cultures, and the seeds sterilized, so that a synthesis of the fungus of *Calluna* was accomplished. It was found that in case this specific fungus did not infect the growing seedling, it did not develop roots, and suffered complete inhibition of growth, remaining alive but rootless for several months. The fungus concerned is said to resemble the genus *Phoma*, and the author proposes that the species should be placed in a new subgenus, for which the name *Phyllophoma* is suggested.—J. M. C.

Development and distribution of Leguminosae.—ANDREWS[7] has brought together all the data dealing with the development and distribution of Leguminosae, and has reached certain conclusions of general interest. His thesis is that "the present distribution of plants and animals is the algebraic sum of the responses made by organisms to their changing environment during the whole of the known geological record, and the present adjustment of the activities involved has been obtained only after ages of development during various geographical changes." This is a problem, therefore, which involves the cooperation of geology, geography, and biology. ANDREWS finds that many uniform types of Leguminosae are widely diffused through the tropics, and that in extra-tropical countries these uniform tropical forms are represented by specialized types, which are mainly xerophytic. The details are fully presented, and it is thought that such study will throw light upon the nature of former land connections. For example, the author thinks that the Leguminosae show that New Zealand was separated from the tropics early in the differentiation of the family, while Australia was cut off at a date considerably later.—J. M. C.

Growth of Nereocystis.—Accurate data concerning the behavior of the large marine algae are much needed, the usual statements of the textbooks being vague and often misleading. This need promises to be supplied by the work of the Puget Sound Marine Station, whose first publication describes the growth of the blades of *Nereocystis Luetkeana*. Miss FALLIS[8] finds that this species grows as well when loosened from its foothold on the rocks, the holdfast serving only to fix the plant Nor is the stipe, including the bulb, necessary for the growth of the blade, small pieces from which can grow independently. The growing region is not at the place of transition, between the blade and stipe, but its basal limit is at the beginning of the flattened part of the

[7] ANDREWS, E. C., The development and distribution of the natural order Leguminosae. Jour Proc. Roy. Soc. N.W. Wales 48:333–407. 1914.

[8] FALLIS, ANNIE L , Growth of the fronds of *Nereocystis Luetkeana*. Puget Sound Marine Station Publ. 1:1–8. 1915

blade, while its terminal limit is difficult to determine, because growth gradually decreases toward the tip of the blade.

A second paper, by Miss SHELDON,[9] deals with the growth of the stipe. It was found that the region of maximum growth in mature plants is 2–4 feet below the blade, the rate decreasing in both directions. In July the rate of growth was found to be about an inch a day.—J. M. C.

Ferns of the oriental tropics.—The richness of the fern vegetation of the tropics is being emphasized by the work of COPELAND.[10] In the papers cited, new species are described in the following genera: *Adiantum, Aglaomorpha* (a new subgenus *Holostachyum*, including two other species, being described), *Angiopteris, Asplenium, Athyrium* (6), *Balantium, Cyathea* (6), *Davallia, Dryopteris* (6), *Leptochilus, Loxogramme* (2), *Marattia, Microlepia, Oleandra, Ophioglossum, Polypodium* (8), *Prosaptia, Pteris, Tectaria* (4), *Trichomanes*, and *Vittaria*. The genera *Diplora* and *Triphlebia* were found to be invalid, because founded upon inconstant and "illusory" features, and should be included in *Phyllitis*, which comprises only three well defined species in the Malay-Polynesian region.—J. M. C.

Sphagnum bogs of Alaska.—RIGG[11] has noted the peculiarities of the flora of some Alaskan peat bogs and finds that while sphagnum occurs in many different habitats in Alaska, only where there is an absence of drainage do bogs accompany it. The peat in the bogs visited had a maximum depth of only 2 5 ft. Aside from the sphagnum, *Empetrum nigrum* is the most abundant and uniform in its occurrence, but *Ledum palustre, Kalmia glauca, Oxycoccus oxycoccus*, and *Drosea rotundifolia* are among other characteristic species. The bogs occur surrounded by treeless areas, by tundras, or by coniferous forests, and vary much in area.—GEO. D. FULLER.

---

[9] SHELDON, SARAH M., Notes on the growth of the stipe of *Nereocystis Luetkeana*. *Ibid.* 1:15–18. 1915.

[10] COPELAND, E. B., The ferns of the Batu Lawi Expedition. Appendix III. Roy. Agric. Soc. (Straits Branch). no. 63. 71–72. 1912.
———, Some ferns of northeastern Mindanao. Leaflets Philipp. Bot. 5:1679–1684. 1913.
———, Notes on some Javan ferns. Philipp. Jour. Sci. 8:139–145. *pls. 2–4.* 1913.
———, On *Phyllitis* in Malaya and the supposed genera *Diplora* and *Triphlebia*. *Ibid.* 8:147–155. *pls. 5–7.* 1913.
———, New Papuan ferns. *Ibid.* 9:1–9. 1914.
———, New Sumatran ferns *Ibid.* 9:227–233. 1914.

[11] RIGG, G. B., Notes on the flora of some Alaskan sphagnum bogs. Plant World 17:176–183. 1914.

By JOHN M. COULTER AND CHARLES J. CHAMBERLAIN

up to that time.   The book was based partly on origi-

the reports of other investigators, and it at once took

revision of the book.   This is now presented to the
scientific world in the belief that it will be no less
useful than the first edition.   Each of the seven great

cusses the problem of phylogeny and points out the

of Chicago Press

JICAL GAZ

Editor: JOHN M. COULTER

SEPTEMBER 1915

VOLUME LX   NUMBER 3

THE
BOTANICAL GAZETTE

SEPTEMBER 1915

IS THE BOX ELDER A MAPLE?
A STUDY OF THE COMPARATIVE ANATOMY OF NEGUNDO

AMON B. PLOWMAN

(WITH PLATES V–X)

The difficulties and uncertainties of classification, upon the basis of purely superficial characters, is perhaps nowhere better exemplified than in the case of the common box elder.

This tree occurs in large numbers over the greater part of its range throughout the northern portion of the United States and Canada, and only less frequently in the Southwest and along the Pacific Coast. It is also very generally cultivated in Europe.

With its diffuse, irregular mode of branching, and its light green, compound leaves, the box elder is one of our most easily identified trees, yet it possesses combinations of characteristics that constitute a troublesome puzzle to the systematist.

The tree habit is manifestly quite different from that of the true maples; the leaves are wholly unlike maple leaves; the sap is similar to that of maples, but the odor of the young twigs is unlike anything ever produced by a maple; the fruit is quite like the maple fruit; yet, unlike the maple, the box elder is anemophilous and strictly dioecious; the wood is maple-like in appearance and structure, though much softer, lighter, and less durable.

These and many other peculiarities of the box elder have engaged the attention of botanists since LINNAEUS named this tree *Acer Negundo* in 1753. To the mind of LINNAEUS, even approximate

agreement in reproductive structures and processes easily out-
weighed in importance the most conspicuous differences in general
morphology.   In 1794 MOENCHHAUSEN set up for the box elder
the new genus *Negundo*, at the same time recognizing the close
relationship to *Acer* in the name *Negundo aceroides*.   Extreme
opposition to the Linnaean rating of the box elder was developed
by KARSTEN, who in 1880 gave the name *Negundo Negundo*.   Each
of these three names is held in favor by a considerable group of
botanists at the present time, though KARSTEN has a much smaller
following in this matter than either LINNAEUS or MOENCHHAUSEN.

To all three of these men, as well as to most of their followers, .
the minute anatomy of plants was of course a sealed book, whose
importance was wholly ignored because unknown and unsuspected.
It remained for DE BARY, VAN TIEGHEM, and a host of more
recent investigators, to discover and demonstrate the fundamental
significance and value of anatomical features in the study of
phyletic relationships.

The following study of the comparative anatomy of the box elder
was undertaken a few years ago, in the hope that some additional
light might be thrown upon the problem involved in the title of
this paper.

General morphology

As is well known, the box elder is usually a rather low-growing,
irregularly and diffusely branching tree, hardly more than a shrub
in many parts of its range, attaining to its maximum size of 50–70
feet in height, with a trunk diameter of 2–4 feet, only in the lower
Ohio Valley.   In a number of characters the box elders show a
remarkable range of variety.

Referring to fig. 1, it appears that the compound leaves are
3–9-foliolate, with the leaflets variously toothed, lobed, and
divided.   The range in leaf types is apparently not so great in the
eastern part of North America as it is in the central part.   Most of
the descriptive works published in America follow the early writers
in stating that the leaves are 3–5-foliolate.   As the result of a careful
statistical study of 1250 box elder trees in southeastern Wisconsin,
it was found that the leaves were predominantly of the 3-foliolate
type on a little more than 5 per cent of the trees; 5-foliolate leaves

were in the majority on more than 32 per cent; while on about
62 per cent of the trees most of the leaves were 7-foliolate. Two
trees out of the entire number showed 9-foliolate leaves in greatest
numbers. No tree bore one .type of leaf exclusively, except a very
few having only the 3-foliolate sort. No 3-foliolate leaves occurred
on the 7 and 9-foliolate trees, neither did the 9-foliolate leaves
appear on 3 and 5-foliolate trees. The higher number of leaflets,
as well as the larger and more deeply lobed leaves, occur on suckers and second growths. In general, the leaflets are more numerous, larger, and more deeply divided on staminate trees. The two
9-foliolate trees mentioned above were both staminate trees.

More than 7 per cent of the leaves studied bore trifoliololate
basal leaflets, such as are shown in no. 6 of fig. 1. Intermediate
forms appear in nos. 3, 4, 5, and 7 of the same figure. It was
impossible to detect any intimate and constant relationship between
leaf types and the quality of the soil in which the trees grew.
Light and shade relations seem to be more effective agents in this
respect, the higher number of leaflets usually occurring on the more
brightly lighted trees and parts of trees. The leaves of box elder
are commonly thicker and of softer texture than are those of true
maples growing in similar situations.

An interesting peculiarity of the box elder is its very common
habit of developing new buds and leaves as long as the growing
season lasts. In other words, its growth is indeterminate. This
characteristic seems to be more pronounced in the northern part
of the range. Here not uncommonly the box elder trees continue to develop new leaves at the tips of the twigs until checked
by the first killing frost. This fact accounts for the diffuse and
angular habit of the trees in the colder parts of the range, since
because of this mode of growth the terminal buds are usually killed,
and the lateral buds carry on the development in the following
season. Indeterminate growth is much more conspicuous in the
male trees than in the female, especially in those years when the
female trees bear a heavy crop of fruit. This in turn explains why
the female trees are so much more symmetrical in form, and hence
better suited for artificial planting and cultivation, than are the
male trees.

In a series of forcing experiments, twigs of box elder and of several species of maples were placed in the hothouse at intervals throughout the winter. All were subjected to uniformly favorable conditions for development. The box elder buds opened in about one-half the time required by the maples. The same quick response is noticeable in the box elder in the first warm days of early spring. By thus utilizing the maximum length of the growing season, the box elder is able to make its remarkably rapid growth. This again is particularly true of the male trees.

Brilliant autumnal colors are not developed by box elder leaves. Toward the close of long or dry growing seasons, all but the very young leaves turn a dull greenish yellow, but do not fall in any considerable numbers until after frost, when the trees may be entirely stripped in a single day. In many of the characteristics here noted, the leaves of the box elder differ strikingly from those of the true maples.

. The young twigs of box elder also show marked differences. Two types are commonly described. In one of these types the twigs are of a pale grayish green color, usually quite slender, and with rather short internodes. In the other type the twigs are of a maroon color, often covered with a white bloom, stout, and with longer internodes. These two forms are said to occur in almost equal numbers in the eastern part of the range, while the green type predominates in the southern part, and the maroon type in the northern part. In the course of these studies, it has been found that green twigs are usually developed by female trees, and maroon twigs on male trees. However, the color of the twigs, in addition to being in a measure a secondary sex character, is subject largely to weather conditions, exposure, etc. Thus we find vigorous young shoots of both sexes pale grayish green in color when growing in protected or poorly lighted situations, while both take on more or less of the maroon tint when exposed to full light and severe weather conditions. Also it is to be noted that the maroon color in all these twigs gives place more or less completely to green in the warm days of spring. In all probability we have here to do with a phenomenon similar to the development of rhodophyll and anthocyanin in autumn leaves, in rosette plants in winter, in alpine types, etc.

In the series of hothouse experiments already mentioned, it was observed that all lateral buds in the box elder are of nearly equal vitality, and that most of them spring into active development at the first favorable opportunity. It was also observed that the box elder twigs were able to develop a much greater amount of new growth from their stored food material than were true maple twigs or even poplar and willow twigs of the same size. In the matter of rhizogenetic capacity, the twigs of box elder compare well with those of black poplar, and while the development is not so rapid as in the case of willow twigs, the number of roots developed is considerably greater, one or more appearing at every lenticel. Thus the propagation of the box elder by cuttings is comparatively easy, while in the case of true maples it is well-nigh impossible.

On the older trunk surfaces the bark of the box elder shows a very characteristic "expanded metal" appearance, with the rather blunt ridges arranged in a fairly regular oblique diamond pattern, as shown in fig. 2. This is wholly unlike most of the true maples, in which the bark is seldom ridged, but in which it usually scales off in larger or smaller thin plates, as shown in fig. 3. The most notable exception to this rule among maples is *Acer platanoides*, which, by the way, seems to be one of the few contact points between the true maples and the box elder. In fact, the trunk surface of the box elder is strikingly similar to that of the white ash, even to the gray or grayish brown color.

Anatomy of the root

The root system of the box elder is very wide-spreading in comparison with the size of the tree. Slender fibrous roots are developed in immense numbers, forming a close-meshed network to an unusual depth in the soil. Where a larger root is uncovered and exposed to the air and light, buds appear and rapidly develop into vigorous shoots, much as in the case of the adventitious buds that form so abundantly on roots of the common locust, when they are exposed or injured. This bud-forming activity of box elder roots is especially marked when the main trunk has been injured in any way. Thus great clumps of second-growth shoots are likely to spring up around the stumps of recently cut trees.

Fig. 7 shows the more prominent features of the anatomy of a young box elder root. The pith is almost all eliminated by growth pressure. Medullary rays are numerous, straight, 1–2-seriate, and expanded considerably in the inner bark. Tracheae are large and very numerous, often in clusters of 5 disposed in radial rows. Tracheids are large, and of thin-walled and thick-walled sorts disposed in irregular groups in such a way as to give a marbled appearance to the section. The cambial zone in growing roots is 8 or 10 cells thick. The bark is comparatively thin, and contains but very few sclerotic, crystallogenous, or tanniniferous cells. The many large sap-storage cells and canals are a conspicuous feature of the bark. The dead bark scales off in small thin plates, leaving the root quite smooth, in striking contrast to the appearance of the older stems.

A comparison of figs. 6 and 7 makes it evident that the maple root is heavily charged with tannin in the older part, the wood is more compact, the bark is thicker, more dense with numerous groups of stone cells, few sap reservoirs, and it scales off in larger plates. The cambial zone is not so extensive, and everything about the structure points to a less vigorous functional activity in the root of the maple than in the root of the box elder. The appearance of the medullary rays in the outer wood and bark of an older root of *Acer rubrum* is shown in figs. 11 and 12. Fig. 12 also shows fairly well the irregular massing of the sclerotic cells of the bark. Assuming the root to be a very conservative structure, it is interesting to compare this figure with figs. 13–16 inclusive, which show corresponding regions in some stems.

Reference may here again be made to the remarkable rhizogenetic powers of box elder stems when covered with water or buried in moist soil. Young shoots may be propagated by cuttings or by "layering," as in the case of many berry canes.

Anatomy of the stem

The young shoots of box elder are commonly of very robust growth, especially in the male plants, and on all second growths. Not infrequently the season's growth may reach a length of 5 or 6 feet, with a basal diameter of five-eighths of an inch or more. The twigs are not quite cylindrical, but more or less elliptical in section, the longer axis lying in the plane of the two leaves borne at the top of the internode. The surface of the young twigs is quite smooth and shining, except in those forms which develop a whitish bloom. Lenticels are not very numerous, but they are of rather large size, long and narrow in the early part of the season, and later becoming broad.

Fig. 19 shows a section of a moderately vigorous young shoot of a male box elder. For comparison a section of a young stem of *Acer saccharinum* is shown in fig. 17. The characteristic 6-sided appearance of the pith in the box elder section is due to the symmetrical arrangement of the 6 large leaf traces, 3 for each leaf. These leaf traces are of course most prominent near the top of the internode, but they are conspicuous even to the lower end. The section was taken from near the middle of an internode 5 inches long. While the leaf traces are prominent in all types of box elder, it is worthy of note that they are relatively much larger in those plants which bear leaves with the higher numbers of division.

The medulla is usually about one-half the diameter of the stem. The pith cells are large, thick-walled, circular, oval, or hexagonal in outline, and usually quite empty, except in the outer amyliferous zone. Fig. 40 shows a portion of this zone, very greatly magnified. While the zone is very irregular in width, it is everywhere at least 3 or 4 cells wide. In the true maples this starch zone is much narrower, even sometimes entirely absent. An average condition of the starch zone of *Acer saccharinum* is shown in the drawing fig. 41. It will be observed, also, that the medullary ray of box elder is expanded at its inner end, thus affording a better connection with the starch zone than in the case of the maple. The box elder's quick response to the first warmth of spring is doubtless made possible, in part at least, by its large starch-storing capacity and its highly efficient medullary rays. These rays are very numerous, mostly uniseriate, and seldom much farther apart than the width of a large trachea. The cambial zone, in the growing season, is made up of a considerable number of layers of actively dividing cells, thin-walled, and filled with protoplasm. The bark is comparatively thin on young twigs, and is composed of small, thin-walled cells for the greater part of its thickness, with a few layers of collenchyma toward the outside, covered by a small-celled epidermis over which is spread a smooth, uniformly thick cuticle. Sclerenchyma tissue occurs as a narrow and frequently interrupted band of bast fibers, with numerous small and irregularly scattered masses of stone cells. The bark contains some starch, and usually a few large empty cells, corresponding to the sap-storage cells of the root. No true phellem is formed in the first season.

Comparing figs. 19 and 17, it will be seen that the leaf traces of the maple are inconspicuous, the medullary starch zone is not prominent, medullary rays are less numerous, and the bark is thicker and more dense. In twigs of *Acer rubrum* the starch zone is thinner; there is less sclerenchyma; no phellem, but many thick-walled tanniniferous cells; cuticle thick and lenticular over a larger-celled epidermis. In *A. saccharum* and *A. platanoides*, the starch zone is thin; few medullary rays; but little sclerenchyma; phellogen and phellem very prominent, with numerous large lenticels; epidermis early tanniniferous; cuticle thin and smooth.

The medullary rays of box elder are continued far out into the cortex, both in young and in older stems. A study of figs. 13–16, which show sections of older stems, will make clear the relative development of box elder in this respect. Further comparison with figs. 11 and 12 indicates that this prominent development of the medullary rays in the cortex of box elder is a primitive ancestral characteristic, as well as another important factor making for rapid growth.

An additional point of interest in this connection is found in the form and number of the pits in the walls of both the ray cells and the cells of the amyliferous zone. While these pits are comparatively small in the box elder, they are far more numerous here than in any of the true maples. Moreover, they are ideally arranged to facilitate the movement of sap to the regions of most active growth. For comparison of sizes of the various cells, and the thickness of the walls, the reader is referred to table I. In general, it may be said that the tissues of the young stem of box elder are much less compact than in the case of the true maples.

TABLE I

Sizes of various tissue elements in some species of *Acer*, as seen in cross-section

Tissue type	Acer Negundo	A. platanoides	A. rubrum	A. saccharum	A. saccharinum
Mla ells	40–60 μ diam.	30–60 μ iam.	40–50 μ diam.	20–40 μ diam.	45–55 μ diam.
Wall	0.2–0.3 μ	0.1–0.2 μ	0 3–0.4 μ	0.2–0 3 μ	0 4–0 5 μ
Amyliferous ells	10–15×25–40	8–10×12–25	10–12×15–25	8–12×15–25	10–12×15–30
Wall thi ness	1–2 μ	0.7 μ	0 6 μ	1–1.5 μ	0.6–0 7 μ
Tracheae	30–40×55–65	20–30 μ diam.	20–40 μ diam.	10–20×15–30	20–40 μ diam.
Wall thi ness	0.6 μ	0 7 μ	0 5–0.6 μ	0 5 μ	0 7–0.8 μ
Tracheids— Thin, summer	12–15 μ	7–10 μ	6–15 μ	8–12 μ idm.	8–12 μ diam.
Thick, "	8–10 μ			–70 μ diam.	
Mary rays, wood	5–7×10–15	4–8× a–15 seriate	3–4×10–14 1–6-seriate	5×10 μ 1–3–(5)-seriate	4–5×12 μ 1–3- iate
Ray cells, t.s.	7–10×25–40	8–12×30–40	8–12×40–75	8–10×35–50	6–10×30–60
					3–4×12–14
Parenchyma, bark	15×35–40				10–14×20–25
Ray cells, bark	8–10×30–45	8–12×15–17	10–15 μ oval	5–8×10–15	10–18×18–25
Bast fibers	8–12×25–30	8–12×25–30	10–18×15–25	8–16×25–30	10–18×18–25
Cavity	–92 μ dim.	10–12 μ diam.	10–17 μ diam.	8–10 μ iam.	1015 μ iam.
	1–2 μ diam.	1–3 μ diam.	1–5 μ diam.	1–2 μ diam.	0 5–1 μ iam.
Stone cells	25–40 μ diam.	20–30 μ diam.	25–35 μ diam.	25–35 μ diam.	20–35 μ diam.
Crystallogenous	10–18×15–25	15–20×25	12–15×15–20	10–18 μ diam.	12–20 μ di.
Phellem cells	7–12×20–35	10–15×25–30	5–7×15–17	3–10×18–22	7–8×15–18

Considering now more specifically the minute anatomy of the older trunk of box elder, we find that the rough "expanded metal" appearance of the trunk surface shown in fig. 2 is almost exactly reproduced in the photomicrograph of a tangential section of the bark (fig. 24). The numerous concentric layers of bast fibers are imbedded in a spongy mass of soft, thin-walled parenchymatous phloem, which is split up into thin radial plates by the medullary rays. Near the cambial zone these layers of hard bast are almost continuous, except where perforated by the thin medullary rays. As the sheet of hard bast is forced outward by growth, the fibers cling tenaciously to each other, and the whole sheet is expanded into an oblique diamond-mesh pattern, as a result of the great tangential tension. At the same time, centripetal pressure is brought to bear upon all of the tissues inside the network of bast fibers. As a result of these conditions, there is more or less radial compression and tangential expansion of all the softer elements of the bark. For the same reasons, there is likely to be a considerable tangential shifting of each layer of hard bast with reference to that next inside. When this shifting is all in the same direction in all the layers, the medullary rays are often bent aside 30° or 40° from the radial line. This is of course a common feature in those plants in which the medullary rays extend far out through a thick bark. The hard bast fibers are quite slender, except occasional isolated specimens, but they are usually several millimeters long, and very firmly united with each other. The walls are very thick, often almost eliminating the cell cavity. Angular, irregularly shaped sclerotic cells are found in fairly large numbers, most frequently in the angles of the bast fiber network. In the young bark the medullary ray cells are oval in form, and occur in loose moniliform rows. The nuclei are here of large size and the protoplasm is especially abundant. The medullary ray cells in the outer part of the live bark are shorter and thicker. As seen in tangential section, these cells are commonly drawn out transversely into an oval shape. In the still older bark the ray cells are empty, crushed and distorted by the pressure of growth. The bast parenchyma begins its development in the form of thin-walled, radially compressed cells about the diameter of autumn wood cells. From these are derived two types of parenchymatous cells: the larger

are 15 $\mu \times$ 35–40 μ in size, becoming crystallogenous in the older
part; the smaller 8–10 $\mu \times$ 30–45 μ in size, and are more numerous,
in the ratio of about 3 to 1 of the larger.   The walls of the larger
are not lignified at any stage, while the walls of the smaller cells
are slightly lignified at maturity.   Sieve tubes are numerous in
the younger bark, with large sieve plates especially conspicuous in
tangential sections.   The outer, dead bark consists of a mass of
crushed, irregular, thin-walled cells, alternating with several zones
of thick-walled phellem.   Each zone of phellem is made up of
3–8 layers of brick-shaped cells, 7–12 μ thick radially, and 20–35 μ
square on the tangential surface.   These cells are very thick-walled
(2–4 μ), lignified, and only the outer ones in each zone are conspicuously suberized.   Progressive parenchymatous degeneration
of the hard bast and stone cells occurs in the outer portion of the
live bark, so that very few of these elements are to be found in the
dead bark of the box elder.

In the radial section of box elder bark (fig. 25) the bast fibers
appear apparently in discontinuous masses, scattered throughout
the live portion of the bark.   As a matter of fact these strands are
connected longitudinally for great distances, but, owing to the fact
that they are tangentially oblique, the section shows only short
portions of each strand.   This type of structure evidently gives
great strength to the bark, while at the same time it secures a
maximum degree of elasticity even where the bark is quite thick,
thus readily permitting the rapid expansion so characteristic of the
growing box elder stem.

In the case of *A. saccharinum*, the bark is thinner than in
box elder, and quite smooth even in older trunks, scaling off from
larger trunks in thin, even plates (fig. 3).   Under the microscope,
the cross-section of a 10-year twig shows the bark to be made up of
parenchyma in alternating narrow zones of radially compressed
cells and round cells, with 2 or 3 nearly continuous zones of hard
bast lying close together near the middle of the bark, and also many
small scattered islands of sclerotic cells, both inside and outside
of the bast fiber zones.

The medullary rays are commonly bent aside from the radial
line 20–30°, to the first hard bast zone, where they are again nearly
radial, but beyond which they quickly disappear.

In radial and tangential views the most striking feature is the incoherence of the hard bast zones. Only rarely do the elements hold together in a diamond-mesh network, such as is so characteristic of the box elder. This is apparently due to the fact that most of the hard bast cells are comparatively short, blunt at the ends, and not interlocked to any great extent. In the outer part of the bark some of the fibers lie in the tangential plane almost at right angles to the axis of the stem, as the result of the great tangential tension in this region. There is extensive parenchymatous atrophy of the sclerotic elements in the outer bark. Only one phellem zone is commonly present. This is made up of 6–10 layers of small cells, quite thin-walled and strongly suberized. Crystallogenous cells are found in small numbers, principally in the middle and outer portions of the bark. Occasionally an older medullary ray cell is found containing tannin.

In *A. rubrum* (fig. 16) the bark is much thinner than in box elder, smooth on young stems, and separating into thin regular plates on older trunks. Hard bast occurs in larger proportion than in *A. saccharinum*, and is distributed quite irregularly throughout the outer three-fourths of the thickness of the bark. The zonation is imperfect, and there is scarcely any tendency to form a diamond-mesh network (fig. 23).

Only the larger medullary rays extend out into the bark. These are only slightly oblique as far as the first hard bast, beyond which they extend radially one-half to two-thirds of the way to the outer surface, becoming increasingly diffuse. Crystallogenous cells are quite numerous, while tanniniferous cells occur less frequently. The parenchyma cells are nearly all much flattened. The hard bast zones and masses are incoherent and the cells are drawn out to oblique and transverse positions in the outer bark. There is a single zone of phellem, 5–12 cells thick, the walls strongly lignified and suberized.

The bark of *A. saccharinum* (fig. 15) is thin, and shows a definite zonation of hard bast in the inner two-thirds of its thickness. The zones are 2 or 3 cells thick, interrupted by the broad medullary rays which usually disappear about half-way out through the bark, beyond which point the sclerotic tissue is very irregularly scattered and progressively atrophied. Most of the parenchyma cells are irregularly flattened. Both tanniniferous and crystallogenous cells are collenchymatous. The single phellem zone is 5–10 layers of cells thick, suberized, but only slightly lignified. Sieve plates are prominent in tangential view of the youngest phloem.

A. platanoides (fig. 14) shows a much smaller proportion of hard bast than do most of the other maples. There are only 2–4 narrow zones, and these are widely interrupted by the broad, irregular medullary rays, which extend about two-thirds of the way to the outer surface of the bark. The rays are not much deflected from a radial course, and the deflection is not at all uniform. The parenchyma is less flattened than in other forms. The bast fiber network is least coherent in this species. Crystallogenous and tanniniferous cells are very numerous. Parenchymatous atrophy is pronounced in the outer part. There is a single zone of phellem, very irregular in thickness. The cells are highly suberized. The collenchyma zone just inside of the phellem is conspicuous and of uniform width around the stem.

These brief studies of box elder and of four species of maples, together with the measurements of elements tabulated on p. 177, show that the bark of the maples is more dense and better able to resist unfavorable conditions and the attack of enemies, but less rapid in growth, less elastic, and hence less perfectly adapted to the needs of a quick-growing tree than is the bark of the box elder.

In figs. 26, 27, and 28 are shown sections in three planes of the wood of A. saccharinum, while the three succeeding figures show corresponding sections of box elder wood. The maple wood is evidently more compact, with somewhat smaller elements arranged with greater regularity than in the case of box elder, thus readily accounting for the fact that the maple wood splits more easily than the box elder wood. The groups of tracheae are larger in the box elder, showing as many as four or five elements in a radially disposed row, while the maple rarely shows more than three in a group. The medullary rays are not so straight in the box elder, hence the radial section does not show such large plates of "silver grain" as in the maple.

Fig. 39 shows a bit of the cross-section of box elder wood highly magnified. The tracheae are elliptical or oval in section, with the longer axis radially disposed. They range in size from 30 μ to 65 μ, with an average size of 40 $\mu \times$ 55 μ. The individual cells of which the tracheae are composed are 150–200 μ long, with their end walls obliquely disposed at an angle of about 45°, the dip in almost every case being radial, so that the end wall seems to be quite transverse as seen in tangential section (fig. 31). The perforation through the end wall is about one-half the size of the end plate, and elliptical in shape. The tracheal walls are marked off in a regular hexagonal pattern, each area of which is about 5 μ across, with a simple pit 0.3 $\mu \times$ 1.5 μ transversely disposed at the center. Where tracheae lie in contact with medullary rays, the pits are circular and 1.5–2.5 μ in diameter. Some smaller tracheae show occasional traces of scalariform and even spiral thickening of the walls. In the acute angle with the oblique end wall, there is sometimes a considerable area of the tracheal wall in which the thickness is uniform and unbroken by pits. In these regions quantities of tannin may be stored.

The tracheids of box elder are of three fairly distinct sorts. (1) The thin-walled summer tracheids are 12–15 μ in diameter, with walls only 0.5–0.8 μ thick. These occur in largest proportion near the beginning of the season's growth, but they are also to be found in small numbers even bordering upon the zone of the thick-walled autumn cells. (2) The thick-walled summer tracheids are 8–10 μ in diameter, and their walls are 1.5–2.5 μ thick. These occur in small groups at the beginning of the annual ring, the groups becoming larger and more numerous as the season's growth progresses. The two kinds of summer tracheids are commonly grouped in such a way as to give a distinctly marbled appearance to the cross-section. All of these cells are angular and very irregular in shape. The majority of them are 400–600 μ in length, and they are firmly interlocked at the ends. (3) The autumn tracheids constitute a dense zone 3–6 cells thick at the close of the season's growth. These cells are much flattened, measuring 10–15 μ tangentially and 5–7 μ radially. The walls are 2 μ or more in thickness and more strongly lignified than other parts of the wood. This zone of

autumn tracheids is frequently divided into two parts by the inter-
polation of a few thick-walled summer tracheids. All tracheids
show a few small circular pits in their walls at points of contact
with medullary rays, while elsewhere their pits are very rare.
Faint oblique striae are occasionally found in the walls of the
thicker-walled sorts.

Wood parenchyma cells in the mature parts of box elder are
very few, small, short, and thin-walled. They are usually found
bordering upon the larger tracheae. Very rarely they become
crystallogenous. The medullary rays are 1–3-seriate, about
100–120 μ apart, and slightly wavy as seen in cross-section of the
wood. In radial and tangential sections the medullary rays show
a breadth of about 200–500 μ, and a thickness of about 20 μ. The
ray cells are of two fairly distinct kinds (figs. 34, 36). (1) The
ray body cells are nearly cylindrical, 7–10 μ in diameter, with
walls 2–2.5 μ thick. Their length is from 25 μ to 40 μ, except in
the region of the autumn growth, where the length hardly exceeds
the diameter, and where the walls are slightly thicker. The end
walls are usually only slightly oblique, with very many minute
simple pits. The pits communicating with the tracheae are large
and numerous, while those connecting with tracheids are small and
few. (2) The ray marginal cells form usually a single, sometimes
a double, row on each edge of the ray. These are a little larger
than the ray body cells, with somewhat thinner walls. They are
triangular in section, and when seen in radial sections they show
marked irregularity in form along the free border (fig. 36). The
pits are larger and more numerous than in the body cells. Much
protoplasm and large nuclei are usually to be seen in the ray body
cells, while the marginal cells are usually almost empty.

Where injuries have been inflicted, the wood of box elder shows
traumatic tissue made up of thin-walled, unlignified cells containing a very little tannin.

As would be expected from the comparison of figs. 27 and 30,
the density of box elder wood is considerably less than that of the
maples. The average for box elder wood is 27 pounds per cubic
foot, while the maples range from 32 to 43 pounds per cubic foot.
In color the wood of box elder is a pale cream or white. Its rather

coarse and uneven texture makes it unfit for the more exacting uses to which maple wood is commonly put.

The wood of *A. saccharinum* is shown in figs. 26, 27, 28, and 38. Here the tracheae are nearly circular in section, except where two or three are crowded together. The diameter is 20–40 μ, and the walls are 0.7–0.8 μ thick. In the older wood the tracheae often contain tannin plugs 250–300 μ in length. The oblique end walls have in most cases a tangential dip of 40–60°. The pit areas are hexagonal and 5–6 μ in diameter. The transverse pits are about 1 $\mu \times$ 2 μ. The thickenings in the walls of the smaller tracheae are sometimes scalariform, but *not* spiral. The tracheids are of two sorts, both rectangular in section in a large proportion of cases. (1) The summer tracheids are about 8–12 μ across, with walls 1–1.5 μ thick, more highly lignified in the neighborhood of the tracheae. (2) The autumn tracheids appear as 1–3 rows of much flattened cells, 4–5 $\mu \times$ 10 μ, with walls 1.5–2 μ thick. There is often a very gradual transition from summer to autumn types of cells. The tracheids are 250–300 μ long. Some wood parenchyma cells occur near the tracheae, but they do not contain either resin or crystals.

The medullary rays are 1–3-seriate, 125–150 μ apart, 200–500 μ broad, and 20–30 μ thick. The body cells are cylindrical, 6–10 μ in diameter and 30–60 μ long, with walls 0.7 μ thick. The marginal cells are a little larger, triangular in section, and quite straight on the outer margin, as in the case of other true maples (figs. 35, 37). The pits are circular, large, and very numerous at points of contact with the tracheae. All ray cells contain much protoplasm and some starch. The end walls as seen in radial section are but slightly oblique, while in the transverse section they stand at an angle of 20–30°. Large parenchymatous masses appear at intervals, connected with the medullary rays.

In *A. platanoides* the tracheae are nearly round in transverse section, occurring in irregular groups. The end walls have a tangential dip of 30–40°. The smaller vessels show well marked spiral thickenings in their walls. The tracheids are of only one general sort, with a very gradual increase in thickness of walls and in lignification through the year's growth. All are very irregular in shape. Some are considerably distended and show numerous

pits. Tannin is found in these larger cells. There are but few
wood parenchyma cells, and these do not contain crystals. Tracheae in the older parts of the wood contain much tannin and
some tyloses. The medullary rays do not differ essentially from
those of *A. saccharinum*, except that the cells are much shorter at
the close of the season's growth, and elsewhere an occasional cell
is shorter, thick-walled, and filled with tannin.

In *A. saccharum* the end walls of tracheal cells are about 45°
oblique, but with no definite direction of dip. The smaller tracheae
show scalariform and imperfect spiral markings. Tracheids are of
three kinds, quite similar to those of box elder. Crystallogenous
wood parenchyma cells are numerous, especially along the sides of
the medullary rays. The ray marginal cells are very irregular in
shape, and show a marked tendency to overlap each other.

In *A. rubrum* the end walls of the tracheal cells have a dip of
30–60°, chiefly in the tangential direction. The smaller vessels
show scalariform but not spiral markings. There are occasional
tannin plugs and diaphragms in the tracheae of the older wood.
Tracheids are of two types. The autumn tracheids are flattened
to a greater extent than in other species, measuring 10–14 μ by
only 3–4 μ in cross-section. The wall is 1.5 μ thick except near the
edges of the flat cells, where it is 2 μ or more. Many older tracheids
contain crystals and tannin, and small simple pits are common in
the walls. The medullary ray cells are often hexagonal in tangential view, and the walls are unusually thick. The rays are
frequently much broader than in other species. The marginal cells
are larger and quite irregular. They are much shorter at close of
the season's growth, and often erect on the edge of the ray. The
most striking feature of this species is found in the much greater
number and larger size of pits in all kinds of elements than in other
species of maple.

Further details in regard to size of elements in the wood of the
various species studied may be found in the table of measurements
on p. 177.

Anatomy of the leaf

It has already been stated that the leaf of box elder is thicker
and of softer texture than are the leaves of the true maples. The
principal features of the minute structure are indicated in fig. 32.

Palisade tissue is developed to an unusual degree, there being at least two well defined rows of these cells, and often as many as four, in which case there is scarcely any spongiophyll and the air spaces are small and few. Where the spongy layer is prominent, there are groups of collecting cells at the lower ends of the deepest palisade cells. Protoplasm is abundant in all palisade and sponge cells, where chlorophyll is also present in large amounts. Crystallogenous cells are found in small numbers. The upper epidermis is made up of medium-sized, lenticular, empty cells, which are regular and even in arrangement, and covered with a moderately thick cuticle. The lower epidermis is composed of cells much less uniform in size and shape, and in consequence the lower surface of the leaf is not so smooth as the upper. A few small hairs may be found widely scattered over the lower surface of the older leaves; hairs are quite numerous on younger leaves. The stomata are small, but very numerous, with the guard cells set flush with the lower surface. The midrib of the leaflet of box elder is very similar to that of *A. rubrum* (fig. 18), except that the crest of spongy tissue on top is even more prominent.

In all of the maple leaves examined, there was found but one rank of palisade cells, and these formed not more, and usually much less, than half the thickness of the lamina (fig. 18). No well defined zone of collecting cells was observed. The spongiophyll contains many good sized air spaces, except in *A. platanoides*. The epidermal cells of both surfaces are quite varied in size, and the cuticle is thin. Stomata are comparatively few and somewhat depressed from the surface. The midrib crest is present in all forms, but not so large as in box elder. Hairs are short and few on all types but *A. platanoides*, where they are quite numerous and long. A section of *A. saccharinum* leaf is shown in fig. 33.

The petiole of box elder presents some interesting structural features (fig. 9). If we follow the three large leaf traces a little way up the petiole from their emergence from the stem, they are found to break up into a considerable number of fibrovascular strands arranged in an interrupted ring around a large medulla. A little farther out, not more than one-fourth of the way to the first pair of leaflets, there will usually appear from 1 to 4 or 5 medullary

fibrovascular strands, which become larger out to the first pair
of leaflets. Beyond that point these strands are smaller, and they
may be fewer. These medullary strands are larger and more
numerous in petioles of leaves having the larger numbers of leaflets.
In these cases the strands are often typically amphivasal, but
with the greater part of the xylem directed upward. Large cells
and ducts appear in the cortex, similar to those found in the cortex
of the root. Neither *A. saccharum* nor *A. platanoides* (fig. 10)
develops medullary strands in the petioles. In *A. rubrum* (fig. 8)
there is a single small centrivasal strand, conspicuous for its dense
and tanniniferous phloem. In *A. saccharinum* there are a few
very small medullary strands clustered near a sclerotic rib that
projects into the upper side of the medulla. In neither of these
cases, however, is there any indication of a true amphivasal condition. The petiole of box elder is larger and less compact than
that of the true maples. Sclerotic tissue is almost entirely wanting,
and the same is true of crystallogenous and tanniniferous cells.
All of the maples possess these three sorts of cells in greater or less
amount. The petiole of *A. rubrum* is particularly dense (fig. 8).

Anatomy of reproductive axis and fruit

The reproductive axis of box elder is characteristically compressed, so that the cross-section is a broad ellipse, with axes
about in the ratio of 3:4 (fig. 5). The surface is ribbed and the
cuticle is very thick. The pith is composed of cells of very unequal
size, some of them quite large. The xylem ring is interrupted
by 10–15 one or two-seriate medullary rays. The hard bast is
prominently developed in a thick continuous zone which is
crowded close upon the phloem. All cell walls are comparatively
thick.

In *A. platanoides* (fig. 4), which is fairly typical of the maples,
the reproductive axis is quite cylindrical, smooth, and covered with
a thin cuticle. The pith cells are uniformly rather small. The
xylem ring is interrupted by 5–8 two to six-seriate rays. The hard
bast zone is but slightly developed, narrow, widely interrupted, and
remote from the young phloem. Compared with the vegetative
stems, the reproductive axis of box elder and the maples shows an

almost complete reversal of structural characteristics, but with far more conspicuous differences.

Fig. 21 shows a section through the fruit (samara) of box elder at about the mid-level of the embryo. The walls of the seed vessel are very thick throughout, with a dense fibrous sclerotic lining of remarkable thickness. The fibrovascular strands are not very numerous, but some of them are quite large, with dense, centrifugally massed pericycle. There is a subepidermal sclerotic zone of usually one layer of large thick-walled cells. The embryo is simply and symmetrically folded, and surrounded by a moderately thick, tanniniferous coat.

Figs. 20 and 22 show sections of the fruit of *A. platanoides*. The wall is very thick at the base, but much thinner around the embryo. There is a fibrous sclerenchyma lining only near the base, and it is not thick or dense even here. Fibrovascular strands are numerous, but all are small. The subepidermal sclerotic zone is composed of small, thin-walled cells. The embryo is very irregularly folded, and covered with a thick, dense, tanniniferous coat.

The box elder fruit is clearly better fitted to withstand unfavorable weather conditions than is the fruit of the maple.

Geological record of box elder

The *Acer* group apparently made its first appearance in the Upper Cretaceous, and became widespread and diversified in species during the Eocene, in which fossil remains of various maples are abundant. However, there is recorded only a single instance of the mention of *Acer Negundo* in a fossil state, and that in the Miocene at Oeningen in Baden (Neues Jahrb. 1835. p. 55). But since this record has not been referred to by any recent authority, it may be set aside as of very doubtful value. Fossil leaves of the *Negundo*-like type are of rather frequent occurrence as far back as the Upper Cretaceous. LESQUEREAUX in 1868 founded the genus *Negundoides* for a single species from the Cretaceous (Dakota Group) of Nebraska, but PAX, in his revision of *Acer*, reduced this genus to *Negundo* (Bot. Jahrb. 6:346. 1885). Under the genus *Negundo* the following fossil species have been described: EOCENE, *N. europaeum* Heer (Switzerland and Oeningen in Baden), *N.*

decurrens Lesq. (Colorado), *N. triloba* Newberry (Ft. Union Beds, North Dakota); OLIGOCENE, *N. bohemica* Menzel (Bohemia); MIOCENE, *N. trifoliata* Braun (Oeningen in Baden), *A. Negundo* (from Oeningen, 1835; doubtful). To these may be added the apparent box elder described by KNOWLTON, under the name of *Rulac crataegifolium*, from the Miocene of the John Day Basin, Oregon (Bull. U.S. Geol. Surv. no. 204. p. 77. *pl. 16. fig. 7*).

Since the Glacial Period the *Negundo* type has been abundant and varied throughout all of the morainal regions. The remarkable elasticity of the type under shifting stress of environment doubtless accounts in large measure for its unusual success.

Theoretical considerations

As a result of the studies and observations outlined in this paper, the writer is of the opinion that the box elder, in its present highly specialized form, is a product of the Glacial Period. The evidence may be stated concisely as follows:

1. *Negundo* characters were but slightly developed before the Pleistocene or Glacial Period. They have been widespread since that time. *Negundo* occurs in greatest abundance in regions of the richest glacial drift, especially upon and below the great terminal moraines.

2. *Negundo* characters were apparently developed rapidly, and partially fixed, through exposure to the inclement conditions along the margins of the great continental ice sheet.

3. *Negundo* was apparently a primitive variant from the ancestral *Acer* stock, possessing peculiarities especially adapted to glacial conditions. These features were greatly emphasized by the glacial experience of the species. The impetus gained from glacial influences is not yet lost. *Negundo* is highly variable, yet irretrievably separated from the true maples. The nearest points of correspondence are found in *A. pennsylvanica, A. spicatum,* and *A. platanoides.*

4. Characters of *Negundo* that would fit it for glacial environment are as follows: (*a*) leaf morphology and anatomy; maximum utilization of light; (*b*) medullary strands in petiole; great capacity for transportation; (*c*) extended insertion of leaf trace into stele;

(*d*) color of twigs; energy absorption; protection; (*e*) indeterminate growth; maximum growing season; (*f*) food storage capacity; amyliferous tissue; (*g*) high vitality of lateral buds; (*h*) vegetative activity of shoots; quick response to warmth and light; rhizogeny; (*i*) medullary rays; marginal cells; extension into bark; (*j*) large and numerous pits in wood elements; (*k*) bark; thick, tough, elastic, persistent; (*l*) unobstructed conduction in roots; (*m*) great extent of root system; (*n*) large number of seeds; (*o*) anemophily; (*p*) extreme protection of embryo; thick, resistant seed coats; (*q*) food storage in embryo; (*r*) fruit long persistent on the tree.

Summary and conclusions

1. *Negundo aceroides* does not appear as an authentic species in the geological record before the Glacial Period.

2. The fundamental *Negundo* characters made their appearance as early as the Upper Cretaceous, but only as minor variations from the *Acer* type.

3. In structure of leaf, efficiency of transporting tissue, capacity of storage organs, and in maximum utilization of light, heat, and growing season, *Negundo* became peculiarly adapted to the rigors of a glacial environment.

4. The impetus acquired by *Negundo* during the strenuous period of its adaptation to glacial conditions is still manifest in the pronounced inconstancy of the *Negundo* type. However, there seems to be no true reversion to the pre-glacial ancestral forms.

5. In practically every particular, except the morphology of the fruit, *Negundo* is now essentially different from the true maples.

6. Upon purely anatomical grounds, it appears that *Negundo* possesses characteristics of generic rank, and while the box elder is undoubtedly a descendant from the ancestral *Acer* stock, it has now reached a stage of differential development that may fairly exclude it from the group of true maples. Thus there seems to be ample justification for the name "*Negundo aceroides* Moench. 1794."

In conclusion, I wish to express my thanks to Professor F. H. KNOWLTON for facts in regard to the geological record of the box

elder; and to Professor WILLIAM TRELEASE for valuable suggestions and for his lively and encouraging interest in the progress
of this work.

WAUKESHA, WIS.

EXPLANATION OF PLATES V–X

FIG. 1.—Photograph showing the commoner types of box elder leaves; ×⅓.

FIG. 2.—Trunk surface of box elder; ×⅓.

FIG. 3.—Trunk surface of *Acer saccharinum;* ×⅓.

FIG. 4.—Transverse section of the reproductive axis of *Acer platanoides;* ×20.

FIG. 5.—Transverse section of the reproductive axis of box elder; ×20.

FIG. 6.—Transverse section of young root of *A. rubrum;* ×20.

FIG. 7.—Transverse section of young root of box elder; ×20.

FIG. 8.—Transverse section of petiole of *A rubrum;* ×20.

FIG. 9.—Transverse section of petiole of box elder; ×20.

FIG. 10.—Tranverse section of petiole of *A. platanoides;* ×20.

FIG. 11.—Radial section of the bark and outer wood of an older root of *A. rubrum;* ×20.

FIG. 12.—Transverse section of the bark and outer wood of an older root of *A. rubrum;* ×20.

FIG. 13.—Transverse section of the wood and bark of a 10-year stem of box elder; ×20.

FIG. 14.—Transverse section of the wood and bark of a 12-year stem of *A. platanoides;* ×20.

FIG. 15.—Transverse section of the wood and bark of an 8-year stem of *A. saccharum;* ×20.

FIG. 16.—Transverse section of the wood and bark of a 20-year stem of *A. rubrum;* ×20.

FIG. 17.—Transverse section of a vigorous 1-year stem of *A. saccharinum;* ×15.

FIG. 18.—Transverse section of the midrib of the leaf of *A. rubrum;* ×25.

FIG. 19.—Transverse section of a vigorous 1-year stem of box elder; ×10.

FIG. 20.—Transverse section of a seed of *A. platanoides;* ×8.

FIG. 21.—Transverse section of a seed of box elder; ×15.

FIG. 22.—Tranverse section through the base of the samara of *A. plata-noides;* ×8.

FIG. 23.—Tangential section of the bark of *A. rubrum;* ×20.

FIG 24.—Tangential section of the bark of box elder; ×20.

FIG. 25.—Radial section of wood and bark of box elder; ×20.

FIG. 26.—Radial section of wood of *A saccharinum;* ×20.

FIG. 27.—Transverse section of wood of *A. saccharinum;* ×20.

FIG. 28.—Tangential section of wood of *A. saccharinum;* ×20.

FIG. 29.—Radial section of wood of box elder; ×20.

FIG. 30.—Transverse section of wood of box elder; ×20.

FIG. 31.—Tangential section of wood of box elder; ×20.

FIG. 32.—Transverse section of leaf of box elder; ×500.

FIG. 33.—Transverse section of leaf of *A. saccharinum;* ×500.

FIG. 34.—Tangential section, medullary ray of box elder; ×500.

FIG. 35.—Tangential section, medullary ray of *A. platanoides;* ×500.

FIG. 36.—Radial section, medullary ray of box elder; ×500.

FIG. 37.—Radial section, medullary ray of *A. rubrum;* ×500.

FIG. 38.—Transverse section of wood of *A. saccharinum;* ×500.

FIG. 39.—Transverse section of wood of box elder; ×500.

FIG. 40.—Transverse section through the amyliferous zone of the medulla of box elder; ×500.

FIG. 41.—Transverse section through the amyliferous zone of the medulla of *A. saccharinum;* ×500.

PLOWMAN on BOX ELDER

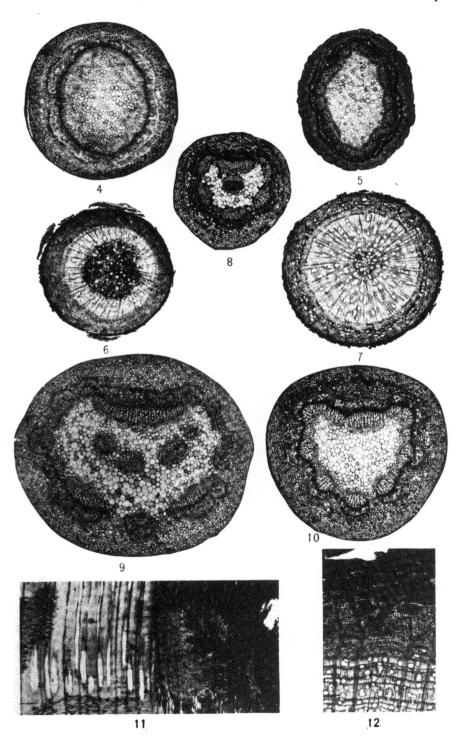

PLOWMAN on BOX ELDER

PLATE VII

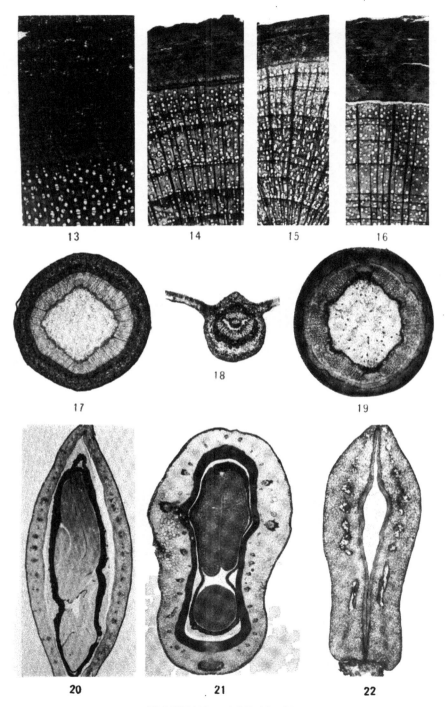

PLOWMAN on BOX ELDER

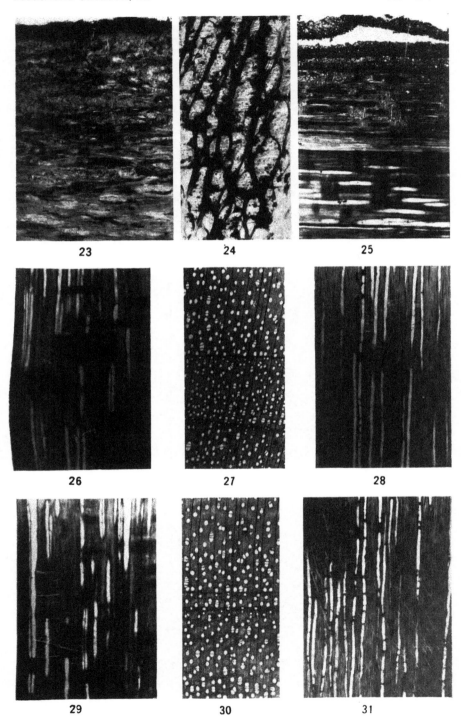

PLATE IX

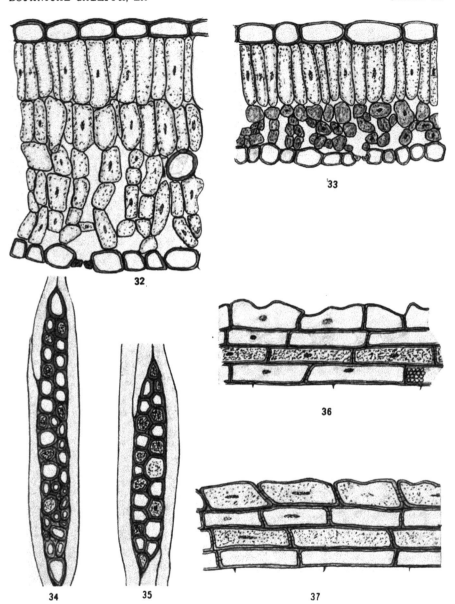

PLOWMAN on BOX ELDER

PLATE X

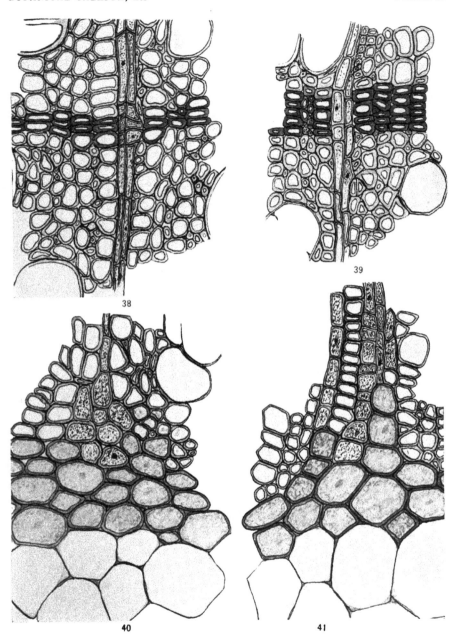

38

39

40

41

PLOWMAN on BOX ELDER

SOME EFFECTS OF ETHYLENE ON THE METABOLISM OF PLANTS

CONTRIBUTIONS FROM THE HULL BOTANICAL LABORATORY 207

EDWARD MARIS HARVEY

(WITH TWO FIGURES)

Introduction

Etiolated pea seedlings develop abnormally when they are grown in the "impure air" of a laboratory. This response to atmospheric impurities has become well known through the work of NELJUBOW (25) and others. Three phases, at least, are usually distinguished in the response of the epicotyl of the seedling: (1) a retardation in the rate of elongation, (2) swelling, and (3) a change from negative geotropism to diageotropism. Furthermore, this characteristic response can be produced by a large number of chemical compounds. However, the three phases mentioned are not induced with equal ease; the third may never appear for a given substance, although that substance readily causes swelling and interferes with the rate of elongation. Likewise, both the second and the third may not appear, although there is a marked retardation of growth. The swelling of developing plant organs in the presence of poisonous substances is a very common response, especially when the concentration of the substance in question is near the lower toxic limit. So frequently does this phenomenon occur that one is perhaps justified in saying that swelling is one of the first superficial indices of a disturbance in the metabolism of a plant.

Of the large number of chemical compounds capable of inducing swelling in the pea seedling, ethylene has been found to be the most effective. According to KNIGHT, ROSE, and CROCKER (19), ethylene will cause swelling of the epicotyl of the sweet pea seedling in dilutions of about 0.00004 per cent (by volume), while to produce similar results with chloroform, for example, the concentration must be about 1 per cent.   This illustrates how sensitive the sweet pea seedling is to traces of ethylene.   Moreover, it seems fairly well demonstrated that for plants in general (but by no means all) ethylene is relatively very toxic.

The remarkable capacity of ethylene to induce swelling naturally suggests the question, What is the effect of ethylene on plant metabolism?   A certain amount of work has been published on the effects of illuminating gas and "laboratory air," which should furnish required data on the question, since in both those gaseous mixtures ethylene has probably been an important factor.   Nevertheless, with regard to the effect of ethylene as such, I have been able to find no literature.   This was largely the reason for undertaking the investigation reported here, the work having as its subject the determination of the changes brought about in plant tissue by ethylene.

Historical

Our knowledge of the changes in metabolism causing and accompanying swelling of plant organs has been gained largely from investigation of the effects of anaesthetics, particularly ether and chloroform.   For a general historical résumé of the literature of the effects of anaesthetics on plants, the reader is referred to an excellent paper by HEMPEL (11).   The following consideration of the literature deals only with the effects of anaesthetics on the chemical composition and the respiratory processes.

JOHANNSEN (16, 17) found that certain concentrations of ether and chloroform caused an increase of soluble sugars and a decomposition of proteins in bulbs of *Crocus* and seeds of pea and barley. But he also noted that very weak ether gave reversed effects, that is to say, favored starch and protein synthesis.   His explanation for the increase in sugars and amino bodies was simply that anaesthetics interfered with the condensation, but not with the hydrolyzing processes.   ZALESKI (36), working with *Lupinus*, found that protein synthesis was favored by ether and hindered by caffein. BUTKEWITSCH (3) and BARTEL (2) both reported an increase of tyrosin in *Lupinus*, as an effect of chloroform.   PRIANISCHNIKOW (28) was able to demonstrate a considerable increase of asparagin in *Lupinus* when the seedlings were grown in an atmosphere containing traces of gaseous impurities. LESCHTSCH (22) studied the effect of turpentine on protein metabolism in bulbs of *Allium*. An acceleration of protein synthesis occurs in wounded bulbs, but the process is further accelerated by small amounts of turpentine and hindered by large amounts. PURIEWITSCH (27) noted that ether interfered with the synthesis of starch. To account for this phenomenon he assumed an increased rate of respiration, whereby the sugars were used up. BUTKEWITSCH (4) reports similar effects for toluol and chloroform. Starch in the bark and wood of *Morus* and *Sophora* was rapidly hydrolyzed. This hydrolysis cannot be explained, he thinks, on the supposition of an increased respiration, since sugars increase concomitantly with the decrease of starch. Also, he points out the analogy between the effects of toluol and chloroform and of low temperatures; both may be explained on the basis of injury to the plastids. REINHARD and SUSCHKOFF (29) determined the effects of several substances upon starch synthesis. Ether seemed to act, not only as a hindrance to starch formation, but also as an accelerator of hydrolysis. Antipyrin, morphine, and caffein hindered, but urea and asparagin favored starch synthesis. Similarly DELEANO (6) observed a rapid destarching of leaves in the presence of chloroform, a result apparently contradictory to that reported by CZAPEK (5). RICHTER (30), working with illuminating gas, laboratory air, xylol, etc., and GRAFE (8), with formaldehyde, have shown that an accumulation of sugar is favored by these substances. ARMSTRONG and ARMSTRONG (1) have demonstrated that toluol, ether, chloroform, etc., cause an increase of glucose and HCN in leaves of *Prunus lauro-cerasus*, due to a rapid splitting of the glucoside present. HEMPEL (11) has made a careful study of the effects of ether on seedlings of *Pisum* and *Lupinus* with particular regard to the CO_2 output, and the changes in the nitrogen compounds and sugars. Her results show that ether effects are dependent upon the concentration. The normal destruction of the proteins in germination was retarded by "weak" doses (up to approximately 0.01 per cent by volume), but the process was accelerated in strong doses. All concentrations interfered with the inversion of sugars. GRAFE and RICHTER (9) have published an article on the effects of acetylene on the chemical composition of several kinds of seeds and shoots. They found that the sugars and amino acids increased in tissues which were naturally high in carbohydrate (*Vicia, Laburnum*, and potato shoots), while in fatty tissue (seeds of squash, mustard, and flax) there was a slight decrease. Also, acetylene caused an increase in the amount of glycerine and fatty acid in the seeds, resulting in a decrease in the amount of fat. Like results were obtained for illuminating gas. They conclude that the condensation processes alone are affected. However, it seems quite possible that anaesthetics sometimes also hasten the hydrolyzing processes. This is further indicated by the recent work of McCool (24), in which he claims that the acceleration of enzymatic activity (of diastase and oxidase) takes place during etherization, although the activity of catalase is depressed.

A preliminary examination of the results referred to above shows a number of inconsistencies; but GRAFE and RICHTER have well pointed out that these inconsistencies are probably not real. Most of them become clear when the effect of anaesthetics, with regard to the general chemical reactions of plants, is expressed as follows: that the condensation processes are favored by "weak" and hindered by "strong" concentrations; but that the effect on the hydrolyzing process is uncertain.

The literature dealing with the effects of anaesthetics on respiration processes uncovers about the same general situation as stated above, since weak doses seem to accelerate and strong doses to retard respiration. This statement is borne out by the results of ELFVING (7), JOHANNSEN (15), MORKOWIN (23), LAURÉN (21), and others. However, MORKOWIN considers that the respiration of carbohydrates cannot be accelerated by ether; that such is possible only with nitrogenous substances. Also LAURÉN found that whether or not respiration could be accelerated by ether depended upon the kind of plant used. Respiration was accelerated in proportion to the dose in *Ricinus* and *Lupinus;* slightly accelerated in limited doses, later depressive, in *Pisum, Phaseolus*, and *Cucurbita;* and there was no acceleration in *Brassica, Hordeum*, and *Zea.* IRVING (12) has shown that for chloroform the effect depended upon the dose. Small doses increased the CO_2 releasal; medium doses

caused an initial outburst, afterward a falling off; and strong
doses caused depression from the beginning. THODAY (33) has
made some careful determinations, both of CO_2 releasal and of
O_2 absorption. Weak concentration of chloroform accelerated
both processes to an equal degree, hence the respiration ratio
remained the same. When the doses were considerably stronger,
respiration was retarded, but the correlation between the two
processes was broken up. In leaves without tannin (*Tropaeolum*)
O_2 absorption was depressed more than the CO_2 output. But in
leaves containing tannin (*Pinus* and *Helianthus*) the situation was
reversed; there was an initial rapid absorption of O_2 which soon
fell to a level somewhat above the CO_2 production.

Material

The sweet pea seedling was chosen as experimental material
for the present study, largely because it is so sensitive to toxic
substances, and on account of the general interest surrounding its
characteristic responses.

Etiolated seedlings were used throughout the experiments.
The seeds were purchased under the trade name Gladys Unwin
(Vaughn's Seed Store). The cultural methods employed have
been described by KNIGHT and CROCKER (20), although some minor
changes were necessary in order to care for large cultures. The
methods are briefly outlined below.

The seeds were scratched with a file (to secure quick and uniform germination), soaked for 12 hours in distilled water, and
germinated on wet filter paper. When the hypocotyls had become
3–7 cm. long the seeds were sowed upon wet absorbent cotton in
large pans ($2 \times 30 \times 48$ cm.) and covered with a layer of wet filter
paper. They were allowed to develop in absolute darkness at a
temperature of 21–24° C., until the epicotyls had reached an average
length of about 2 cm. The filter paper was then taken off the
seedlings and the culture equally divided into two portions, one
for treatment with ethylene, the other for control. The entire
culture usually consisted of 12 pans, each containing about 250
seedlings. The portion for ethylene treatment was transferred
to a galvanized iron box of 225 liters' capacity; the lid sealed gas-

tight; and enough ethylene[1] admitted to make the concentration about 0.0001 per cent by volume. Both control and treated portions were then allowed to continue development for 72 hours under the same conditions (that is, of moisture, temperature, and darkness), except for the ethylene in one. At the end of this period the epicotyls were collected for experimentation. The total culture period was 9 days.

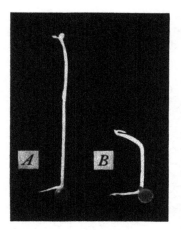

FIG. 1.—Etiolated seedlings of the sweet pea, showing the stage of development at which the epicotyls were taken for experimentation: *A*, normal; *B*, ethylene treated, showing the "triple response"; ×⅓.

The epicotyls of the control seedlings, at the time of collection, were 8–11 cm. in length and vertical and straight. But the ethylene treated seedlings showed the well known "horizontal nutation" (NELJUBOW 25) or "triple response" (KNIGHT and CROCKER); that is, the epicotyls were only 3–5 cm. long, swollen, and had assumed horizontal or nearly horizontal positions. The difference in appearance between the ethylene treated and the untreated seedlings is shown in fig. 1. Certain histological differences are shown by the drawings of fig. 2.

Methods and experimentation

The present attack of the problem on the effects of ethylene has been made through a study of the following questions: (1) chemical composition, (2) acidity, (3) osmotic pressure and permeability, and (4) respiration. A decided emphasis has been laid upon the chemical phase of the problem.

[1] The ethylene used in these experiments was prepared by dropping ethyl alcohol into syrupy phosphoric acid at a temperature of about 215° C. The final dilutions of ethylene were made from a stock ethylene-air mixture containing 2.5 per cent ethylene.

A. CHEMICAL ANALYSIS

1. *General procedure*

At the close of the culture period, the epicotyls of the seedlings were collected, their wet weight determined, and preserved in

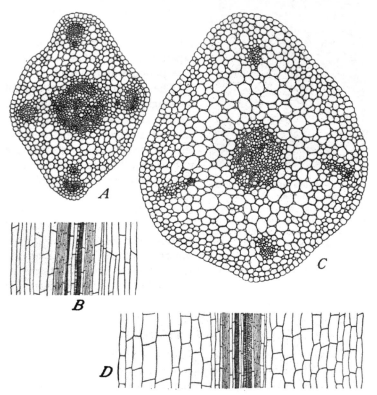

FIG. 2.—Sections of the sweet pea epicotyl at the stage corresponding to fig. 1: *A* and *B*, normal; *C* and *D*, ethylene treated, ×50.

85 per cent alcohol[2] (redistilled). Both the ethylene treated and the untreated samples usually weighed 100–150 gm. Soon after

[2] Enough 95 per cent alcohol was added to make the final concentration 85 per cent, allowing for water in the tissue, which was about 92 per cent.

preserving, the tissue was heated in a water bath for one hour at 70° C. and set aside for at least one week before proceeding with the analysis. Later the epicotyls were cut with scissors into 2–5 mm. lengths, transferred to ashless filter paper extraction cups, and the preserving liquid filtered through the cups. Two extractions followed, one with hot 95 per cent alcohol for 4 hours and the other with hot ether for 2 hours. Then the tissue was powdered in a mortar and finally extracted for 12 hours more with hot 95 per cent alcohol. This procedure separated the sample into alcohol-ether soluble and insoluble portions. The dry weight of each was determined by methods which are described below. That of the insoluble fraction was found simply by drying to constant weight in an oven at 104° C. But the soluble fraction was concentrated upon a water bath to about 450 cc., transferred to a 500 cc. volumetric flask, and made up to the mark. Then an aliquot part (usually 100 cc.) was taken for dry weight determination, and later for ashing. The final drying was carried out in a vacuum desiccator over $CaCl_2$.

Analysis of the alcohol-ether soluble fraction was carried out on the remaining 400 cc. This was evaporated to small volume to free it from alcohol, taken up with water, and transferred to a 500 cc. volumetric flask. Fats and lipoids were precipitated from solution by the addition of 3–5 cc. of chloroform and 10–15 cc. of 5 N HC. After shaking, the solution was made up to volume and set aside in an ice box for 24 hours to settle. The clear supernatant liquid was then decanted and filtered. No determinations were made upon this fat and lipoid residue. The water solution was divided into portions for the following determinations: (a) carbohydrates, (b) total nitrogen, and (c) ammonia and alpha-NH_2 nitrogen.

Analyses on the alcohol-ether insoluble fraction included the following: (a) reducing sugars after acid hydrolysis, (b) total nitrogen of the fraction, (c) total nitrogen of the portion rendered soluble by acid hydrolysis, (d) "crude fiber" (dry weight only), which was that portion remaining insoluble after acid hydrolysis, and (e) weight of ash. The "crude fiber" would largely be cellulose plus protein which was yet undissolved by acid hydrolysis, hence the results from $(d)-[(b)-(c)]$ should give an approximate figure for the cellulose present.

2. *Carbohydrates*

Reducing sugars, before and after hydrolysis, were determined for the alcohol-ether soluble fraction. The methods employed were those of MUNSON and WALKER (**26**, pp. 241–251). The reduced copper was determined by the volumetric permanganate method described in the above-named bulletin (p. 52). Similarly, the reducing sugars were determined for the insoluble fraction after 2.5 hours of hydrolysis with 3.5 per cent HCl.

3. *Nitrogenous substances*

Total nitrogen was determined by the KJELDAHL method and the amino acids and amids by both the formol titration (as described by JESSEN-HANSEN, **14**, and VAN SLYKE's methods, **34**). While employing the latter method, estimations were made of ammonia and àlpha-NH_2 groups both before and after a 24-hour hydrolysis with 20 per cent HCl at 98–99° C. (see **35**).

4. *Fats*

Only the dry weight of the ether extract and the free fatty acid value were estimated. Separate samples were used for fat determinations. The tissue was collected and preserved as previously described, but instead of the alcohol extraction being used as before, the tissue was spread out to dry in a current of air, absolute alcohol being added from time to time to hasten the drying process. The dried tissue was then extracted for 12 hours with absolute ether. Also, the preserving liquid was evaporated, the residue dried *in vacuo*, and taken up with absolute ether. The ether extracts were combined, dried *in vacuo*, and the weight determined. Free fatty acids of this residue were estimated by titration in the usual manner.

5. *Results*

The following statements are made in explanation of the tables of results. Whenever percentages are given, the figures are always in terms of the *total dry* weight of the samples. Also, the numbers of the samples are given in order to facilitate proper comparison between ethylene treated and untreated tissue; for, as has been stated, a given culture was not divided into the two portions until 72 hours before the end of the cultural period. Considering the unavoidable variations which must enter into different cultures, the treated and untreated pair of samples from one culture should therefore be more comparable than any other combination; for example, the treated sample of one culture and the untreated of another. Samples I and II, afterward IV and V, VI and VII, etc., are directly comparable.

TABLE I

TOTAL ALCOHOL-ETHER SOLUBLE SUBSTANCES IN THE ETHYLENE TREATED AND UNTREATED TISSUE

UNTREATED				TREATED			
No. of sample	Percentage of alcohol-ether soluble	Percentage of alcohol-ether insoluble	Percentage of total solids	No. of sample	Percentage of alcohol-ether soluble	Percentage of alcohol-ether insoluble	Percentage of total solids
I........	59.79	40.21	7	II.	67 63	32.37	8.54
V........	59.28	40.72	8 03	IV......	68 80	31.20	7.82
VII... .	59.04	40.96	7 82	VI . .	67 43	32.57	8.32
XIII. .	59.40	40.60	8.50	XII	64 98	35.02	8.60
XV. ..	60.00	40.00	7.80	XIV	67 60	32.40	8.30
XIX. .	59.60	40.40	8	XVIII (a)	68 64	31.36	8.90
XXI . .	59.70	40.30	8	XVIII (b)	68 76	31.34	8.94
XXIII .	59.66	40.37	8.36				
Mean	59 55	40 45	8.27	Mean	67.70	32 30	8.48

Table I gives the results of the separation of the samples into alcohol-ether soluble and insoluble portions. Two facts are noteworthy: first, that the soluble substances are more abundant in the ethylene treated than in the untreated tissue, a difference amounting to about 8 per cent; and, secondly, the water content of the two tissues is practically the same, being about 91.5 per cent. To what substances this difference is due should become clearer from the tables to follow.

Table II shows the amount of reducing sugars in the alcohol-ether soluble fraction before and after hydrolysis, and also the reducing power of the insoluble fraction after hydrolysis. These latter data cannot be expressed, even approximately, in percentages, on account of the probability of a great variety of hydrolyzable polysaccharides present.

The amount of reducing sugars in the soluble fraction is considerably greater in the treated tissue, although the results of sample II are inconsistent. Again, after hydrolysis, the treated tissue still shows more reducing sugar, but the difference is less pronounced, which means that the higher soluble carbohydrates, such as the disaccharides, are really less in this tissue than in the untreated. One seems justified in saying that the ethylene treated tissue has about 11 per cent more of the lower, and about 3 per cent

TABLE II

CARBOHYDRATES

		ALCOHOL-ETHER SOLUBLE FRACTION			ALCOHOL-ETHER INSOLUBLE FRACTION
	No. of sample	Percentage of reducing sugars before hydrolysis	Percentage of reducing sugars after hydrolysis	Increase by hydrolysis	Mg. cu. reduced for each gm of material after 2.5 hours' hydrolysis
Untreated tissue	I	....	....	....	153.9
	V	....	....	....	162.2
	VII	10 19	15 92	5.83	143.4
	XIII (a)	....	....	....	131.4
	XIII (b)	....	....	....	145.2*
	XV	12.30	15.70	3.40	140.1
Ethylene treated tissue.......	II	13.84	15.15	1.31	120.7
	IV	23.18	25.56	2.38	124.6
	VI	23.70	25.30	1.60	109.7
	XIV	20.39	21.51	1.12	118.8

* Hydrolysis 5 hours

less of the higher soluble sugars than the untreated. The reducing power of the alcohol-ether insoluble fraction after hydrolysis is clearly less in the treated tissue. The polysaccharides, which are likely to be present and which are capable of yielding reducing sugars by this acid hydrolysis, are starch, ligno-celluloses, galactans, pectins, etc. Microchemical tests show that very little starch is present in either tissue. The reducing sugars, therefore, are largely from other polysaccharides. An examination of the drawings of fig. 2 will aid in interpreting the differences found. Around the four leaf traces, mechanical tissue is considerably more abundantly developed in the untreated epicotyl. This anatomical difference agrees with the findings of KAUFMANN (18) for lupine seedlings treated with ether. From a chemical viewpoint the difference seems sufficient to account for the greater power of reduction of the untreated tissue after hydrolysis, inasmuch as mechanical tissue contains a large proportion of rather easily hydrolyzable carbohydrates. That these polysaccharides are not completely hydrolyzed by the 2.5 hours' hydrolysis is shown by comparing samples XIII (a) and XIII (b), in which the hydrolyzing time of the latter was doubled.

TABLE III

	No. of sample	Percentage of "crude fiber"	Approximate percentage of cellulose
·Untreated tissue......	V	15.09	9.2
	VII	16.18	11.4
	XIII (a)	15.00	....
	XIII (b)	13.94*	....
	XV	15.05	....
Ethylene treated tissue......	II	9.22	....
	IV	11.12	6.22
	VI	11.28	8.3
	XIV	11.66	....

* Hydrolysis 5 hours.

Table III gives the results of "crude fiber" determinations. "Crude fiber" obviously is a mixture of a large number of substances, such as unhydrolyzed protein, cellulose and other polysaccharides, etc. However, this fiber must be largely cellulose and protein. Less crude fiber was found in the treated tissue. The approximate percentage of cellulose was estimated by subtracting from "crude fiber" the total protein present before hydrolysis, minus the protein rendered soluble by the acid hydrolysis. By this method relative differences, at least, should be shown, and the results indicate that the treated tissue had about 3 per cent less cellulose than the control.

In table IV are shown the total nitrogen and the ammonia and amino nitrogen, before and after hydrolysis, in the alcohol-ether soluble fraction, and the total nitrogen in the alcohol-ether insoluble fraction.

The total nitrogen figures for the former fraction do not show a difference between the treated and untreated tissues. Both the formol titration and the VAN SLYKE methods show a higher amino nitrogen content in the treated tissue before hydrolysis, but after hydrolysis the difference does not appear. (Determinations after hydrolysis were made only by the VAN SLYKE method.) Since the increase of alpha-NH_2 nitrogen, by hydrolysis, is somewhat more in the untreated tissue, it suggests a less amount of polypeptides in the treated tissue, a situation for the soluble nitrogenous substances corresponding to that of the soluble carbohydrates of this fraction. The ammonia-nitrogen results are very variable and no conclusions can be drawn from them.

TABLE IV

NITROGEN

	No of sample	ALCOHOL-ETHER SOLUBLE FRACTION						ALCOHOL-ETHER INSOLUBLE FRACTION	
		Total nitrogen	NH₃ before hydrolysis	Amino-nitrogen before hydrolysis	NH₃ after hydrolysis	Amino-nitrogen after hydrolysis	Total nitrogen	Protein
Untreated tissue.....	I	5.18	.....	.. .	. ..	......	2.19	13.68
	V	5.41	.. .	. ...	.....	.. . ..	1.91	11.93
	VII	5.07	. . .	. . .	. . ..	. . . .	1.50	9.38
	XIII	....	....	1.62*	. . .	. ..	....	... .
	XV	....	....	1.71*	. .	. .	... ..	. ...
	XIX	. . ...	0.32	1.96	... .	....	.. ....	. .
	XXI	....	.	1.93	...	2.89	.. ....	. ....
	XXIII	....	0.76	2.02	1.30	2.78	.. ..	.....
Ethylene treated tissue.....	II	5.20	....	....	....	....	1.82	11.37
	IV	5.62	....	....	....	....	1.57	9.81
	VI	5.20	....	....	....	....	0.95	5.93
	XII	....	....	2.25*	.....	.. .	, ...	....
	XIV	....	....	2.34*	....	....	....
	XVIII (a)	. ...	0.83	2.54	0.93	3 22	... .	...
	XVIII (b)	.. .	0 40	2.41	1.25	2 98	. ....	. .. .

*Obtained by the formal-titration method.

The total nitrogen of the alcohol-ether insoluble fraction is less in the treated tissue. By employing the factor 6.25 to the nitrogen figures, it is found that the proteins are about 3 per cent less in the treated tissue. Although the results at first seem somewhat inconsistent, they no longer appear so when treated and untreated samples of the same culture are compared.

Table V gives the percentage of ash of the soluble and insoluble fractions. No marked difference appears between the two tissues.

TABLE V

Asᴴ

	No. of sample	Percentage of ash of the alcohol-ether extract	Percentage of ash of the alcohol-ether insoluble residue
Untreated tissue . {	I	....	1.58
	V	....	1.73
	VII	3.08	1.35
Ethylene treated tissue............{	II	2.76	1.54
	IV	3.15	1.41
	VI	3.22	1.47

Table VI shows the amount of ether soluble substance and the free fatty acid value. The figures in parentheses were not determined, but calculated on the assumption that the percentage of

TABLE VI

Fᴀᴛꜱ

	No of sample	Wet wt of tissue gm	Dry wt. of tissue gm	Dry wt of ether-extract gm	Percentage of ether-extractives	cc. of N/10 NaOH to neutralize free fatty acid	cc. of N/10 NaOH to neutralize free fatty acid in 1 gm. of ether extract
Untreated tissue . {	XXVII	119 57	10 043	(0 220)	(2 2)	3 00	(13.63)
	XXIX	171 64	14 417	0 313	2.17	3 86	12.33
Ethylene treat-{ ed tissue {	XXVI	139 45	11 713	(0 140)	(1.2)	1.92	(13.71)
	XXVIII	264 55	22 222	0 271	1.22	3 56	13.13

ether-extractives was the same in these samples as in others. One thing is rather clear from the table, namely that less fat is present in the treated tissue, a fact which agrees with the effects of acetylene in oily seeds as studied by Grafe and Richter (9). The free acid value is of particular interest on account of the claim of Iwanow (13) that the free acid value is predetermined by the degree of saturation of the fatty acids involved in the fat in question. However, the free acid value of the fats in the two tissues was not found to differ. On the foregoing assumption one may therefore say the nature of the fats in the treated and untreated tissues is the same as regards degree of saturation.

B. ACIDITY

For acidity determinations the epicotyls were collected as described for the chemical analysis. The wet weight of the sample was determined, but instead of being preserved in alcohol, the tissue was directly triturated with water in a mortar. More water was added to bring the mixture up to a definite volume, and finally the free acids present were titrated with N/10 NaOH, using phenolphthalein as indicator. The entire procedure, from the cutting of the seedlings to the end of the titration, required about one hour. The foregoing method is rather unsatisfactory; in addition to the fact that in this way one estimates only the surplus H-ions, other objections may be offered. However, any marked relative difference can be caught by this method.

TABLE VII

ACIDITY

	Wet. wt of tissue in gm	cc of N/10 NaOH to neutralise 1 gm. wet wt of tissue
Untreated. ...	87.25	{ 0.7241 0.7482
	25.65	{ 0.8457 0.8730
Ethylene treated...	79.65	{ 0.7706 0.7313
	29.55	{ 0.8288 0.8118

The results obtained are found in table VII, expressed in terms of N/10 NaOH required to neutralize 1 gm. wet weight of the tissue. No consistent difference is evident between the treated and control tissues.

C. OSMOTIC PRESSURE AND PERMEABILITY

Osmotic pressure was estimated by two methods, freezing point and plasmolysis. For the former method, the juice was expressed by means of a hand press giving about 300 kgm. per sq. cm., the tissue having been coarsely cut up with scissors and wrapped in a single layer of art canvas. The freezing point was determined with the BECKMAN apparatus, following the directions given by HAMBURGER (10). Both osmotic pressure and permeability were investigated by the plasmolytic method. Plasmolysis was observed in the cortical cells just underlying the epidermis at the base of the second leaf scale (that is, second from cotyledons). Plasmolytic agents employed were sucrose, glucose, KNO$_3$, and glycerine. A solution was considered isotonic with the cell sap if it just caused plasmolysis after 30 minutes. The temperature was 20–24° C.

TABLE VIII

OSMOTIC PRESSURE BY FREEZING POINT

	No of sample	Δ	Pressure in atmospheres	Mean pressure
Untreated..	V	0.610	7.33	.....
	VII	0.632	7.60	. .
	XI	0.703	8.46	. . .
		....	....	7.79
	IV	0.755	9.08	....
	VI	0.818	9.84	....
Ethylene treated .	VIII	0.827	9.94	. ...
	X	0.782	9.41	.. . .
	XIV	0.821	9.87	. . .
		....	....	9.63

In table VIII are the results by the freezing point method. It is evident that the juice of the treated tissue has a higher osmotic pressure than that of the control, a difference of about two atmospheres.

Similarly, table IX gives the results by the plasmolytic method. The figures show that the same relative difference of about two atmospheres exists between the treated and untreated tissues, although the pressures themselves are somewhat higher. RICHTER (31) and others have assumed a rise of osmotic pressure in tissues under the influence of anaesthetics. This assumption is based upon the fact that sugars and other osmotically active substances were known to increase. However, no previous measurements of osmotic pressure under such conditions have apparently been made.

Results with KNO_3 and glycerine indicate probably two things: first, that neither the treated nor the untreated tissues are very

TABLE IX

OSMOTIC PRESSURE AND PERMEABILITY

	Plasmolyzing agent	Concentration gm -mol	20–24° C Osmotic-pressure in atmospheres	Difference in gm.-mol. between treated and untreated
Untreated. ..	Sucrose .	0.37	9 38–9 52	.. ..
	Glucose	0.37	....	....
	KNO	0.21	....	....
	Glycerine	0.43	....	....
Ethylene treated .	Sucrose	0.46	11 70–11.84	0.09
	Glucose	0.47	....	0.10
	KNO	0.29	....	0.08
	Glycerine	0.56	....	0.13

permeable; and, secondly, that the treated tissue is slightly more permeable than the control.

D.   RESPIRATORY CHANGES

For the study of respiration, the cultural methods differed in some respects from those already described.   When the epicotyls had become 2–3 cm. long (that is, ready for the usual 72 hours' exposure period), the entire seedlings or the epicotyls only were taken from the pans and placed in test tubes of 20 cc. capacity, graduated for 15 cc.   The condition in the test tubes were as follows: They were filled with mercury and inverted over a dish of the same.   The mercury in the tubes was displaced to the 15 cc. mark, either with pure air or an ethylene-air mixture containing 0 0002 per cent ethylene.   Three entire seedlings or four epicotyls were introduced from below into the various tubes.   The experimental periods were 3, 6, 12, 24, 48, and 72 hours.   At the close of a period the seedling or epicotyls were withdrawn by means of a hooked wire and the gas present preserved for analysis.   The BONNIER and MANGIN apparatus was employed for the gas analysis, following practically the procedure suggested by THODAY (32).

In table X are the results of the analyses. Each set of figures represents the average of a number of analyses; for example, the 12-hour cultures are from 16 analyses upon gas of 3 different cultures. The results plainly show a general depression of respiration by ethylene, both in the CO_2 production and the O_2 absorption. The respiratory ratio gradually increases with the time in both tissues, an increase which probably is due to the lowering of the

TABLE X

RESPIRATORY CHANGES

No. of hours		UNTREATED TISSUE			ETHYLENE TREATED			
		cc CO_2	cc O_2	$\frac{CO_2}{O_2}$	Concentration of gas used, percentage (by vol.)	cc. CO_2	cc O_2	$\frac{CO_2}{O_2}$
3	Epicotyls only	0 2185	0 3664	0 66	0 0002	0 2449	0 3030	81
6	" "	0 2255	0 3419	0 66	"	0 1703	0 2898	61
6	Entire seedling.	0.4138	0 6069	0 69	"	0 3780	0 5744	66
12	" " ..	0 3758	0 5345	0.70	"	0 3197	0 4668	69
24	" " .	0 3587	0 4190	0.84	"	0 3127	0 3873	81
24	Epicotyls only	0 3145	0 4025	0.78	0 01	0 3380	0 4086	84
48	Entire seedling.	0 2501	0 2897	0 86	0 0002	0 2404	0 2869	85
72	" " ..	0 3210	0 2360	1 39	"	0 2610	0 2212	04

oxygen pressure. In the 3-hour culture with ethylene the ratio is very large. The result, as it stands, comes from an excessive production of carbon dioxide. This ratio of 0 81 seems extremely high in consideration of the 0 66 ratio of the control, and particularly of the 0 61 ratio of the 6-hour ethylene culture. However, IRVING (12) in her study of the effects of chloroform on barley leaves found that "medium" doses cause a large initial outburst of CO_2 quickly followed by a depression.

Conclusion

The results of the present study seem to indicate that the general effect of ethylene on plant metabolism is exactly comparable to the effects of the common anaesthetics, chloroform, ether, etc., as reported by other workers. Also, that the 0.0001 per cent ethylene concentration used is equivalent in its physiological effects to the so-called "strong" concentrations of those anaesthetics, concentrations which are in reality, for example in the case of chloroform or ether, thousands of times stronger. Probably most of the ether concentrations employed by HEMPEL (11) were many times weaker, physiologically, than the ethylene concentrations of the above-described experiments. Such an assumption would account for the difference between some of her results and those reported in this paper. It seems probable that ethylene, also, would favor condensation processes if used in "weak" or "medium" concentrations.

In the presence of ethylene the simple soluble substances increase at the expense of the higher soluble and insoluble forms; direct reducing sugars against soluble non-reducing sugars and insoluble polysaccharides; amino acids and amids against proteins; and probably fatty acids and glycerine against fats, seeing that the latter were found to diminish. Accordingly, ethylene appears to affect the balance of the general chemical reactions of the plant in favor of the simpler substances. The experimental work offers no evidence as to whether or not this result is accomplished through an acceleration of the hydrolytic as well as through a retardation of the condensation processes, since all the substances present in the tissue examined (epicotyls) had, within a relatively few hours, arrived, in simple translocation forms, from the cotyledons.

The accumulation of soluble substances in the tissue changes the osmotic relations of the cells and may have much to do with the observed swelling of plant organs in the presence of ethylene, for example in the characteristic "horizontal nutation" or "triple response" of the pea epicotyl. Also, the observed retardation of the rate of elongation may partly be accounted for by the fact that the gas interferes with the synthesis of complex substances, that is to say, perhaps with tissue formation.

Summary

1. Ethylene was found to be very effective in producing changes in the general processes of plant metabolism.

2. Chemical analyses showed that ethylene caused the simple soluble substances to increase at the expense of the higher soluble and insoluble forms.

a) The hot alcohol-ether soluble substances (sugars, amino acids, amids, polypeptides, lipoids, etc.) increased by 8–9 per cent, while the insoluble substances (proteins, starch, cellulose, ligno-celluloses, etc.) were correspondingly diminished. The water content of the ethylene treated and control tissues was the same.

b) The lower soluble sugars (by direct reduction) were about 11 per cent more and the higher soluble sugars (by reduction after hydrolysis) about 3 per cent less. The reducing power of the alcohol-ether insoluble residue, after hydrolysis, was decidedly less for the ethylene treated tissue; also, the cellulose content was diminished by about 3 per cent.

c) Amino acid plus amids were more, and the polypeptides apparently less in the ethylene treated tissue. The protein content also was about 3 per cent less.

d) Fats were much less abundant in the treated tissue. The free fatty acid value was unchanged.

3. The acidity of the ethylene treated tissue was not found to be changed.

4. Ethylene caused an increase of osmotic pressure, as measured both by the freezing point and plasmolytic methods.

5. The permeability was not sharply affected by ethylene, although it was somewhat increased.

6. Ethylene affected respiration, retarding both the CO_2 production and the O_2 absorption, but the respiratory ratio remained practically the same. An exception to the preceding statement was found in the case of the shortest exposure period (3 hours), in which there occurred, apparently, an excessive production of CO_2, thereby increasing the ratio.

I wish to acknowledge the many valuable suggestions of Drs. WILLIAM CROCKER, of the Department of Botany, and FREDERICK C. KOCH, of the Department of Physiological Chemistry, under whom the foregoing investigation was undertaken.

UNIVERSITY OF CHICAGO

LITERATURE CITED

1. ARMSTRONG, H. E., and ARMSTRONG, F. E., The origin of osmotic effects. II. The function of hormones in stimulating enzymatic change in relation to narcosis and the phenomena of degenerative and regenerative change. Proc. Roy. Soc. London. B 82:588–602. 1910.

2. BARTEL, R., Über Tyrosinabbau in Keimpflanzen. Ber. Deutsch. Bot. Gesells. 20:454–463. 1902.

3. BUTKEWITSCH, W., Über das Vorkommen proteolytisches Enzyme in gekeimten Samen und über ihre Wirkung. *Ibid.* 18:358–364. 1900.

4. ———, Zur Frage über die Umwandlung der Stärke in den Pflanzen und über den Nachweis der amylolitischen Enzyme. Biochem. Zeitschr. 10:314–344. 1908.

5. CZAPEK, FR., Über die Leitungswege der organischen Baustoffe im Pflanzenkörper. Sitz. Ber. Akad. Wien Kl. 106:117–170. 1897.

6. DELEANO, N. T., Über die Ableitung der assimilate durch die intakten, die chloroformierten und die plasmolytierten, Blattstiele der Laubblätter. Jahrb. Wiss. Bot. 49:129–286. 1911.

7. ELFVING, F., Über die Einwirkung von Aether und Chloroform auf die Pflanzen. Abst. in Bot. Jahresb. 14:167. 1886.

8. GRAFE, V., Untersuchungen über das verhalten grüner Pflanzen zu gasformigen Formaldehyde. Ber. Deutsch. Bot. Gesells. 29:19–36. 1911.

9. ———, und RICHTER, O., Über den Einfluss der narkotika auf die chemische Zusammensetzung von Pflanzen. I. Das chemische verhalten pflanzlicher objekte in einen Acetylenatmosphäre. Sitz. Ber. Akad. Wien Kl. 120:1187–1229. 1912.

10. HAMBURGER, H. J., Bestimmung der Gefrierpunktniederungen nach Beckmann. Osmoticher Druck u. Ionenlehre 1:89–97. 1902.

11. HEMPEL, JENNY, Researches into the effect of etherization on plant metabolism. Kgl. Danske. Vidensk. Selsk. Skrift. VII. 6:215–277. 1911.

12. IRVING, ANNA A., The effects of chloroform upon respiration and assimilation. Ann. Botany 25:1077–1099. 1911.

13. IWANOW, S., Über die Umwandlung des Oels in der Pflanze. Jahrb. Wiss. Bot. 50:375–386. 1912.

14. JESSEN-HANSEN, H., Die Formoltitration. Abderhalden's Handb. d. Biochem. Arbeitsmethochem. 6:262–277. 1912.

15. JOHANNSEN, W., Aether- und Chloroform-Narkose und deren Nachwirkung. Bot. Centralbl. 68:337–338. 1896.

16. ———, Studier over Plantemes periodiske Livs yttringer. I. Om antagonistiske Virksomheder i Stofshiftet särlig under Modning og Hvile. Abst. in Bot. Jahresber. 25:143–144. 1897.

17. ———, Über Rausch und Betaubung der Pflanzen. Naturw. Wochenschrig. Neue Folge 2:97–101, 109–113. 1903.

18. KAUFMANN, C., Über der Anaestetica auf Protoplasma und dessen biol.-physio. Eigenschaften. Inaug. diss. Erlangen. 1899.

19. KNIGHT, L. I., ROSE, R. C., and CROCKER, WM., Effect of various gases and vapors upon the etiolated seedling of the sweet pea. Science N.S. 31:635–636. 1910.

20. KNIGHT, L. I., and CROCKER, WM., The toxicity of smoke. BOT. GAZ. 55:337–371. 1913.

21. LAURÉN, W., Om inverkan of eteråga på groddplantors andning. Inaug. Diss. Helsingfors. 1891. Abst. in Bot. Jahresber. 20:92–93. 1892.

22. LESCHTSCH, MARIE, Über den Einfluss des Terpentinöls auf die Verwandlung der Eiweisstoffe in den Pflanzen. Ber. Deutsch. Bot. Gesells. 21:425–431. 1903.

23. MORKOWIN, M. N., Récherches sur l'influence des anesthésiques sur la réspiration des plantes. Rev. Gén. Botanique 11:289–303. 1899.

24. McCOOL, M. M., The influence of etherization on certain enzymatic activities of bulbs and tubers. Science N.S. 39:261. 1914.

25. NELJUBOW, D., Über die horizontale Nutation der Stengel von *Pisum sativum* und einiger anderen Pflanzen. Beih. Bot. Centralbl. 10:128–139. 1901.

26. Official and provisional methods of analysis. U.S. Dept. Agric. Bur. Chem. Bull. no. 107. 1912.

27. PURIEWITSCH, K., Zur Frage über Verwandlung des Stärke in der Pflanzenzelle. (Russian) Kiev. 1898 (cited by BUTKEWITSCH 4, p. 320).

28. PRIANISCHNIKOW, N., Zur Frage der Asparaginbildung. Ber. Deutsch. Bot. Gesells. 22:35–43. 1904.

29. REINHARD und SUSCHKOFF, Beiträge zur Starkebildung in der Pflanzen. Beih. Bot. Centralbl. 18:133–146. 1905.

30. RICHTER, O., Über den Einfluss der Narkotika auf die Anatomie und die chemische Zusammensetzung von Keimlingen. Ver. Gesells. Deutsch. Naturf. und Aertze 80:189–190. 1908.

31. ————, Über Turgorsteigerung in der Atmosphäre von Narkotika. Lotos 56:106–107. 1908.

32. THODAY, D., On the capillary eudiometric apparatus of BONNIER and MENGIN for the analysis of air in investigating the gaseous exchanges of plants. Ann. Botany 27:565–573. 1913.

33. ————, The effects of chloroform on the respiratory exchanges of leaves. *Ibid.* 27:697–717. 1913.

34. VAN SLYKE, D. D., The quantitative determination of aliphatic amino groups. II. Jour. Biol. Chem. 12:275–284. 1912.

35. ————, The conditions for complete hydrolysis of proteins. *Ibid.* 12:295–299. 1912.

36. ZALESKI, W., Zur Atherwirkung auf die Stoffumwandlung in den Pflanzen. Ber. Deutsch. Bot. Gesells. 18:292–296. 1900.

ON THE CUTICLES OF SOME INDIAN CONIFERS

RUTH HOLDEN

(WITH PLATE XI)

In no branch of science has the improved technic of the last few decades brought about a greater increase of knowledge than in paleobotany. The old purely systematic work based on impressions alone has been supplanted, or at least supplemented, by a · microscopic examination of structurally preserved material. The results have been valuable along both geological and botanical lines; the former by insuring the reliability of stratigraphical correlations through more accurate diagnoses, and the latter by indicating the relationships between living groups of plants through more extensive information regarding their extinct ancestors. Recently the examination of epidermal tissues has opened up a new line of attack. This method has been especially fruitful among the Cycadales, and our ideas of the affinities of fossil cycads have been materially altered. The next group to be attacked is obviously the Coniferales. Such genera as include both fertile and sterile shoots have, as a rule, certain definite diagnostic characters, but where a knowledge of the reproductive parts is lacking, chaos reigns supreme. In a few cases structurally preserved material has been examined, and the results have shown in a very striking manner the futility of attempting to classify according to impressions alone, and the folly of affiliating specimens with the living genera which they may simulate in external appearance. When we consider that *Thuyitis cretacea* and *Widdringtonitis Reichii*, both formerly included in the Cupressineae, have been proved to be araucarians (14), the truth of this statement is evident. In cases where the state of preservation precludes the possibility of a satisfactory investigation of the internal anatomy, the next best thing is to examine their cuticles. Such work has been undertaken in a few instances, as for example, ZEILLER (31) and BERRY (3) on *Frenelopsis*, SCHENK (21) and NATHORST (19) on *Palissya*, THOMPSON (30) on *Frenelopsis, Androvettia*, and

215]

Brachyphyllum, and THOMAS (28) on *Taxites*.   In addition to these, there are many scattered references to the distribution of stomata in the description of coniferous leaves, as by JEFFREY (14), BERRY (3), STOPES (26), and STOPES and KERSHAW (27), but too often there has been no attempt at correlation with living forms.

In the structure of the epidermis there are certain features which are constant, and certain others which seem to vary, not only within the genus or species, but even in different individuals. For example, in the case of flattened dorsiventral leaves of the *Taxites* type, the stomata are always abundant on the lower surface, but may or may not be present on the upper as well.   Thus MALHERT (16) states that in *Sequoia sempervirens*, *Abies*, and *Pseudotsuga* they are present on both surfaces, but I have found many instances where they were completely lacking above.   As regards numbers, such a character must obviously vary, and even a hasty examination shows flaws in the elaborate keys of BERTRAND (6), where *Araucaria excelsa* is described as having 3 nerves, and 4 groups of stomata of 5 rows each; *A. Balansae*, with 5 nerves, and 4 groups of stomata of 8 rows each, etc.   This is commented on by SEWARD and FORD (25): "The veins vary in number in the leaves of a species according to the part of the lamina examined and the age of the leaf.   The rows of stomata exhibit similar varieties; for example, *Araucaria imbricata*, said by BERTRAND to have 70 rows, may have any number from 60 to 80."

As regards arrangement of the stomata with reference to each other, there seems to be remarkable constancy.   Thus in practically all leaves of the *Taxites* type, they are in regular rows on each side of the midrib, with the long axis of the stoma parallel with the edge of the leaf.   This seems to hold irrespective of the family to which the specimens belong: *Pseudotsuga*, *Tsuga*, *Abies*, and *Keteleeria* in the Abietineae; *Widdringtonia* in the Cupressineae; *Cunninghamia* and *Sequoia sempervirens* in the Taxodineae; *Taxus*, *Cephalotaxus*, and *Torreya* in the Taxineae; and *Prumnopitys* and *Saxegothea* in the Podocarpineae.

The character of the epidermal cells has not been fully described by most investigators, but there is every reason to regard it as fixed. Such is undoubtedly the case with cycads; for example, *Ptillophyllum hirsutum* (29) from Marske is identical both with specimens from Whitby in the Sedgwick Museum, Cambridge, England, and with others collected by the writer from Navidale, Sutherland, Scotland. Since there is such specific constancy in the cycads, it is probable that the same holds true for the conifers, although too little work has been done on the latter to speak dogmatically.

On the other hand, so far as uniformity within large groups is concerned, there seems to be less in the case of the conifers than in the cycads. To illustrate: the Bennettitales group may be marked off from the Nillsoniales by the sinuous-walled epidermal cells of the former as contrasted with the straight walls of the latter; while within the Abietineae, sinuous walls have been observed only in some species of *Abies* and *Keteleeria;* in the Taxodineae, only *Cunninghamia;* in the Podocarpineae, *Saxegothea* and *Podocarpus;* and in the Araucarineae, *Araucaria.* Furthermore, this tendency toward diversity even within a single genus is much more marked among the conifers than among the cycads. Thus, *Dictyozamites Johnstrupi* Nath. (18) from Bornholm is substantially like *D. Hawelli* Seward (23) from Marske, and the three species of *Ptillophyllum* (*P. hirsutum* from Yorkshire and Sutherland, *P. pecten* from Yorkshire, and *P. acutifolium* from India) are very similar, both in the sinuous walls and in arrangement of stomata. Within the genus *Araucaria*, on the other hand, members of § COLYMBEA have the long axis of all the stomata parallel to the leaf margin, while in § EUTACTA there is no uniform angle; or even in § EUTACTA, *A. Cunninghami* has distinctly sinuous-walled epidermal cells, while in *A. Cookii* and *A. elegans* they are straight. A similar state of affairs exists for the genus *Podocarpus*, where within § EUPODOCARPUS, *P. macrophylla* and *P. totara* have sinuous walls, whereas in *P. alpina* and *P. elatus* they are straight.

Another feature which seems to be fairly constant is the presence of characteristic thickenings of the cuticle, either on the accessory cells of the stomata or on those of the general epidermis. These may constitute teeth projecting into the cavity of the stomatal opening, as in *Frenelopsis* (30), or they may make a rim around the opening itself, as in species of *Taxus, Juniperus, Thuja, Libocedrus*, etc. In other cases, all the epidermal cells, at least in the vicinity of the stomata, may have peculiar knoblike projections of cuticle. These last seem to be diagnostic in their constancy; they are present in *Taxus baccata*, including vars. *erecta, fastigiata*, and *variegata*.

As a last feature, we may refer to the shape and extent of the lignified lamellae of the guard cells. These have been found to be absolutely constant in the case of the cycads, and the investigations of the writer on the conifers indicate a considerable uniformity not only within the species, but even within the genus or family. Thus in all the members of the Araucarineae examined, the ventral thickenings are relatively larger and overlap the dorsal to a greater extent than in any other family. Unfortunately, however, the employment of this feature in examining fossil conifers is usually rendered impossible through imperfect preservation, although it has been found to be of considerable value for the cycads.

From this discussion of the significant points in structure of the epidermis, two conclusions may be drawn. The first is that the number and general distribution of stomata (upper or lower surface of leaf) is variable within the species, but that the character of the walls of the epidermal cells (uniformity of thickness, cuticular projections, straight or sinuous), arrangement of stomata (whether or not in rows, angle of long axis with reference to leaf margin), and extent of lignified lamellae of guard cells, are remarkably constant. The second relates to the phylogenetic value of these features and seems to indicate that it is small. For example, all leaves of the *Taxites* type, no matter to what family they belong, have long rows of stomata on each side of the midrib of the under surface of the leaf all with their long axes parallel to the margin. Similarly, leaves of the *Thuyites* type, as exemplified by *Dacrydium, Arthrotaxis, Tetraclinus, Thuya, Juniperus, Cupressus, Libocedrus*, etc., have the stomata scattered indiscriminately, usually avoiding the midrib, but with no sign of rows or constant angle. It seems obvious, accordingly, that the investigation of cuticles is of importance to the systematic botanist engaged in accurate specific diagnoses, but that it is of little interest to one concerned with the broader problem of the evolutionary history of the Coniferales.

Palissya

Among the numerous specimens of the genus *Palissya* sent by the Director of the Indian Geological Survey to Professor SEWARD, there were but two with cuticle preserved. One of these, from the Umia group of Thrombow, is labeled *Palissya* sp.; the other, from the Jabalpur group, is called *P. indica*, Fstm., both being Jurassic. OLDHAM and MORRIS had previously described specimens of this character as *Taxitis indicus*, but FEISTMANTEL transferred them to the genus *Palissya* Endl., believing them to be very near to the European *P. Brauni* (9). The two specimens differ but slightly in external appearance, and the structure of the cuticle indicates that they are specifically identical. The general habit is shown in fig. 1, and it is evident that in the spirally arranged, linear, and decurrent leaves, they resemble closely typical specimens of *Taxitis* or *Palissya*. The only discrepancy is the absence of a midrib, a feature noticed also by FEISTMANTEL. A general view of the epidermis is given in fig. 9. Toward the left are represented the cells of the upper surface, angular in shape, with straight walls and no stomata; toward the right, those of the lower, showing the stomata scattered with no semblance of regularity, but with their long axes more or less parallel to the margin of the leaf (see also fig. 11). Details of a single stoma are shown in fig. 4. The accessory cells are usually 6 in number, though not rarely 4 or 5; the guard cells are deeply sunken and often lacking; occasionally, however, the thickenings persist. Sometimes, as in fig. 4, the dorsal lamellae remain, while the ventral at one or both ends disappear. Not infrequently there are a few stomata on the upper surfaces of the leaf, and on the stem itself there are usually a few, due probably to the decurrent nature of the leaf bases. In no case, however, is there the slightest indication of the central astomatic region which would normally cover a midrib. The fact that the midrib is indistinguishable either in gross specimens or in detached cuticles, suggests a doubt as to the propriety of referring these specimens to either *Taxitis* or *Palissya*. The former is always described as "uninerva" (see SCHIMPER **22**, UNGER **32**, etc.), and THOMAS (28) has pointed out that the stomata, at least in *Taxitis zamioides*, occur in two bands, one on either side of the midrib. *Palissya* is also "uninerva" in Endlicher's original description; and in *P. sphenolepsis* (**19**) and *P. Brauni* (**21**), the stomata have the same distribution as in *Taxitis*. It seems clear, accordingly, that these Indian specimens do not conform to the *Taxitis* type, as represented either by the living *Tsuga*, *Abies*, *Taxus*, etc., or by the fossil *Taxitis* or *Palissya*. Among other existent forms the absence of parallel veins separates them from the Araucarineae and from the genus *Podocarpus*. A possible affiliation would be with *Dacrydium* or *Arthrotaxis*, but in both these genera, although there is no obvious midrib in gross specimens, preparations of the cuticle show a marked astomatic path running down the center of the lower surface of the leaf. By the process of elimination, we are driven to the only other flat-leaved conifers, namely Cupressineae of the *Retinospora* type, such as are found in seedlings of *Thuya* or *Juniperus*. Here, also, the stomata are scattered irregularly, sometimes on the upper surface, sometimes on the lower, sometimes avoiding the midrib, but often disregarding its presence; furthermore, the leaves are not constricted at the base, and they are markedly decurrent. The only objection to referring them to that family is the spiral phyllotaxis, but when the diversity existing in nearly related forms, both living and fossil, is considered, it seems doubtful whether this point is of much importance. For example, the leaves of *Podocarpus* are spiral except § NAGEIA, where they are decussately opposite; again, in all the Araucarineae they are spiral except in the two fossil forms *Thuyites cretacea* and *Androvettia* (**14**). The different types of phyllotaxis sometimes found in the same specimen add further evidence in the same direction. Thus, the *Retinospora*-like seedling leaves of *Thuya* and *Juniperus* are occasionally arranged in a spiral fashion, which soon gives way to characteristic verticils. DAGUILLON (**7**) has described an *Abies* seedling with whorled instead of spiral leaves, and MASTERS (**17**) one of *Cephalotaxus* with first a pair of opposite leaves and then a whorl of four. Moreover, although the foliage leaves of *Microcachrys* are decussately opposite, the sporophylls, both microsporangiate and megasporangiate, are in spirals.

From these considerations it seems evident that this so-called *Palissya* presents a type of leaf new to paleobotany, and to indicate its similarity to living forms, it may advantageously be called **Retinosporitis indica.** It should be emphasized, however, that this name is not intended to signify that it is necessarily closely related to *Retinospora*, or indeed that it belongs to the Cupressineae at all; but merely that in external appearance and epidermal structures it has certain features in common with that genus.

Echinostrobus expansus

The next specimen to be described has been referred by FEISTMANTEL (8) to *Echinostrobus expansus*, with the statement that it is identical with *Thuyitis expansus* L. & H. (15). Fig. 2 shows the general decussately opposite disposition of the leaves; other figures are given by FEISTMANTEL (*loc. cit. pl. 9010*); and fig. 6 shows the epidermis of a single detached leaflet. The dark crescent in the upper part may correspond to what was originally the free end of the leaf. Above it is a rim of fairly regularly arranged cells which is probably the "marginal depression" mentioned by LINDLEY and HUTTON, while the astomatic part to the left may have been overlapped by the leaf adjacent to it. The epidermal cells are exceedingly irregular in shape, though below the midrib they tend to become somewhat elongated. The stomata are scattered without definite order beyond the fact that there is an astomatic area down the center, and that they are more abundant near the margins, where they might have been partially shaded by the adjacent leaves. The accessory cells (fig. 5) are almost invariably 4 in number, beneath them is at least one intercalary layer, and then the guard cells, which have practically disappeared. The depth to which the stomata are sunk is probably correlated with their relative abundance and direct exposure to the sun's rays, for in living conifers of similar habit they are often less deeply sunken, but are usually confined either to the under surface of dorsiventrally flattened shoots (for example, *Thujopsis dolabrata*), or to the depressions where one leaf overlaps another (for example, *Libocedrus decurrens, Thuja gigantea, Frenella* sp., etc.).

The sytematic position of branches of this type has long been a disputed point. They were first referred by LINDLEY and

HUTTON (15) to *Thuyitis* because of the verticillate arrangement of the leaves. SCHIMPER (22) then transferred them to STERNBERG's taxodineous genus *Echinostrobus*, where the leaves are sometimes spiral and sometimes whorled. SAPORTA (20), however, pointed out the inadvisability of this step and put them back into the Cupressineae, this time as *Palaeocyparis expansa*. SEWARD (24) refers to the difficulty of distinguishing between the whorled leaves of *Thuyitis* and the spiral ones of *Brachyphyllum*, and suggests (23) that, at least so far as the specimens of LINDLEY and HUTTON are concerned, the two genera are identical. The structure of the cuticle is not without bearing on this question, for in the species of the latter which have been examined—*B. macrocarpum* NEWBURY (14), *B. Münsteri*, and *B. affine* (21)—the stomata are in long rows alternating with strands of sclerenchyma. This condition, of course, is entirely different from that of *Thuyitis expansus*, though it is singularly like *T. Schlonbachi* SCHENK (21) and the living podocarpineous genus *Microcachrys*. For a parallel, we are driven to the Taxodineae (*Arthrotaxis*) or the Cupressineae (*Thuja*, etc.), and though, as suggested above in the case of *Palissya*, phyllotaxy is not an invariable test for affinities, still in view of the fact that, as a whole, the leaves of the Taxodineae are in spirals and those of the Cupressineae in whorls, it would seem to be advisable to retain the original name, and, at least until an examination of their internal structure settles the question of affinities, to continue to call shoots of this type *Thuyitis expansus*.

Taxitis tenerrimus

The next specimen to be described has been referred by FEISTMANTEL (10) to *Taxitis tenerrimus*, and the spiral arrangement of the linear, uninerved, and decurrent leaves, shown in fig. 3, indicates the correctness of this identification. The cuticle of the upper surface is entirely devoid of stomata; that of the lower is represented in fig. 8. The epidermal cells are irregular in shape, with a slight tendency to become elongated below the midrib. The stomata are scattered without definite arrangement, but the indifferent state of preservation prevents any detailed description. In general, however, there are 4–6 accessory cells, and the opening

is parallel to the leaf margin. This distribution is quite unlike
that of *Taxitis zamioides* (28), where there are two narrow rows of
stomata, one each side of the midrib, and warrants at least a specific
distinction. As to its affinities, it is impossible to go farther than
to state that it is totally unlike *Taxus* or any other living member of
the Taxineae.

Podozamites lanceolatus

The last specimen to be described was referred by FEISTMANTEL
to *Podozamites lanceolatus* (11). Isolated leaves were found fairly
commonly in the Jabalpur group of South Rewah (Jurassic), but
there were none attached to the rachis. Various specimens showing
the characteristic shape are represented in *figs. 2–5. pl. 1 (loc. cit.)*,
and their resemblance to the type specimen of LINDLEY and HUTTON
is obvious. There is no difference between the epidermis of the
upper and lower surfaces; these cells (fig. 10) are all straight-
walled, more or less elongated over the veins, while the stomata are
confined to the area between the veins, with their long axes parallel
to the margin of the leaf. The structure of a single stoma is shown
in fig. 7. There are usually 6 accessory cells, rarely 4 or 5. The
character of the guard cells is unfortunately difficult to determine,
but there seems to be a double rim of cuticle around the opening.
This appearance is constant in the best preserved specimens, but
its interpretation is doubtful. Probably there was at least one
row of cells intercalated between the accessory cells and the guard
cells, and the rims referred to may represent cuticular projections
on these intercalary cells, such as are characteristic of certain
living conifers and cycads. The lignified lamellae of the guard
cells have invariably disappeared.

The resemblance of the cuticle of this Indian specimen to that
of *Zamites distans* Prestl., as described by SCHENK (21), is very close;
both have straight-walled epidermal cells with stomata between the
veins. In the latter, however, there are no stomata on the upper
surface. The difference between *Podozamites* and *Zamites* is
rather obscure, BRONGNIART including the former as a subsection
of the latter. The cuticles, however, show them to be entirely
distinct, for *Zamites* (29) has all the bennettitalean characters,
sinuous-walled cells and long axis of stomata at right angles to

leaf margin; while if the affinities of *Podozamites* are cycadean at all, they are with the Nillsoniales. It seems entirely probable, however, that they are coniferous. SEWARD (23) has stated the pros and cons of the situation, and in view of the spiral phyllotaxis and bud scales at the base of the petiole reaches that conclusion. SCHENK (21), on the other hand, compares these scales to those found in the living *Cycas*, and argues that *Podozamites* cannot be related to the conifers as exemplified by *Dammara orientalis* for three reasons: (1) the leaves are not opposite; (2) the vascular tissue in the petiole is not like that of *Dammara;* and (3) the epidermal structure is different. The first reason may hold for *Zamites distans* and *Dammara orientalis*, but it does not hold for *Zamites lanceolatus*, for the original description of this species by LINDLEY and HUTTON (15) states that the pinnae are "sometimes opposite and sometimes alternate"; nor does it hold for other species of *Dammara*, where the leaves are spiral. The second reason seems equally questionable, for of the two vascular strands figured by SCHENK, one shows protoxylem rings and the other has the crowded hexagonal pitting characteristic of both cycads and araucarians. As regards the third reason, the difference in epidermal structure is slight; in both *Podozamites* and *Dammara* the stomata are in rows between the veins, but in the former the long axis is parallel to the edge of the leaf, while in the latter it is at right angles. The resemblance to *Araucaria* § COLYMBEA, however, seems to be very close. The phyllotaxis is the same, and both have rows of stomata with their long axes parallel to the leaf margin. Another possibility is presented by § NAGEIA of the genus *Podocarpus*. It is not suggested that *Podozamites* can be identified specifically with any living conifer; for example, the sinuous walls of *P. Nageia* and *A. brasiliana* bar them out, as do the heavily pitted epidermal cells of *A. imbricata* and *A. Bidwillii;* but it does seem fairly clear that *Podozamites* is nearer to the conifers than to the cyc ads.

Summary and conclusions

1. A comparative study of living and fossil conifers indicates that epidermal structures are of great value for accurate specific diagnoses, but of relatively little importance for indicating affinities.

2. On account of the character of its cuticle, the so-called *Palissya indica* of FEISTMANTEL cannot properly be referred to that or any other fossil genus; and to point out its resemblance to the living *Retinospora*, it is suggested that it be called *Retinosporitis indica*.

3. *Echinostrobus expansus* closely resembles many living members of the Cupressineae, both in epidermis and in phyllotaxy; accordingly it would seem better to retain the old name of LINDLEY and HUTTON, *Thuyitis expansus*.

4. *Taxitis tenerrimus* has a type of cuticle common to many extant conifers, and its affinities cannot be decided.

5. The epidermal structure of *Podozamites lanceolatus* constitutes another reason for referring that genus to the conifers rather than to the cycads.

In conclusion, I wish to thank Professor SEWARD for this opportunity to study the fossil conifers sent by the Director of the Indian Geological Survey, and to compare them with the living ones in the collections at the Botany School, and for valuable suggestions in regard to this work.

BOTANY SCHOOL
CAMBRIDGE, ENGLAND

LITERATURE CITED

1. BERRY, E. W., Some araucarian remains from the Atlantic Coastal Plain. Bull. Torr. Bot. Club **35**:249–260. 1908.
2. ———, Contributions to the mesozoic flora of the Atlantic Coastal Plain. V. N. Carolina. Bull. Torr. Bot. Club **37**:19–29. *pl. 8.* 1910.
3. ———, Epidermal character of *Frenelopsis ramosissima*. BOT. GAZ. **50**:305–309. *figs. 2.* 1910.
4. ———, Revision of several genera of gymnospermous plants from the Potomac group in Maryland and Virginia. Contrib. U.S. Nat. Museum **40**:289–318. 1911.
5. ———, Maryland Geol. Survey, Lower Cretaceous. Baltimore. 1911.
6. BERTRAND, C. E., Anatomie comparée chez les Gnet. et Conif. Ann. Sci. Nat. Bot. V. **20**:5–201. *pl. 12.* 1874.
7. DAGUILLON, A., Sur le polymorphisme foliaire des Abietinées. Compt. Rend. pp. 108–110. 1889.

8. FEISTMANTEL, O., Fossil flora of the Gondwana System. Mem. Geol. Surv. Ind. 2^1: Jurassic (Oolitic) flora of Kach. 1876.

9. ———, *ibid.* 1^2: Jurassic (Liassic) flora of the Rajmahal group in the Rajmahal Hills. 1877.

10. ———, *ibid.* 2^2: Flora of the Vabalpur group. 1877.

11. ———, *ibid.* 4^1: Fossil flora of the South Rewah Gondwana Basin. 1883.

12. HALLE, T. G., Mesozoic flora of "Graham Land." Schwed. Süd Polar Exped. 3:14. 1913.

13. HILDEBRAND, F., Bau d. Spaltöffenungen d. Coniferen. Bot. Zeit. 18:17. 1860.

14. HOLLICK, A., and JEFFREY, E. C., Cretaceous coniferous remains from Kreischerville, N.Y. Mem. N.Y. Bot. Gard. 3:viii+138. *pls. 29.* 1909.

15. LINDLEY, J., and HUTTON, W., Fossil flora of Great Britain. 1836.

16. MALHERT, A., Beiträge zur Kenntnis der Anatomie der Laubblätter der Coniferen, mit besorderer Berücksichtigung des Spaltöffenungs-Apparates. Bot. Centralbl. 24:54. 1885.

17. MASTERS, M. T., Review of some points in the comparative morphology anatomy, and life history of the Coniferae. Jour. Linn. Soc. Bot. 27:226. 1891.

18. NATHORST, A. G., Palaeobotanische Mitteilungen. 2. Die Kutikula von *Dictyozamites Johnstrupi* Nath. Kungl. Svensk. Handl. 42: no. 5. 1907.

19. ———, *ibid.* 7. Über *Palissya, Stachyotaxus,* and *Palaeotaxus. Ibid.* 43: no. 8. 1908.

20. SAPORTA, G. DE, Paléontologie française. II. Végétaux plantes Jurassiques. 3: "Conifères." Paris.

21. SCHENK, A., Die Fossile Flora der Grenzschichten des Keupers und Lias Frankens. Wiesbaden. 1867.

22. SCHIMPER, W. P., Traité de paléontologie végétale 2:1870–1872.

23. SEWARD, A. C., Catalogue of the mesozoic plants in the Department of Geology, British Museum (Nat. Hist.). The Jurassic flora. Part 1. 1900.

24. ———, *ibid.* Part 2. 1904.

25. SEWARD, A. C., and FORD, S. O., The Araucarieae, recent and extinct. Phil. Trans. Roy. Soc. London B 198:305–411. 1906.

26. STOPES, M. C., and FUJII, K., Studies on structure and affinities of cretaceous plants. Phil. Trans. Roy. Soc. London B 201:1–90. 1910.

27. STOPES, M. C., and KERSHAW, E. M., Anatomy of cretaceous pine leaves. Ann. Botany 24:395–402. *pls. 27, 28.* 1910.

28. THOMAS, H. HAMSHAW, The fossil flora of the Cleveland District of Yorkshire. I. The flora of the Marske Quarry. Quart. Jour. Geol. Soc. 69: 223. 1913.

29. THOMAS, H. HAMSHAW, and BANCROFT, N., On the cuticles of some recent and fossil cycadean fronds. Trans. Linn. Soc. London II. Bot. 8:155–204. *pls. 17–20.* 1913.

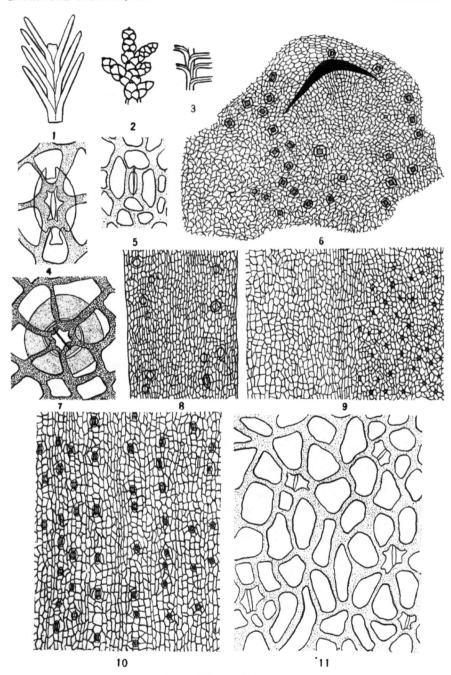

30. THOMPSON, W. P., The structure of the stomata of certain cretaceous conifers. BOT. GAZ. **54**:63–88. *pls. 5, 6.* 1912.

31. ZEILLER, R., Observations sur quelques cuticules fossiles. Ann. Sci. Nat. Bot. VI. **13**:217–238. 1882.

32. UNGER, F., Genera et Species Plantarum Fossilium. Vindobonae. 1850.

EXPLANATION OF PLATE XI

FIG. 1.—*Palissya indica:* showing general habit; linear leaves, with decurrent bases, in spiral phyllotaxis.

FIG. 2.—*Echinostrobus expansus:* showing general habit; small, closely imbricated, decussately opposite leaves.

FIG. 3.—*Taxitis tenerrimus:* showing general habit; linear leaves, with decurrent bases, in spiral phyllotaxis, with distinct midrib.

FIG. 4.—*Palissya indica:* single stoma, showing 6 accessory cells, and thickened lamellae of guard cells.

FIG. 5.—*Echinostrobus expansus:* single stoma showing 4 accessory cells.

FIG. 6.—Same: cuticle of single leaf, showing irregular shape of epidermal cells and scattered stomata.

FIG. 7.—*Podozamites lanceolatus:* single stoma, showing 6 accessory cells, and cuticular rims, probably on intercalary cells.

FIG. 8.—*Taxitis tenerrimus;* epidermis of lower surface, showing stomata scattered on each side of the midrib.

FIG. 9.—*Palissya indica:* cuticle from leaf; toward the left is the upper surface; toward the right, the lower.

FIG. 10.—*Podozamites lanceolatus:* cuticle showing rows of stomata, with their long axes parallel to the side of the leaf.

FIG. 11.—*Palissya indica:* cuticle of lower surface, showing irregular disposition of stomata and shape of epidermal cells.

THE DETERMINATION OF ADDITIVE EFFECTS

W. J. V. OSTERHOUT

(WITH FOUR FIGURES)

It was pointed out in previous papers[1] that in measuring antagonism it is of importance to determine the additive effect; this is the effect produced by dissolved substances in a mixture when each substance acts independently of all the others. It was also stated that when two equally toxic solutions are mixed the additive effect may be predicted, since it will be equal to that of one of the pure solutions. In this discussion it was assumed that if two solutions are equally toxic they will not become unequally toxic when both are diluted to the same degree. This is true (either completely or with negligible error only) for cases which have hitherto come under the writer's observation, but other cases might possibly occur to which it would not apply, and it seems desirable to discuss briefly the treatment of such cases.

As an example of this we may consider the influence of dilution on the effects of two solutions, A and B. These may be mixtures, but for the sake of simplicity we may assume that they are pure solutions of two salts, A and B, and that 100 cc. of solution A, or of solution B, diluted to 200 cc. will permit the same amount of growth to take place, as shown in fig. 1. In this figure the abscissas represent growth, while the ordinates represent the number of cc. which are taken and diluted to make 200 cc. of the culture solution. Thus on the curve A, A, the abscissa at 60 represents the growth in a culture solution made by taking 60 cc. of solution A and adding water to make 200 cc. Similarly on the curve B, B, the abscissa at 40 represents the growth in a culture solution made by taking 40 cc. of solution B and adding water to make 200 cc.

Ordinarily we should expect these curves to be almost or quite identical, but we may imagine cases in which they diverge, as shown in fig. 1. It is apparent from the figure that while 100 cc. of either

[1] BOT. GAZ. **58**:178, 272. 1914.

solution (diluted to 200 cc.) produces exactly the same effect, 50 cc. of solution A (diluted to 200 cc.) produces a different effect from 50 cc. of B (diluted to 200 cc.).

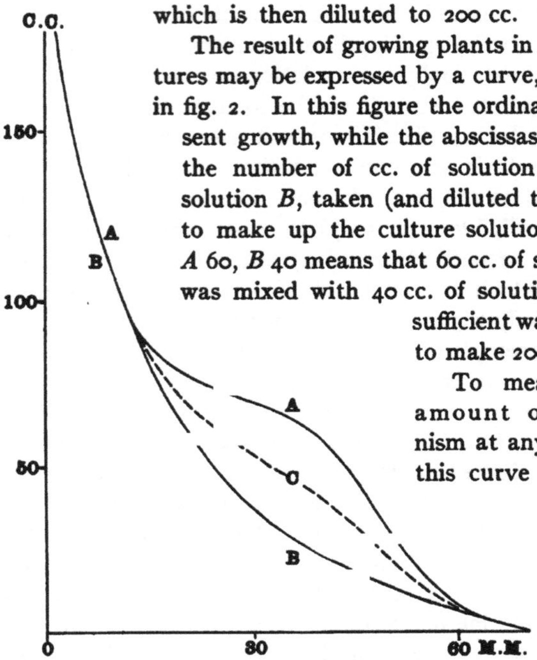

Let us now consider an antagonism curve obtained by growing plants in a culture solution made by mixing the two solutions, A and B, so as to make 100 cc. of mixture, which is then diluted to 200 cc.

The result of growing plants in such mixtures may be expressed by a curve, as shown in fig. 2. In this figure the ordinates represent growth, while the abscissas represent the number of cc. of solution A, or of solution B, taken (and diluted to 200 cc.) to make up the culture solution. Thus, A 60, B 40 means that 60 cc. of solution A was mixed with 40 cc. of solution B and sufficient water added to make 200 cc.

To measure the amount of antagonism at any point on this curve according

FIG. 1.—Curves showing growth in various dilutions of two solutions of salts, A and B: the abscissas represent growth; the ordinates represent the number of cc. of the salt solution which are mixed with water to form 200 cc. of the culture solution in which the plants were grown; the two salt solutions are equally toxic at certain concentrations but not at others; the curve C is drawn by taking points half-way between A and B (measured vertically), it serves as a basis of comparison in computing additive effects.

to the method outlined in previous papers,[2] we must first determine the additive effect. To ascertain this at any point,

[2] BOT. GAZ. 58:178, 272. 1914.

as for example at *A* 60, *B* 40, in fig. 2, we must answer the question: What is the effect of 60 cc. of *A* + 40 cc. of *B* + 100 cc. of water, when each salt acts independently of the other? It is obvious that

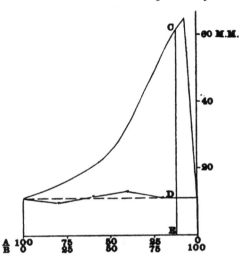

we cannot answer this by merely adding together the abscissas at these points on the curves[3] in fig. 1. Since this cannot be done, it might be thought feasible to express the effect of *A* in terms of *B*, or vice versa. If the curves *A*, *A*, and *B*, *B*, in fig. 1 were identical, this would be very simple, since the additive effect of (60 *A* + 40 *B*) would equal the additive effect of (60 *A* + 40 *A*) or the effect of 100 *A*, which is shown by the curve to be 11. Proceeding in this way, we should find the additive effect at any point on the antagonism curve to be exactly the same (that is, 11), and the additive effect could therefore be represented by a straight horizontal line, as is done in fig. 2 (dotted line).

FIG. 2.—Antagonism curve showing growth in various mixtures of solutions of the two salts, *A* and *B*, the dilution curves of which are shown in fig 1. the ordinates represent growth; the abscissas represent the number of cc. of the salt solutions which are mixed with water to make 200 cc. of the culture solution in which the plants were grown; thus *A* 75, *B* 25 signifies that 75 cc. of solution *A* were added to 25 cc. of solution *B*, the whole being then diluted to form 200 cc.; the additive effect (calculated by the method explained in the text) is shown as an unbroken line, the additive effect which would be obtained if the two dilution curves in fig. 1 did not diverge is shown as a horizontal dotted line; the antagonism at the point *C* is *CD* ÷ *DE*.

But when the curves diverge, as in fig. 1, we cannot consider the effect of 40 *B* as equal to that of 40 *A;* we see by inspection

---

[3] This is evident, for example, from the fact that the abscissa at 50 on curve *AA* is not equal to exactly twice the abscissa at 100 on curve *AA;* the abscissa at 30 is not equal to exactly twice the abscissa at 60, etc.

of the figure that the growth in 40 A is 47, while that in 40 B is 28.5 (this is equivalent to the growth in 71 A). The additive effect of (60 A + 40 B) is therefore equal to the effect of (60+71=) 131 A, which gives (as read from the curve in fig. 1) the additive effect 6.

If we calculate the additive effect of the same mixture in terms of B, we find that the effect of 60 A is 39, which is equal to the effect of 24 B. Hence the additive effect of (40 B + 60 A) equals the effect of (40 B + 24 B=) 64 B, which gives as the additive effect 19.5.

TABLE I

ADDITIVE EFFECT WHEN THE EFFECT OF M = THE
EFFECT OF 2 N

Mixture which is diluted to 200 cc to make the culture solution	Additive effect
100 cc. M	1.0
80 " M ⎫ = 90 M . . . . . 20 " N ⎭	2 0
60 " M ⎫ = 80 M . . . . 40 " N ⎭	3.3
40 " M ⎫ = 70 M . . 60 " N ⎭	5.0
20 " M ⎫ = 60 M . . . . 80 " N ⎭	7.5
100 " N = 50 M	11 0

We have in this case, therefore, two values for the additive effect, namely 6 and 19.5. One is undoubtedly too high, the other too low. Instead of taking the mean (or the weighted mean) of these two values, it seems desirable to avoid this complication altogether by calculating A and B in terms of a third curve, C. This may be obtained by taking points midway between the two curves A and B (the distance being measured vertically) and drawing a line through them, giving the dotted line C. The curve C could be drawn in any convenient manner (it might, for example, be a parabola or a hyperbola), but it should have two points in common with each of the other curves. This might be arranged by multiplying or dividing the ordinates or abscissas so as to make these curves coincide at the origin and at the half-way point with the arbitrary standard curve.

By this method we find that the effect of 60 A is equal to the effect of 42 C, while the effect of 40 B is equal to the effect of 56 C. The additive effect of (60 A+40 B) is therefore equal to the additive effect of (42 C+56 C=) 98 C, which is seen from the figure to be 11.5.

The values of the additive effect thus obtained are plotted in fig. 2. It will be seen that these values do not differ greatly from the value of an additive effect which is constant (and equal to the effect of 100 A or 100 B). Unless, therefore, the two dilution curves (as plotted in fig. 1) diverge widely, there will be no great error in regarding the additive effect as constant (and equal to the effect of 100 A or 100 B); this error will in fact ordinarily be less than the experimental error.

In case there are several salt solutions to be mixed, we may draw the corresponding dilution curves and average the ordinates of these curves at various elevations to obtain points through

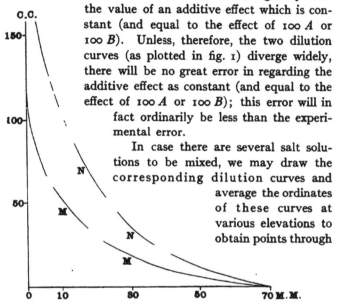

FIG. 3.—Curves showing growth in various dilutions of two unequally toxic solutions of salts, M and N: the abscissas represent growth; the ordinates represent the number of cc. of the salt solution which are taken and diluted to 200 cc. to make the culture solution in which the plants are grown.

which a curve may be drawn which shall serve the same purpose as the curve C in fig. 1; or an arbitrary curve (for example, a parabola or hyperbola) may be drawn for this purpose.

When the two salt solutions are not equally toxic, we often find cases in which a constant relation exists between the amounts of the two solutions which (diluted to 200 cc.) produce the same

effect. For example, it will be seen in fig. 3 that a growth of
20 mm. may be found at 30 M or 60 N; or a growth of 15 mm. may
be found at 40 M or 80 N. If this relation holds throughout (even
approximately), we may consider the effect of M as equal to the
effect of 2 N. If we grow plants in culture solutions made by taking sufficient of $M+N$ to make
100 cc. (and then diluting this mixture to 200 cc.),
we shall get an additive effect which is not constant but which will always be the same for any
given mixture, whether calculated as M or as N.
The additive effect obtained under these
conditions is shown as a curved and dotted
line in fig. 4 (cf. table I). This procedure,
as will readily be seen, is the same as
determining the additive effect of M
mixed with another solution of M
which has been diluted
to a definite degree with
water (each 100 cc. of
the mixture of $M+[M$
diluted] being itself di-
luted to 200 cc.). This
is in fact the method
suggested in a previous
paper.[4]

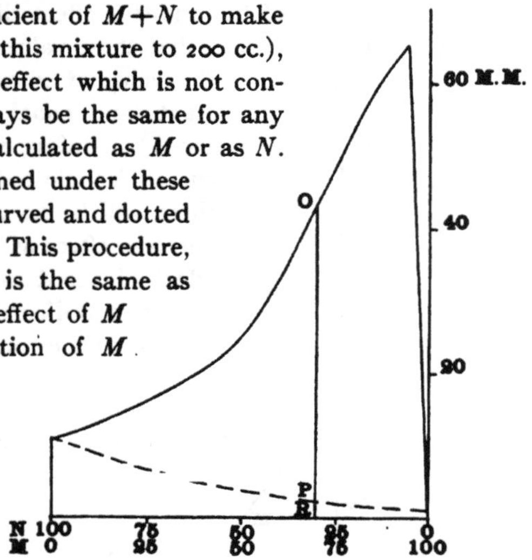

A relation such that
the effect of M = the
effect of $\times N$ will be
found to hold (at least
approximately) in most
cases. If it does not,
the additive effect may

Fig. 4.—Antagonism curve showing growth in
various mixtures of two solutions of salts, M and N,
of which the dilution curves are shown in fig. 3: the
ordinates represent growth; the abscissas represent
the number of cc. of the solutions of salts which
are taken and diluted to 200 cc. to make the culture
solution in which the plants were grown; thus N 75,
M 25 signifies that 75 cc. of N were added to 25 cc.
of M and the whole diluted to 200 cc.; the additive
effect is shown by the curved dotted line; the
antagonism at the point O is $OP+PR$.

be calculated in terms of a third curve drawn arbitrarily or by
taking points midway between the two (measured vertically), as
previously explained.

4 Bot. Gaz. 58:178. 1914.

When the solution contains more than two components, we may follow a method similar to that already outlined for equally toxic solutions containing more than two components.

In most cases the calculations described above may be dispensed with, as the error (if any) in proceeding by the method described in the two preceding papers[5] is so small as to be negligible. Calculations such as are here discussed will be necessary only in those cases in which equally toxic solutions acquire, when diluted to the same degree, a very marked difference in toxicity, or in cases where a mixture is made of unequally toxic solutions which have dilution curves of very dissimilar character.

Summary

In most cases two solutions which are equally toxic remain so (at least approximately) when both are diluted to the same degree; this allows the additive effect to be easily determined. But in exceptional cases, where this does not hold, a value may be assigned to the additive effect.

Similar considerations apply to unequally toxic solutions.

LABORATORY OF PLANT PHYSIOLOGY
    HARVARD UNIVERSITY

[5] *Op cit.*

BRIEFER ARTICLES

TISSUE TENSION IN AMORPHOPHALLUS

(WITH ONE FIGURE)

Large monocotyledonous plants which do not develop woody mechanical tissue afford excellent opportunity to demonstrate the tension to which cortical and central tissues are subjected. The combined result of the stretching of the cortex and the compression of the central tissues is the rigidity of the organ.

FIG. 1.—Leaf of *Amorphophallus campanulatus*

Such a plant in the Philippine Islands is *Amorphophallus campanulatus* (Roxb.) Blume (Araceae). The decompound blade may often be more than a meter above the ground and occasionally spreads out over an area 1 5–2 meters in diameter. The stemlike petiole, which is often 10 cm. in thickness, is the part utilized in studying tissue tension.

Two sets of experiments were performed. In the first set, medium sized petioles were collected in the middle of a hot day, when the plant was partly flaccid, brought into the laboratory, and there experimented upon during the following hour. In each case a piece 30 cm. in length and as nearly uniform in thickness as possible was selected from the petiole. A strip of the cortex, 1 cm. wide, was taken from one side of the petiole and laid on a table. The rest of the cortex was peeled off and the core laid next to the strip. When the elongation of the core and the contraction of the strip had ceased, both were measured and the results assembled. For the second set of experiments, petioles were collected in the early morning and soaked for about 40 minutes in tap water to make sure that they were turgid. In the experiment three lengths were employed: 20, 25, and 30 cm., with each following the same methods as above. The results of both sets, summarized in tabular form, are shown in the accompanying table.

TABLE I

TABLE OF TISSUE TENSION IN *Amorphophallus*

NUMBER OF PLANTS	LENGTH			PERCENTAGE INCREASE IN CORE	PERCENTAGE DECREASE IN STRIP
	Original	Core	Strip		
	Petioles partly flaccid				
1 . . 69....	30.0	31.6	29.4	5 3 per cent	2 0 per cent
	Petioles turgid				
2a ..12....	20.0	21.7	19.8	8.5	1.0
2b...12....	24.7	26.8	24.5	8.4	0.8
2c...12....	30.0	32.5	29.7	8.3	1.0
Average of turgid petioles.......				8.4	0.9

As shown in the table, the core is capable of greater elongation when turgid, but the cortex will contract more when flaccid. The possibility of change in length in the core is greater than in the cortex because the former is more vascular. In the petiole, in nature, under turgid conditions, the sum of the possibilities of change in length of the tension-producing elements is greater than when the organ is flaccid, consequently greater tissue tension is present under turgid conditions.—FRANK C GATES, *Los Baños, P.I.*

CURRENT LITERATURE

BOOK REVIEWS
A montane rain-forest

As a result of several visits to Jamaica, affording in all about 11 months' residence upon the island, SHREVE[1] has contributed a study of a tropical rain-forest that is particularly valuable, since it represents practically the first attempt to determine quantitatively the factors concerned in the production of such vegetation. The forest investigated covers the Blue Mountains above an altitude of 1372 meters. This ridge, having an altitude of 1500–2265 meters, runs lengthwise of the island and, therefore, is nearly at right angles to the direction of the trade winds. The northern or windward side is fog-covered for 70 per cent of daylight hours during 9 months of the year and for 30 per cent of daylight hours during the remaining months, while the rainfall approximates 425 cm. annually. The southern slope is drier, with less than half the amount of fog, and with an annual precipitation of about 265 cm. A further indication of the humidity is to be found in the evaporation measurements upon the leeward slope, giving maximum and minimum daily average losses of 17 9 cc. and 1 8 cc. (compared with approximately 45 cc. and 10 cc. respectively for the Chicago region) as shown by the standard atmometer placed in the open, with half of these amounts for the forests of the ravines. These low amounts seem to be four or five times as great as for corresponding situations upon the more humid northern slopes.

Temperature has an annual range of 23° C., with the monthly mean from 15° to 17° C., the maximum seldom exceeding 22° C. or the minimum 7° C., while frosts are unknown. Detailed temperature and humidity data for limited periods, as well as soil temperature records, are given.

The mountain soil is very subject to erosion, and this instability may account, in part, for the absence of large trees; still, the forest is very continuous, reaching its best development, however, in ravines, where trees 20 meters in height are seen, with trunks 80 cm. in diameter in such species as *Solanum punctulatum* and *Gilibertia arborea*. There seems to be no great complexity of floristic composition, the forest resembling in this respect temperate forests rather than those of tropical lowlands. This is indicated by the fact that *Clethra occidentalis*, *Vaccinium meridionale*, and *Podocarpus Urbanii* form about 50 per cent of the stand, and that an additional 35 per cent is made

---

[1] SHREVE, FORREST, A montane rain-forest. A contribution to the physiological plant geography of Jamaica. Washington. 8vo. pp. 110. *pls 29. figs. 18.* Carnegie Institution of Washington. Publication 199. 1914. $1 50.

up of 10 other species. The comparatively small number of lianas, the paucity
of floral display, the slow rate of growth, and the absence of plank buttresses
and cauliflora, all show rather unexpected variations from the rain-forests of
other lands. Epiphytes are abundant, however, but limited to comparatively
few species of bromeliads and orchids, a profusion of the Hymenophyllaceae,
and a great bulk of mosses and hepatics.

The richer forests of the windward ravines show a distinct stratification
of vegetation, passing from the more or less continuous canopy of large trees,
through under-trees and shrubs, to the herbaceous carpet, the lower levels being
festooned with mosses (species of *Phyllogonium* and *Meteorium*), while the
larger limbs and leaning trunks are crowded with orchids and ferns. Among
the notable genera, represented by several species each, are tree ferns (*Cyathea*)
found among the under-trees, *Piper* occurring among the shrubs, and the
herbaceous species of *Peperomia* and *Pilea*. Upon the slopes the trees are
smaller, the stratification is less marked, and epiphytes are less abundant,
changes that are more accentuated upon the ridges, where a low open canopy
results. Here thickets of ferns and of climbing bamboo (*Chusquea abietifolia*)
are common. The distinct type of forest occupying the various habitats are all
carefully described, and SHREVE concludes that by no possible physiographic
change could any one of these habitats occupy all or nearly all the region;
hence there is no means to fix on any one of the types as the climax of the region.

The great uniformity of temperature, with no pronounced dry season,
affords excellent opportunities for studying the seasonal behavior of the various
species. It is quite interesting, therefore, to note that although the months
from October to January have a maximum rainfall it is a season of relative rest,
during which a few tree species, such as those belonging to *Rhamnus*, *Clethra*,
and *Viburnum*, allied to north temperate forms, shed all or a part of their leaves,
while others make little growth. This may be accounted for by the comparative lack of sunlight. Other trees, while evergreen, flower in the spring and
complete their growth by October; but in still others, growth and leaf formation is continuous, but flowering periodical, while not a few show continuous
growth and blooming. Among the large trees belonging to the last category
are *Ilex montana* and *Solanum punctulatum*, while under-trees and shrubs
include species of *Piper*, *Boemeria*, *Malvaviscus*, and *Datura*   Quite as interesting is the behavior of introduced species, *Quercus robur* and *Liquidambar
Styraciflua* becoming evergreen, while *Liriodendron Tulipifera* and *Taxodium
distichum* retain their deciduous habits.

Rates of relative transpiration in the rain-forest and in the desert of
Arizona are found to be of the same order of magnitude, that is, they are
proportional to evaporation, but in its own climate the desert plant loses far
more water per unit area than the plant of the rain-forest. The foliage of the
rain-forest, however, shows great diversity of structure, corresponding to
the different strata of vegetation, from sclerophyllous leaves of the trees, the

coriaceous succulent leaves of the epiphytes, and the mesophytic shade leaves of the herbaceous plants, to filmy ferns with leaves composed of a single layer of cells. The effect of rainfall upon the foliage is reported elsewhere,[2] and it is seen that, with a single exception, all adult leaves are wettable. In spite of this, aside from the Gramineae and Cyperaceae, only four seed plants and two ferns have functioning hydathodes, and dripping points are neither abundant nor conspicuous. Injected leaves were seen only once, and that after five days of violent rains, when those with as well as those without hydathodes alike recovered, showing no evidence of injury. The wettable leaf surface reduces the water intake from the shoot, but does not reduce the temperature to an extent sufficient materially to affect transpiration. Epiphyllous plants are common in the ravines, and in other situations of maximum humidity. They are favored by the wet leaf surfaces, nor do the dripping points, when developed, so greatly promote drying as to reduce the probability of the leaves being thus overgrown. The epiphyllae are mostly Hepaticae of the genus *Lejunea*.

Most of the conclusions arrived at in these two publications are the result of experiment, and are supported by qualitative data, thus marking a new era in the investigation of tropical vegetation, and necessitating a readjustment of many generalizations resting upon less definite evidence. Taken together, they form one of the most notable of recent contributions to our knowledge of rain-forest phenomena.

The importance of the Cinchona Botanical Station, the headquarters from which these investigations were carried on, has been emphasized quite recently by JOHNSON,[3] who draws attention to its many advantages for the student who would become familiar with a great variety of tropical conditions or who would undertake the solution of some of the many ecological and physiological problems of tropical vegetation. Situated upon the slopes of the Blue Mountains at an altitude of about 1500 meters, it has a climate agreeable to workers from temperate zones, a supply of pure water, freedom from tropical diseases, and yet from it as a center easy access may be had both to the higher parts of the forest-covered mountains, now reserved by the government of the island as a watershed, and to the more torrid plains below. It has an equipment of residence, laboratories, and greenhouses capable of affording accommodations for eight or ten workers, and gardens and grounds planted with species from other tropical and temperate lands. In addition, two botanical gardens situated in the lowlands can be used as substations of the main laboratory. One at Castleton has an altitude of 150 meters and a rainfall of 355 cm., while the other at Hope at a similar altitude has less than half as much precipitation.

---

[2] SHREVE, FORREST, The direct effects of rainfall on hygrophilous vegetation. Jour. Ecol. 2:82–98. 1914.

[3] JOHNSON, D. S , The Cinchona Botanical Station. Pop. Sci. Monthly 85:512–530. 1914; 86:33-48. 1915.

Not less important than the natural advantages of the Station in climate and vegetation is its accessibility, and the fact that it is located in an English-speaking country with a stable government and reliable sanitary control. For the past ten years it has been a station of the New York Botanical Garden, but it is now to be maintained under the auspices of the British Association for the Advancement of Science with the cooperation of the Jamaican government.—GEO. D. FULLER.

NOTES FOR STUDENTS

Cultures of the Uredineae.—In the review covering the cultural work with the Uredineae for 1912,[4] the following results of TREBOUX and of LONG should have been included. TREBOUX[5] in two papers from Nowotscherkask, reports the following cultures. Teleutospores of *Uromyces Festucae* Syd. from *Festuca ovina* L. produced aecidia on *Ranunculus illyricus* L. (The reverse culture has previously been reported.)[6] Aecidiospores from *Allium decipiens* Fisch., *A. moschatum* L., *A. rotundum* L., and *A. sphaerocephalum* L. produced uredospores and teleutospores (*Puccinia permixta* Syd.) on *Diplachne serotina* Lk. The reverse infection on 3 of these and 13 other species of *Allium* was successful also. Teleutospores of *Puccinia stipina* Tranzsch. from *Stipa capillata* L. infected 5 native species of *Salvia* and 15 others grown from seed, and also *Origanum vulgare* L., *Lamium amplexicaule* L., *Glechoma hederacea* L., *Lallemantia iberica* F. et M., *Leonurus cardiaca* L , and *Stachys recta* L. This rust shows very little selection among the Labiatae. *Puccinia littoralis* Rostr. from *Juncus Gerardi* Lois. produced aecidia on *Cichorium Intybus* L. (the reverse culture has been previously reported). Aecidiospores of *Puccinia Polygoni-amphibii* Pers from *Geranium collinum* Steph. infected *Polygonum amphibium* L. but not *P. lapathifolium* L. The reverse infection was successful on *Geranium collinum* L., *G. pratense* L., *G. divaricatum* Ehrh., *G. columbinum* L., and *G. rotundifolium* L. Aecidiospores of the autoecious form *P. ambigua* Alb et Schw on *Galium aparine* L. produced successive generations of aecidia when sown on that host. Aecidiospores of *Puccinia Agropyri* Ell. et Ev. from *Clematis pseudo-flammula* Schmalh. infected *Agropyrum repens* P.B. The uredospores from this culture infected *Agropyrum cristatum* Bess. and *A. prostratum* Eichw. Aecidiospores of *Puccinia bromina* Erikss. from *Lithospermum arvense* L. infected *Bromus tectorum* L. and *B. squarrosus* L Similarly aecidiospores from *Myosotis silvatica* Hoffm. infected *B. tectorum*. The two aecidia belong to the same rust Aecidiospores and uredospores of *Uromyces Limonii* (DC.) from *Statice latifolia* Sm infected *Statice Gmelini* Willd. also. Aecidiospores of an unnamed species of *Puccinia* from *Centaurea trichocephala*

4 BOT. GAZ. **56**:233-239. 1913.

5 TREBOUX, O., Infektionsversuche mit parasitischen Pilzen II. Ann. Mycol. **10**:303-306. 1912; and *idem* III. *Ibid.* 557-563. 1912.

6 Rev. BOT. GAZ. **56**:239. 1913.

M.B. produced uredospores and teleutospores on *Carex stenophylla* Wahlenb.
Aecidiospores from *Euphorbia virgata* W.K. infected *Caragana arborescens* L.,
Trifolium agrarium L., and *Lotus corniculatus* L., but none of the species
usually inhabited by *U. Genistae-tinctoriae* Pers. This result seems to indicate
that the species of *Uromyces* on *Caragana* is a biological form distinct from
U Genistae-tinctoriae. With aecidiospores from the same host from another
locality, *Medicago minima* Bart., *M. murex* Willd., and *Trifolium arvense* L.
were infected. This form proved to be *Uromyces striatus* Schroeter. A third
form with its aecidia likewise on *Euphorbia virgata* produced uredospores and
teleutospores (*Uromyces Astragali* Opiz.) on *Astragalus criticus* Lam. and *A.
sanguinolentus* M.B. Aecidiospores of *Uromyces caryophyllinus* (Schrank)
Wint. from *Euphorbia Gerardiana* Jacq. infected *Dianthus arenarius* L., *D.
campestris* M.B., *D. capitatus* DC., *D. caryophyllus*, and *D. pseudomeria* M.B.
Aecidiospores from other plants of *Euphorbia Gerardiana* produced an abundance of uredospores and teleutospores of *Uromyces Schroeteri* De Toni on
Silene otites Sm Sowings of aecidiospores (*Puccinia coronifera* Kleb.) from
Rhamnus cathartica L. and reverse cultures and cross-sowing of the aecidiospores thus derived seem to show that the specialization of forms in this
species of crown rust is not so well marked as former experiments appeared to
indicate.

LONG'S[7] experiments dealt with three species of rusts on members of the
genus *Andropogon*. He reports the following successful cultures. Teleutospores of the type of *Puccinia Andropogonis* Schw. from *Andropogon furcatus*
Muhl. collected in Texas produced aecidia on *Oxalis corniculata* L. Teleutospores of *Puccinia Ellisiana* Thüm. from *Andropogon virginicus* L. collected in
Virginia produced aecidia on *Viola fimbriatula* Sw., *V. hirsutula* Brainard, and
V. papilionacea Pursh. *Viola sagittata* L., which was not infected, had been
successfully infected in a former experiment. Aecidiospores from *V. sagittata*
and *V. papilionacea* reinfected *Andropogon virginicus*. In 1910 the author had
sent material of this same type and from the same locality to ARTHUR, who
made successful sowings of teleutospores on *Penstemon*. These two results
seem to indicate that two forms of *Puccinia* occur on *Andropogon virginicus*.
Finally, teleutospores of *Uromyces Andropogonis* Tracy from *Andropogon
virginicus* L. collected in Virginia infected *Viola primulifolia* L. and *V. cucullata*
Ait. but none of the other violets. Aecidiospores from *V. primulifolia* reinfected *Andropogon virginicus*. In conclusion, the author points out the close
similarity between *Puccinia Ellisiana* and *Uromyces Andropogonis*, which
differ from each other only in the number of cells of the teleutospore.

During 1913 no very extensive series of cultures has been reported, but
small additions to our knowledge of the biological relations of hitherto isolated
forms come from many sources and include studies in several genera. In this

---

[7] LONG, W. H., Notes on three species of rusts on *Andropogon*. Phytopathology
2:164–171. 1912.

connection the past work of FRASER[8] on the fern rusts has been of special interest. This author now reports[9] a further series of cultures supplementing and confirming former work which was not regarded by him as entirely conclusive The work was done at Pictou, Nova Scotia. Successful infections on *Abies balsamea* (L.) Mill. were made with teleutospores of *Uredinopsis Struthiopteridis* Störmer from *Onoclea Struthiopteris* (L.) Hoff., *U. Osmundae* Magn. from *Osmunda Claytoniana* L., *U. Phegopteridis* Arthur from *Phegopteris Dryopteris* (L.) Fée, and *U. mirabilis* Magn. from *Onoclea sensibilis* L. The aecidial form on *Abies* is *Peridermium balsameum* Peck. Successful infections with aecidiospores of this form were made on *Onoclea Struthiopteris*, *O. sensibilis*, and *Aspidium Thelypteris* Sw. The telial phase on the last is known as *Uredinopsis Atkinsonii* Magn.

Cultures with three other species not belonging to the fern rusts are reported also in confirmation of former work. Teleutospores of *Pucciniastrum Myrtilli* (Schum.) Arthur from *Vaccinium canadense* Kalm produced a *Peridermium* of the type of *P. Peckii* Thüm. on *Tsuga canadensis* (L.) Carr. Teleutospores of *Melampsora Medusae* Thüm. from *Populus grandidentata* Michx. produced *Caeoma Abietis-canadensis* Farl. on *Tsuga canadensis* (L.) Carr. Teleutospores of *Melampsora arctica* Rostr. from *Salix* sp. produced aecidia (*Caeoma* sp.) on *Abies balsamea* (L.) Mill.

Further work on the *Peridermium* rusts is reported by MEINECKE, SPAULDING, and by HEDGCOCK and LONG.

MEINECKE[10] infected *Castilleja miniata* Dougl. with aecidiospores of *Peridermium stalactiforme* Arthur and Kern from *Pinus contorta* Loud. The resulting telial stage is *Cronartium coleosporioides* (Dietel and Holway) Arthur.

HEDGCOCK and LONG[11] report the following results. Aecidiospores of *Peridermium inconspicuum* Long from *Pinus virginiana* Mill. produced uredinia (*Coleosporium inconspicuum* (Long) H. and L.) on *Coreopsis verticillata* L.; aecidiospores of *P. delicatulum* Arthur and Kern from *Pinus rigida* Mill. produced uredinia (*Coleosporium delicatulum* (Arthur and Kern) H. and L.) on *Solidago lanceolata* L.; and aecidiospores of *P. stalactiforme* Arthur and Kern from *Pinus contorta* Loud. produced uredinia and telia on *Castilleja linearis* Rydb. The last result is regarded as confirming the work of MEINECKE mentioned above. For the synonomy of these forms see the work of ARTHUR and KERN mentioned below.

In 1907 CLINTON showed that *Peridermium pyriforme* Peck (*ex* ARTHUR and KERN) on *Pinus silvestris* L. is the aecidial form of *Cronartium Comptoniae*

---

[8] Rev. BOT GAZ. **56**:234. 1913.

[9] FRASER, W P., Further cultures with heteroecious rusts. Mycologia **5**:233–239. 1913.

[10] MEINECKE, E P , Notes on *Cronartium coleosporioides* Arthur and *Cronartium filamentosum*. Phytopathology **3**:167–168. 1913.

[11] HEDGCOCK, G. C., and LONG, W. H., Notes on cultures of three species of *Peridermium*. Phytopathology **3**:251–252. 1913.

Arthur on *Comptonia asplenifolia* Banks. SPAULDING[12] has now succeeded in infecting this host with aecidiospores of *Peridermium pyriforme* from *Pinus silvestris* (confirming CLINTON's work), *P. ponderosa* Dougl., *P. Taeda* L, and *P. austriaca* Link. On account of its increasing frequency in nurseries, this fungus is becoming economically important.

JACOB[13] in a short note reports that teleutospores of *Puccinia Poiygoni-amphibii* Pers. from *Polygonum amphibium* L. infected *Geranium pratense* L. (confirming the result of KLEBAHN), *G. pusillum* Burm., and *G. pyrenaicum* Burm. Among the species not infected was *Geranium phaeum* L., which KLEBAHN had successfully infected with this rust. The aecidiospores from these cultures were sown on various species of *Polygonum*, but only *P amphibium* was infected. Teleutospores of *Puccinia Polygoni* Alb. amd Schw. from *Polygonum Convolvulus* L. infected only *Geranium columbinum* L., with a doubtful infection on *G. molle* L. which was successfully infected by KLEBAHN. Aecidiospores from *G. molle* reinfected only *Polygonum Convolvulus*. Uredospores from *Uromyces Kabatianus* Bubák from *G. pyrenaicum* infected *G. pyrenaicum*, *G. maculatum* L., and *G. pusillum* Burm., but not *G. silvaticum* L. which is the principal host of *Uromyces Geranii*. These cultures furnish further evidence in justification of BUBÁK's separation of *U. Kabatianus* from *U. Geranii*. CRUCHET[14] reports that teleutospores from *Polygonum Bistorta* L. infected *Peucedanum Ostruthium* Koch, and that the aecidiospores (*Aec. Imperatoriae* Cruchet) derived from the culture reinfected *Polygonum Bistorta*. The rust is described as *Puccinia Imperatoriae-mamillata*. CRUCHET was led to suspect this connection by the fact that *Peucedanum Ostruthium* bears, in addition to *Aecidium Imperatoriae*, a micropuccinnia whose teleutospores resemble those of *Puccinia mamillata* Schröter on *Polygonum*.

FISCHER[15] in two short papers reports further experiments with *Urmoyces caryophyllinus* (Shrank) Winter and *Puccinia Pulsatillae* Kalchb. which is a micropuccinia of the type of *P. Anemonis-virginianae* Schwein. inhabiting members of the Ranunculaceae. In his former work[16] FISCHER found that aecidio-

---

[12] SPAULDING, P., Notes on *Cronartium Comptoniae*. Phytopathology 3:62, 308–310. 1913.

[13] JACOB, G., Zur Biologie *Geranium*-bewohnender Uredineen. Mycol. Centralbl. 3:158–159. 1913.

[14] CRUCHET, P., Contribution à l'étude des Urédinées. Étude biologique et description de *Puccinia Imperatoriae-mamillata*, nov. sp. Mycol. Centralbl. 3:209–214. 1913.

[15] FISCHER, ED., Beiträge zur Biologie der Uredineen. 4. Weitere Versuche über die Specialisation des *Uromyces caryophyllinus* (Schrank) Winter Mycol. Centralbl. 3:145–149. 1913.

———, idem. 5. *Puccinia Pulsatillae* Kalchb (Syn. *P. de Baryana* Thüm.) und Theoretisches über die Specialisation. *Ibid.* 214–220. 1913.

[16] Rev. BOT. GAZ 56:237 1913.

spores of *Uromyces caryophyllinus* from *Euphorbia Gerardiana* Jacq. collected near Heidelberg infected *Tunica prolifera* (L.) Scop. and rarely *Saponaria ocymoides* L., while aecidiospores from the same host collected in the Wallis, Switzerland, infected *Saponaria ocymoides*. The relations of the last form to *Tunica prolifera* were not determined. Further cultures have now shown that the form from the Wallis infects both *Saponaria ocymoides* and *Tunica prolifera* and to some extent also *T. Saxifraga* (L.) Scop. Uredospores from *Saponaria ocymoides* or from *Tunica prolifera* infect either of those hosts indifferently.

The cultures with *Puccinia Pulsatillae* Kalchb. from *Anemone montana* Hoppe showed that this form infects, besides *A. montana*, *A. vernalis* L., *A. pratensis* L., and *A. Pulsatilla* L., but not *A. alpine* L., *A. sylvestris* L., and *Atragene alpina* L. Comparing the specialization of these forms with that of *Uromyces caryophyllinus*, FISCHER distinguishes two types. The first is correlated with the geographical distribution of the hosts and is illustrated by *Uromyces caryophyllinus*. In the Wallis, where both *Saponaria* and *Tunica* are common, this rust occurs on both of these plants, whereas in Baden, where *Saponaria ocymoides* does not occur, the fungus has become adapted to *Tunica prolifera* to such an extent that it scarcely infects *Saponaria*. Specialized races of this type show no distinctive morphological characteristics by which they might be separated from each other. The second type of specialization is correlated with the degree of affinity of the host plants. The forms of rusts showing this type of specialization each inhabit groups of closely related species of host plants and do not cross readily from one group to another. The races showing this type of specialization usually have slight morphological differences, besides their biological behavior, by which they can be distinguished.

ITO[17] reports the successful infection of *Pourthiaea villosa* Decne. with teleutospores of *Gymnosporangium Photiniae* (P. Henn.) Kern (*G. japonicum* Syd.) from stems of *Juniperus chinensis* L. *Pyrus sinensis* Lindl., *P. Malus* L., and *Amelanchier asiatica* Koch were not infected. These cultures show that the stem-inhabiting form of *Gymnosporangium* on *Juniperus chinensis* is connected with *Roestelia Photiniae* P. Henn. and is distinct from the leaf-inhabiting form which the author identifies with *Gymnosporangium Haraeanum* Syd. (*G. asiaticum* Miyabe) which, according to the experiments of SHIRAI and those of HARA cited by the author, belongs to *Roestelia koreaensis* P. Henn. on leaves of *Pyrus sinensis*. SHIRAI does not state whether he used the leaf-inhabiting form or the stem-form in his experiments.

The following papers were published in 1914.

FROMME[18] successfully infected *Myrica cerifera* L. with teleutospores of *Gymnosporangium Ellisii* (Berk.) Farlow from *Chamaecyparis thyoides* L. This

---

[17] ITO, S., Kleine Notizen über parasitische Pilze Japans. Bot. Mag. Tokyo 27:217–223. 1913.

[18] FROMME, F. D., A new gymnosporangial connection. Mycologia 6:226–230. 1914.

result is of special interest since the aecidial host (*Myrica*) belongs to a family far removed from the Pomaceae.   The aecidia are of the cupulate type.

TRANZSCHEL[19] reports the results of cultures carried out from 1911 to 1913. During that time the connections of 4 species of *Puccinia* with their aecidia were established for the first time, and confirmatory cultures were made with 12 other species whose aecidia were known.   In each case sowings were made on a number of plants besides those infected.   Only the new connections are given here.   *Puccinia simplex* (Körn) Eriks. and Henn. from *Hordeum vulgare* L. produced aecidia (*Asc. ornithogalum* Bubák) on *Ornithogalum umbellatum* L. and *O. narbonense* L.   Aecidiospores from this culture produced uredospores and teleutospores on *Hordeum vulgare*.   *Puccinia Hemerocallidis* Thüm. from *Hemerocallis minor* Mill. produced aecidia (*Aec. Patrinae* P. Henn.) on *Patrinia rupestris* Juss. and *P. scabiosifolia* Link.   *Puccinia nitidula* Tranzsch. from *Polygonum alpinum* All. produced aecidia on *Heracleum sibiricum* L.   *Puccinia Stipae-sibiricae* Tranzsch. from *Stipa sibirica* L. produced aecidia (*Aec. Sedi-Aizoontis* Tranzsch.) on *Sedum Aizoon* L.

KLEBAHN[20] reports new hosts for *Cronartium asclepiadeum* (Willd.) Fr. and various species of *Coleosporium*.   Aecidiospores of *Cronartium asclepiadeum* (*Peridermium Cornui* Rostr. and Kleb.) were successfully sown on *Vincetoxicum officinale* Moench (the usual host), *V. fuscatum* Reichenb., *V. laxum* Koch, *Tropaeolum minus* L., *T. majus* L., *T. canariensis* Hort. (*T. peregrinum* L.), *T. Lobbianum* Hort., *Impatiens Balsamina* L., and *Pedicularis palustris* L. Uredospores obtained from the cultures on *Vincetoxicum officinale* infected *Impatiens Balsamina* and *Pedicularis palustris*.   Aecidiospores of *Peridermium Pini* (Willd.) Kleb. failed to infect *Tropaeolum minus*, *Pedicularis palustris*, *Vincetoxicum officinale* and *Schizanthus Grahami* Gill.   The results of cultures on *Pedicularis* show that the *Cronartium* on that plant belongs to *Peridermium Cornui* Rostr. and Kleb. and not, as SIRO had erroneously supposed, to *P. Pini* Chev.   *P. Pini* remains an isolated aecidium.   The ooservation that the Chilean species, *Schizanthus Grahami* Gill. growing in Brandenburg was infected with a *Coleosporium* led the author to make sowings of a number of European species of *Coleosporium*, with the surprising result that not one but several of the European forms infected *Schizanthus*.   At the same time, cultures were made upon another exotic plant, *Tropaeolum minus* L., upon which *Coleosporium* had been observed.   *Schizanthus Grahami* was infected by uredospores of the following forms: *Coleosporium Euphrasiae* (Schum.) Wint. from *Alectrolophus major* Reichenb. and *A. minor* Wimm. and Grab.; *C. Melampyri* (Rahenh.) Kleb. from *Melampyrum pratense* L.; *C. Campanulae* f. *rapunculoides* Kleb. from *Campanula rapunculoides* L.; *C. Campanulae* f.

[19] TRANZSCHEL, W., Culturversuche mit Uredineen in den Jahren 1911-1913. Mycol. Centralbl. 4:70-71. 1914.

[20] KLEBAHN, H., Kulturversuche mit Rostpilzen. Zeitschr. Pflanzenkrank 24:1-32. 1914.

rotundifoliae Kleb. from *Campanula rotundifolia* L.; *C. Campanulae* f. *Trachelii* Kleb. from *Campanula patula* L. and *C. Trachelium* L.; *C. Tussilaginis* (Pers.) Kleb. from *Tussilago Farfara* L.; *C. Senecionis* (Pers.) Fr. from *Senecio sylvaticus* L. and *S. vulgaris* L.; and *C. Sonchi* (Pers.) Lév. from *Sonchus arvensis*. *Tropaeolum* was infected with all these forms except *C. Euphrasiae, C. Melampyri,* and *C. Sonchi.*

FRASER[21] gives a short account of cultures confirming work that has been previously reported. The cultures with *Uredinopsis mirabilis* Magn. deserve mention since they show that this form, which in common with a number of other species of *Uredmopsis* has its aecidia on *Abies balsamea* (L.) Mill., produces its uredospores and teleutospores only on *Onoclea sensibilis* L. and does not infect *Osmunda Claytoniana* L., *O. regalis* L., *Aspidium Thelyteris* (L.) Sw., *Asplenium Felix-foemina* (L.) Bernh., and *Phegopteris Dryopteris* (L.) Fée. These experiments show that *Uredinopsis mirabilis* is a distinct species.

In their revision of the North American species of *Peridermium* on pine, ARTHUR and KERN[22] mention cultures establishing the connection between *Peridermium cerebrum* Peck. (*P. fusiforme* Arthur and Kern) from *Pinus Taeda* and *Cronartium Quercus* Arthur on *Quercus rubra* L. and *Q. Phellos* L. This species of *Peridermium* occurs also on many other species of pine.

As a result of cultural experiments with *Gymnosporangium Blasdaleanum* (Dietel and Holway) Kern. (*G. Libocedri* (P. Henn) Kern.) and *Libocedrus decurrens* Torr., JACKSON[23] is able to add *Cydonia vulgaris* L., *Pyrus communis* L., *P. rivularis* Dougl., *Amelanchier alnifolia* Nutt., and *Crataegus* Lindl. to the list of hosts upon which the aecidial generation of this fungus has been grown. The aecidium, which has distinctive characteristics, has been found occurring also in nature upon a number of other plants.

In continuation of his work on the rusts of Southeastern Russia, TREBOUX[24] reports the following connections. Aecidiospores from *Ranunculus flammula* L. produced uredospores and teleutospores (*Uromyces Festucae* Syd.) on *Festuca rubra* L. With teleutospores of *Puccinia Magnusiana* Körn. from *Phragmites communis* Trin., which is known to have its aecidia on *Ranunculus repens* L , *R. chaerophyllos* L., *R. creticus* L., *R. illyricus* L., *R. Kotschyi* Boiss. and *R. sardous* Crantz were also infected. The successful infection of *Berberis vulgaris* L. with teleutospores of a *Puccinia* on *Sesleria caerulea* Ard. shows that this rust, which had been described by FISCHER as *P. Sesleriae-caeruleae,* is

---

[21] FRASER, W. P., Notes on *Uredinopsis mirabilis* and other rusts. Mycologia 6:25–28. 1914

[22] ARTHUR, J. C., and KERN, F. D , North American species of *Peridermium* on pine Mycologia 6:109–138. 1914

[23] JACKSON, H S , A new pomaceous rust of economic importance, *Gymnosporangium Blasdaleanum.* Phytopathology 4:261–270. pls. 2. 1914.

[24] TREBOUX, O., Infectionsversuchen mit parasitischen Pilzen IV Ann. Mycol. 12:480–483. 1914.

P. graminis. Sowings of teleutospores of *Puccinia Phragmitis* (Schum)
Körn. showed that *Rumex aquaticus* L., *R. confertus* Willd., *R. maritimus* L.,
R. patientia L., *R. arifolius* All., *R. bucephalophorus* L., *R. fennicus* Murb.,
R. thyrsiflorus Fingerh., *Rheum palmatum* L., *R. undulatum* L., *R. compactum* L.,
and *R. tartaricum* L. are additional aecidial hosts for this rust. Further experi-
ments with the crown rusts lead the author to doubt the validity of the species
Puccinia coronifera which KLEBAHN separated from *P. coronata* (Corda) Kleb.
This doubt is founded on the one hand on the infection of a number of typical
P. coronata hosts with aecidiospores from *Rhamnus cathartica* L., and on the
other hand on the infection of *Avena sativa* L., a *P. coronifera* host, by aecidiospores from *R. Frangula* L., the aecidial host for *P. coronata.*—H. HASSELBRING.

Origin of herbaceous angiosperms.—The question of the relative antiquity
of herbaceous and woody angiosperms has been considered at some length by
SINNOTT and BAILEY.[25] It has frequently been assumed, although definite
statements of the view are rare, that herbaceous plants preceded the woody, and
such a view was likely to be held as long as the monocotyledons were believed
to be the older angiosperms. The authors deal with evidence from four
sources: paleobotany, anatomy, phylogeny, and phytogeography, and reach
a conclusion entirely at variance with the prevailing theory. Under the
first head it is pointed out that the ancient club mosses and horsetails
were arborescent, but it is admitted that the evidence is not conclusive.
The anatomical evidence hinges on the question whether the primary wood was
originally a continuous layer or a series of bundles. Examination of various
groups of plants leads to the inference that the cambium was originally a complete ring, and that its segregation into "fascicular" and "interfascicular"
cambium is a relatively recent occurrence. In explaining how this may have
come about, JEFFREY attaches importance to the leaf traces, but from this
view our authors dissent; they attribute the production of discrete bundles to
a simple decrease in activity of the cambium. In connection with phylogeny,
a survey of the families of angiosperms shows that the primitive types are
much more woody than the recent ones. In more than half of the families of
dicotyledons there are no herbaceous species, and exclusively herbaceous
families consist of insectivores, parasites, or other recent forms. Under the
heading of phytogeography a large array of facts is gathered, leading to the
conclusion that angiosperms made their appearance in the tropics as woody
plants, and spread into the north temperate zone, where gradual stunting
occurred, largely as a consequence of lowered temperature, resulting finally in
the production of annuals. Such herbaceous plants have subsequently
spread to all parts of the earth's surface Insular and other endemic flora

[25] SINNOTT, E. W., and BAILEY, I. W., Investigations on the phylogeny of the
angiosperms 4. The origin and dispersal of herbaceous angiosperms. Ann. Botany
28:547–600. *pls. 39, 40.* 1914.

are examined, and the influence of glacial periods is considered.—M. A. CHRYSLER.

Some abnormal pines.—BOODLE[26] has described an abnormality obtained from a specimen of *Pinus Laricio* growing in the Kew Gardens. Most of the foliage of the tree is normal, but pairs of concrescent leaves are produced every year in considerable number. The fusion of the two leaves seems to be very much as has been described for the double needles of *Sciadopitys*.

WORSDELL[27] has described a remarkable shoot of *Pinus Thunbergii* grown in England. Some of the scale leaves bear ordinary axillary spur shoots with two needles, but a majority of them subtend a very different axillary structure, the most frequent form being "a swollen fleshly foliar organ arching outwards over or against the subtending scale leaf." Another form which the axillary shoot assumes is that of a pair of transversely placed fleshy leaves. The phenomenon of the recurved leaf and its origin by the uniting of the first two leaves of an axillary shoot by their adaxial margins is additional proof of the accepted character of the ovuliferous scale of the Abietineae.—J. M. C.

Death camas.—This name is applied to species of *Zygadenus* to distinguish them from *Quamasia* and *Calochortus*, which were also known as camas, and were much used for food by the Indians. Reports of the poisoning of stock from eating the roots and leaves of the species of *Zygadenus* led to its investigation by MARSH and CLAWSON.[28] It seems that *Zygadenus* grows abundantly on many of the stock ranges of the west, and is one of the most important sources of loss to sheepmen. All the species are poisonous, through the whole season of their growth. The toxicity of the bulbs and tops is about the same, while the seeds are much more toxic than any other part of the plant. The poisonous principle is an alkaloid or alkaloids allied to veratrin and cevadin. Sheep, cattle, and horses are poisoned by the plant, but the fatalities are almost entirely confined to sheep.—J. M. C.

Thelephoraceae.—BURT[29] has begun the publication of a monograph of the North American Thelephoraceae. The first three papers contain a general discussion of the limitations of the family, a key to the genera, 23 of which are recognized, and a presentation of three genera. The genera presented are *Thelephora*, with 23 species, 3 of which are new; *Craterellus*, with 18 species, 6 of which are new; and *Cyphella*, with 21 species, 5 of which are new.—J. M. C.

[26] BOODLE, L. A., Concrescent and solitary foliage leaves in *Pinus*. New Phytol. 14:19–22. *figs. 4.* 1915.

[27] WORSDELL, W. C., An abnormal shoot of *Pinus Thunbergii* Parl. New Phytol. 14:23–26. *figs. 5.* 1915.

[28] MARSH, C. D., CLAWSON, A. B., and MARSH, H., *Zygadenus*, or death camas. Bull. U.S. Dept. Agric. no. 125. pp. 46. 1915.

[29] BURT, E A., The Thelephoraceae of North America. I, II, III. Annals Mo. Bot. Gard. 1:185–228, 327–350, 357–382. *pls. 4, 5, 15–17, 19.* 1914.

OF HICAGO·PRES

d books on subjects in **philosophy**, psycholog
ive philology, classical **archaeology**, Semit
ges, Slavic, political economy, **political** scien
gion and theology, botany, **physiology**, zoölo
nistry, geology, geography, astronomy, astr
aphy. *A complete catalogue will be sent*

AGO PRESS, CHICAGO, ILLINOI

NICAL GA

Editor: JOHN M. COULTER

OCTOBER 1915

Mechanism of Inhibition and Correl
eration of Bryophyllum calycinum

natomy of the Megasporophylls of Conife

of Foliar Transpiring Power as an
ent Wilting in Plants

icles

VOLUME LX   NUMBER 4

THE
BOTANICAL GAZETTE

OCTOBER 1915

RULES AND MECHANISM OF INHIBITION AND CORRE-
LATION IN THE REGENERATION OF
BRYOPHYLLUM CALYCINUM

JACQUES LOEB

(WITH FORTY-ONE FIGURES)

I. Introduction

In the phenomena of regeneration the problem of correlation appears, that is, the influence of the whole on the part.   A part cut out from a whole organism may regenerate, while no such regeneration will occur so long as the part is not separated from the whole. What are the forces inherent in the whole which exercise the control over the part resulting in the prevention of regeneration?

We cannot form a definite idea of this inhibitory mechanism until we know the laws or rules underlying this prevention of regeneration or growth in the normal plant.   Only if we succeed in finding such rules and if they are sufficiently simple can we with any hope of success begin to draw conclusions concerning the nature of the mechanism underlying these phenomena of inhibition and correlation.   The reason why it is difficult to find such laws lies in the fact that the phenomena of regeneration in most organisms are too complicated or too indeterminate for such a purpose, and we are compelled to look for an organism which is especially favorable for such a purpose.   The tropical plant *Bryophyllum calycinum* is apparently such an organism, and the writer has succeeded in finding some rules governing the phenomena of inhibition and correlation of growth. These rules are so simple and transparent that they form, in the opinion of the writer, a securer basis for hypothesis than is offered by most former experiments in this direction which have not led to such simple rules.

The advantage of this plant for the study of the problem of regeneration lies in the fact that shoots can grow out only from definitely located buds in the stem and in the notches of the leaf. The "Anlagen" of roots are not so definitely located, and roots may grow out apparently from practically any spot on the stem of the plant; they are, therefore, not so appropriate for the establishment of definite and simple rules of inhibition, and their growth will not be considered in this paper.

One bud is located in each notch of a leaf of *Bryophyllum calycinum;* when such a notch begins to grow it forms first roots and later shoots. It is well known that if the leaves of this plant are cut off and put on moist soil (or suspended in moist air) they will form roots and shoots from their notches. This is the mode of propagation of this plant. The question is: Why does a leaf not form roots and shoots in its notches so long as it is in connection with a healthy plant? The buds in the notches of the leaf are not the only ones which are inhibited from growing when forming parts of the whole; the buds on the stem, one of which is found in the axilla of each of the two leaves in each node, are in the same condition, and the same question may be raised, namely: Why do not these buds grow out as long as they form part of a plant, while if isolated they may grow into shoots?

A very few words will suffice to show that the stimulus of the wound is not responsible for the growing out of buds, though the conditions at the edge of the wound are responsible for the healing or covering of the wound by the spreading of epithelial cells over the area laid free by the wound; and they may possibly be directly responsible for the callus formation in the case of plants. When we break off a leaf of *Bryophyllum*, the notches of the leaf will grow out into roots and shoots, but these notches are far away from the cut end of the stalk of the leaf. Moreover, as a rule, the notches in the middle of the leaf will grow out first, and not those nearest the wound caused by the cutting or breaking off of the leaf. It is plainly impossible, therefore, to connect in any way the growth of the notches of a leaf with the "stimulus" of the wound. The same may be said for the growth of roots in the main stem of the plant, which may take place several inches away from the seat of injury. We need not dwell on this point any further, since this is generally conceded. It is chiefly in animals that we find regeneration localized at the wound; but this is apparently due to the fact that in such animals any cells may give rise to new growth, while in *Bryophyllum calycinum* the power of giving rise to shoots is restricted to buds located in definite places in the plant.

II. Isolation as the cause of regeneration

It is generally stated that "isolation" is responsible for regeneration, inasmuch as isolation would release the leaf from the inhibiting influence which the whole has on each part.[1] It is obvious, however, that isolation is an abstract term and that it cannot help us, therefore, in visualizing the forces inhibiting the growth of the buds while the plant is intact. We will show in a simple example that the conception of isolation, while it may fit some cases, will not fit others.

The following experiment was often repeated during the winter months. Three leaves of the same plant of *Bryophyllum* were suspended in an aquarium saturated with water vapor, so that the tips of the leaves (or about one-half of each leaf) were submersed in water at the bottom of the aquarium (figs. 1, 2, 3). Leaf 1 was completely isolated from the stem; leaf 2 had a piece of a stem of the plant attached; and leaf 3 had in addition to a piece of the stem of the plant also the opposite leaf attached. The drawings show the condition of the three leaves after 11 days. Leaf 1 formed roots in a few days, and soon after shoots at the notches of the submersed part of the leaf. In leaf 2, as a rule, all growth from the notches was inhibited, but the bud of the stem opposite the leaf grew out very rapidly into a shoot (fig. 2, *S*). The submersed part of leaf 3 again formed roots and stems in its notches, not quite but almost as quickly as leaf 1. Experiments showed that the

[1] CHILD, C. M., Die physiologische Isolation von Teilen des Organismus, etc. Roux's Vorträge und Aufsätze. Leipzig. 1911.

result is the same if both leaves of specimen 3 are partly submersed in water; both form roots and shoots in that case.

According to the idea that isolation is the cause of regeneration, we should say that leaf 1 formed new roots and shoots because it was completely isolated; that leaf 2 did not do so (for a long time at least) because, being connected with a piece of a stem, it was less isolated. But leaf 3, which was still less isolated than leaf 2, inasmuch as it had another leaf attached to the stem, formed roots and shoots much more quickly than leaf 2 and often almost but not quite as quickly as leaf 1. The idea that isolation is the cause of regeneration is obviously inadequate in this case.

Figs. 4 and 5 are a repetition of this experiment. The two leaves had been submersed in water for 5 weeks. The leaf in fig. 4, with a piece of stem attached, had formed no roots or shoots in its notches; instead it had formed a long shoot (S) from the bud of the stem opposite the leaf. The leaf in fig. 5, with a piece of stem and the opposite leaf, had formed four shoots from the submersed notches, while the stem had formed one tiny shoot (S) from a bud in the axilla of the lower leaf, and roots (R) at the under side of the basal end of the stem.

FIGS. 1–3

When we modify this experiment and suspend the three leaves entirely in moist air (instead of submersing them partly in water), leaf 1 (entirely isolated) will again form roots and shoots in its notches; leaf 2 will as a rule show no growth, but from the opposite bud of its stem a shoot will grow (S in figs. 2 and 4); and in the leaves 3 tiny roots may begin to grow from the notches which, however, usually dry up after some time; and no shoots are formed if the leaf is suspended entirely in moist air.

III. Inhibition of growth of leaves by growth of buds on stem

The question now arises: Why does the presence of the piece of main stem in fig. 2 inhibit or retard the formation of roots and shoots in the notches of the leaf, and why does the same piece of stem cease to inhibit (or why does it inhibit considerably less) when, as in fig. 3, in addition to the stem another leaf is left with it? Each node has two buds, one in the axilla of each leaf. When we use a specimen, as fig. 2, a shoot (S) will grow out in a few days from that bud of the stem where the leaf is removed; and this is the first growth which will occur in this specimen. The bud in the axilla of the leaf which is preserved will as a rule not grow out. In fig. 3, where both leaves are preserved, neither bud of the stem will grow out in winter.[2] Hence we

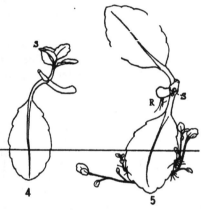

FIGS. 4 AND 5

notice that where a shoot grows out very rapidly from the bud of the stem, as in fig. 2, the leaf in contact with the stem is prevented or delayed in forming roots and shoots, but when no such shoots grow out from the bud of the stem (as in fig. 3), the notches of the leaf (if submersed in water) will form roots and shoots rather quickly.

  [2] In the spring this is not so strictly true, but all these experiments were made in a greenhouse during the winter months. The greenhouse had a temperature of 70° F.

The stalk of an isolated leaf without any piece of the stem is not capable of giving rise to any regeneration. Such a leaf will form adventitious roots and shoots in its notches very rapidly.

All these facts make it appear as if the growth of the buds on a piece of the stem might have an inhibiting influence on the growth of the adventitious roots or shoots of the leaf.

In order to estimate properly such an influence, an extensive series of experiments was made, in which leaves with a piece of stem attached were submersed with their tips in water, while the rest of the specimen was in moist air. In a number of stems both buds were removed (fig. 7), in another only the upper buds were

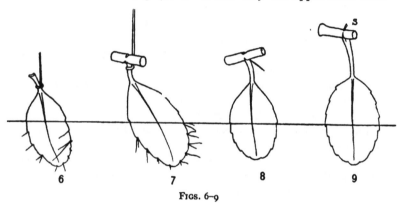

FIGS. 6–9

removed (fig. 8), while in the rest of the stems none of the buds were removed (fig. 9). If the inhibiting effect of the stem were *exclusively* due to the growth of the buds on the stem, the latter should lose its inhibiting effect entirely if these buds were removed; and the leaf connected with such a "debudded" stem should form adventitious roots or shoots as fast as a leaf without any stem. This was, however, not entirely the case. While a leaf connected with a "debudded" stem (fig. 7) formed as a rule its adventitious roots more quickly than a leaf with a normal stem (fig. 9), the leaves connected with the "debudded" stem formed their adventitious roots not quite so quickly as the completely isolated leaves (fig. 6).

Figs. 6–10 give the average results of such experiments. The drawings were made May 10, seven days after the operation. The isolated leaves (fig. 6) had all formed their adventitious roots, and so had some but not all the leaves with "debudded" stems (fig. 7). The leaves which had lost only the upper bud had not formed roots as fast as the leaves with entirely "debudded" stems (fig. 8). The leaves with normal stems (fig. 9) had not yet formed any adventitious roots, but the shoot (S) on the stem where the leaf had been removed had begun to grow out.

The following record of an experiment performed May 1 had yielded on May 10 the following result:

1. Completely isolated leaves (fig. 6). All ten leaves had formed adventitious roots and tiny shoots.

2. Eighteen leaves each attached to a completely "debudded" stem (fig. 7). Eleven leaves had formed adventitious roots and one also adventitious shoots.

3. Ten leaves with a stem whose upper bud was removed (fig. 8). Four leaves had formed adventitious roots or shoots.

4. Ten leaves with a normal stem (fig. 9). All these stems formed shoots from the upper bud. No leaf has formed adventitious roots or shoots.

It is, therefore, obvious first that a stem whose buds are removed has still an inhibiting influence upon the formation of roots in the notches of a leaf; and second, that if the buds of the stem are not removed the growth of the bud opposite the leaf enhances this inhibiting effect of the stem upon the leaf considerably.

Since the growth of this bud of the stem is as a rule also inhibited when the opposite leaf is not removed, as in fig. 3, we understand why the non-removal of this leaf favors the growth of the adventitious roots from the notches of the other leaf.

We have seen that isolated leaves when suspended *in moist air* will form roots and shoots from their notches even if they are not submersed in water; while if a leaf is connected with a stem, the formation of roots and shoots in the notches will be permanently inhibited *in moist air*. It should be added, that the leaves attached to a "debudded" stem may form ·very short adventitious roots when suspended in moist air (instead of in water), but will not form long roots or shoots as will the completely isolated leaf. The analogy between the effect of the non-removal of the opposite leaf from the stem and of the removal of the opposite bud seems thus pretty complete.

IV. Continuation of these experiments

We have thus seen that the growth of buds on the stem is one factor which inhibits or delays the growth of the notches in the oppo-

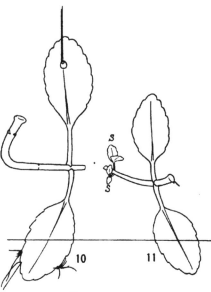

site leaf. We intend to show the influence of this factor in some further observations.

In all previous experiments we had cut out from a plant a piece of stem with only one node. If we cut out a piece of a stem containing two or three nodes (figs. 10, 11, 12, 13) and preserve one pair of leaves, the behavior of these leaves will be different if they are left in the apical or in the basal node of the piece. Figs. 10–13 illustrate this difference. In all cases one or both leaves are partly sub-mersed in water, while the

FIGS. 10 AND 11

rest of the preparation is suspended in moist air. In such cases new shoots (SS) were formed in a few days from the two apical buds of the stem in fig. 11, where the apical leaves had been removed and only the basal leaves left; while in specimens like fig. 10, where the apical leaves were left,.the buds on the stem either formed no new shoots or formed them with some delay. As a consequence, we notice that in fig. 11 the submersed leaf formed at first no shoots, while the submersed leaf in fig. 10 in about 50 per cent of the cases formed roots` and shoots in its notches rather rapidly. This happened often, but not always, when the formation of shoots on the stem itself was long delayed. Ultimately all the leaves may form adventitious roots and shoots in the notches that are under water or near the edge of the water.

This experiment, therefore, supports the conclusion that if the buds of the stem grow out very rapidly, their growth inhibits or

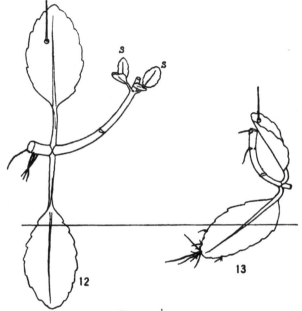

Figs. 12 and 13

delays the growth of roots and shoots in the notches of the leaf attached to the stem.

Figs. 12 and 13 are a repetition of the same experiment. The drawing was made 17 days after the beginning of the experiment. The stem in fig. 12 formed rapidly two shoots (SS) from its apical buds and this inhibited the growth of roots and shoots in the submersed leaf; in fig. 13 the stem formed no shoots and the submersed leaf could form roots. The root formation in both stems

was about equal.   These experiments have been repeated so often
that they can be asserted to form reliable demonstration experiments, during the winter months at least.

We have already stated that in a completely isolated leaf (as in
fig. 1) the roots in the notches in the leaf do not begin to grow until
a few days after the bud in a stem (*S* in fig. 2) has begun to grow.
It would seem, therefore, that we might weaken the inhibiting
influence of a piece of stem, as shown in the experiment in fig. 2, if

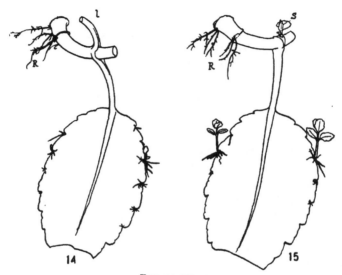

FIGS. 14 AND 15

we inhibit or retard the shoot formation of the bud on the stem.
This can be done, as we have already stated, by not removing the
other leaf on the stem, as in fig. 3.   It is not necessary, however, to
leave the whole leaf attached to the stem; it suffices if we leave a
piece of the stalk of a leaf attached as in fig. 14.   In this case a
leaf (with a piece of stem and a piece of the stalk *l* of the other leaf)
were suspended November 12 in moist air.   The bud in the axilla
of the stalk *l* was by the presence of the latter prevented from
growing out, and after some time roots (*R*) were formed at the basal

end of the stem. Still later, roots and shoots began to grow out from the notches of the leaf (although this was not submersed in water). Fig. 14 was drawn January 18, therefore 9 weeks after the beginning of the experiment. About a week after the drawing was made, the stalk *l* which had wilted fell off and now the bud in the axilla of the stalk *l* was able to grow and the shoot *S* was formed (fig. 14). The drawing (fig. 15) was made when the shoot was one week old. The experiment shows also incidentally that the *root* formation on the stem does not (under the conditions of this experiment) inhibit the formation of roots and shoots on the leaf. We shall return to this fact later.

In this experiment the leaf was merely suspended in moist air and yet shoots developed from the leaf although it was attached to a piece of stem. This is unusual, since in order to obtain such a result with certainty it is necessary to submerse part of the leaf in water.

V. Inhibiting influence of roots on the growth of the notches of a leaf

A piece of stem when cut from a whole plant of *Bryophyllum* is not only able to form shoots but it also forms roots, and it is now our intention to consider the influence which the root formation of the stem has on the growth of the notches of a leaf. WAKKER, DEVRIES,[3] and GOEBEL[4] all have reached the conclusion that it is the presence of the main root or the regenerated roots on the stem which prevent the growth of adventitious roots or shoots on the leaf. If we break or cut off a leaf of *Bryophyllum calycinum* from the stem, neither the stalk nor the base of the isolated leaf has the power of forming roots, and this inability of root formation is considered by WAKKER and DEVRIES to be the cause of the growth of the notches. "According to WAKKER the organic separation of the leaf from the rooted part of the plant acts as a stimulus upon the leaf and induces the growth in the notches."[5]

[3] DEVRIES, HUGO, Jahrb. Wiss. Bot. 22:35. 1890.

[4] GOEBEL, K., Einführung in die Morphologie der Pflanzen. Leipzig, 1908 (pp. 142–149).

[5] DEVRIES, HUGO, *loc. cit.* The writer was not able to obtain WAKKER's monograph.

DeVries describes a very striking experiment which supports the idea of Wakker that the root of the main plant is the factor which inhibits the growth of the notches in the normal plant.

The apices of six plants were cut off beneath the most vigorous adult leaf and planted in soil. After strong roots had been formed (in the soil) their stems were cut above the lowest pair of leaves, the apices removed, and this lowest pair selected for the experiment. Both leaves were put flatly on moist sand, the one after having been removed from the stem, while the other remained connected with the roots. The axillary buds were destroyed. After three weeks the isolated leaves had formed numerous young plants on their margin. The leaves which had remained connected with the rooted piece of stem had formed no plants in their notches (and did not form any afterward), although they had otherwise been exposed to the same conditions as the isolated leaves.

This experiment is in harmony with the view that the normal roots of a stem (if they are under normal conditions) inhibit the growth of notches in the leaves. DeVries reports a second experiment in favor of the view of Wakker. He cut the stem of a plant in its internodes and thus isolated seven pairs of leaves.

From each pair one leaf was broken off; all axillary buds were destroyed. The leaves were now put on moist sand. After a month the seven stems had formed roots. The isolated leaves[6] had formed in their notches rooted plantlets, varying from 10 to 26 in number. The leaves whose stems had formed roots behaved differently. One leaf had formed no trace of growth in its notches; it was the one whose stem had formed roots first. The rest of the leaves had formed only a few plants whose number varied between 2 and 6. They reached only a few mm. in length, while those of the isolated leaves measured from 0.5 to 2 cm.

We see here that if a root is formed on a stem before the roots in the notches of a leaf can grow out, the root (under proper conditions) may inhibit the growth of the notches of a leaf.

While these facts leave no doubt that the root (under proper conditions) can inhibit the growth of the notches in the leaves of *Bryophyllum*, the experiments mentioned on the previous pages of this paper show that this is not the only factor. A piece of stem, even if it does not form any roots but only a shoot, will inhibit or greatly delay the growth of the notches of a leaf connected with it.

---

[6] That is, those broken off from the stem.

VI. Influence of root formation and of root pressure

DE VRIES assumes with WAKKER that it is not the root formation in itself by which the stem or main plant inhibits the growth
in the notches of a leaf, but the root pressure. GOEBEL is inclined
to think that it is the root formation in itself, regardless of the root
pressure (or flow of water caused by it), which in-
hibits the growth in the notches of the leaf. The
writer has made a number of observations which
indicate that of the two views that of WAKKER
and DEVRIES is better supported by the facts.

As an illustration, we may take figs. 14 and
15, in which a leaf with a piece of stem was
suspended in moist air. The basal end of the
stem formed a mass of roots, yet this did not
prevent the growth of roots and shoots from the
notches of the leaf. This contradicts GOEBEL'S
assumption, but is in harmony with the view of
DE VRIES, since these roots in the air were not
able to give rise to "root pressure."

As a further support, we may give the drawings (figs. 16 and 17). In these cases the leaves
were cut off with only a fragment of the stem
attached, which was a little larger in fig. 17
than in fig. 16. The axillary bud was not
removed. The leaves were suspended in moist
air. Although the remnants of the stems formed
roots, yet the leaves formed also roots and
shoots (although they were in moist air). The
root formation on the remnant of the stem preceded the root formation on the leaf but did not
prevent the latter. The experiment shows again
that mere root formation in a stem suspended in

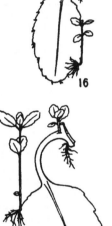

FIGS. 16 AND 17

moist air does not prevent the formation of roots and shoots in a
leaf of *Bryophyllum*. It should be pointed out, however, that the
shoots grew out from the axillary bud of the leaf; it is the growth of
the opposite bud which has the inhibitory power on the leaf
mentioned in the third and fourth sections.

While the root formation on the stem does not inhibit the shoot formation on the leaf if the root is exposed to moist air, the result is different if the roots are in water. If one leaf with a piece of stem (fig. 18) is put into a Petri dish, the bottom of which contains a thin layer of water, the stem will form enormous roots (R_1) at its basal end from the callus, and R_2 from the basal end of the shoot (S) which grows out from the bud of the stem where the leaf was removed. On the other hand, the leaf has formed only a few small roots (R_3) at one notch. (As a rule the notches of the leaf formed no roots in such an experiment.) In this case the roots of the stem which were functioning, and probably established the

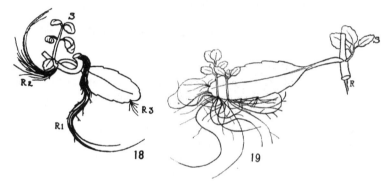

FIGS. 18 AND 19

usual root pressure, inhibited for a long time and in most cases permanently the regeneration in the leaf.

If, however, we do not put the leaf immediately after it is cut from the plant into the Petri dish but suspend it first in moist air, the stem will form a shoot (S in fig. 19). Roots (R) may or may not be formed on the stem, but they will always be formed considerably later than the shoot, at least in winter. If after a month we put the leaf into a Petri dish, while the stem remains in moist air, the leaf (fig. 19) will rapidly form roots and shoots. The contrast between the behavior of the leaves in this case and the one mentioned before is very striking. In the experiment represented in fig. 19, the roots (R) at the base of the stem could not establish a flow of water in the stem and could not inhibit the growth of the shoots in the notches of the leaf; and by the time the leaf was put in water they were obviously not in a position to produce a flow or a root pressure.

Roots formed on the stem have as a rule, therefore, an inhibiting effect on the growth from a leaf if they can produce a root pressure, that is, if they are in water. The formation of a shoot from the bud of a stem can produce such an inhibiting effect upon the opposite leaf if the shoot is in moist air and if no root is formed. This influence of roots on the growth of the notches of a leaf was discussed only for the sake of completeness, since it does not strictly belong to our problem, which deals only with the growth of shoots.

VII. The conditions inhibiting and accelerating the growth of the axillary buds

Each node of *Bryophyllum* has one pair of leaves, and in the axilla of each leaf is found a bud which in the normal life never grows out, but which may grow out as a consequence of a mutilation of the plant.

If we cut through two successive internodes of a stem and isolate a single node, and if we remove the two leaves, the two buds on the stem will grow out rapidly (if we provide the necessary water supply or if the node was cut out from near the base of the stem).

If we remove only one instead of both leaves, only one bud will as a rule begin to grow, namely the one whose leaf is removed. This suggests the idea that the leaf, while favoring the growth of the opposite bud, inhibits the growth of its own axillary bud. If we remove neither of the two leaves, in many cases (especially in winter) neither bud will grow out, a fact which harmonizes with the assumption that each leaf suppresses the growth of its own axillary bud.

The following experiment, however, restricts this last assumption that each leaf will inhibit the growth of its axillary bud. If we isolate a node with its two leaves (which we do *not* remove), and if we split the piece of stem longitudinally, we obtain two leaves, each attached to a half of a node containing the axillary bud of the leaf (fig. 20). In this case the axillary bud will grow out, although often with some delay. Hence the leaf in this case does not prevent the growth of its own axillary bud, and if we speak of an inhibition in the previously mentioned cases, we have to add the remark that this inhibition only exists if the other leaf or the opposite bud are in connection with the first leaf. A comparison of figs. 20 and 21 is of interest.

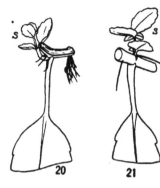

In both cases leaves with a piece of stem attached were suspended in moist air on February 20. They were drawn on April 1. In fig. 20 the shoot (S) in the axilla of the leaf left attached to a longitudinally split piece of stem grew out. In fig. 21, one leaf with a whole, non-split piece of stem attached gave rise to the growth of the shoot (S) on the upper side of the stem where the leaf was removed, while the bud in the axilla of the leaf was prevented from growing.

Fig. 22 is a case similar to fig. 14, one leaf with a piece of stem and the stalk *l* of the opposite leaf. In this case the bud (S) in the axilla of the intact leaf grew out. This is not the most common experience. More often in winter neither of the two axillary buds of the stem will grow out under such conditions. The experiment in which the piece of stem is split longitudinally (fig. 20), however, generally succeeds.

The following observation is also of some significance. If we cut out a node, remove one leaf and its bud, but preserve one leaf and the bud in its axilla,

FIGS. 20–22

the latter will grow out into a shoot after some delay. Hence the removal of the opposite bud removes the inhibiting influences which this bud naturally has on the growth of the bud in the axilla of a leaf. We can accelerate the growth of this latter bud, however, when in addition to the removal of the opposite bud and leaf we make an incision or cut out a piece from the rind apically to the axillary bud whose leaf is not removed. In this case the bud in the axilla of a leaf which is not removed will grow out rather rapidly.

We may anticipate that all these experiments indicate that the growth of the bud depends upon the flow of certain substances from the leaf to the bud. That bud which receives these substances first will grow out first, and thereby prevent the flow to the other bud whose growth is thereby "inhibited." The apparent inhibition of growth in one place is simply due to the fact that under the conditions of the experiment the substances required for growth flow to some other place and are retained there, and the removal of the inhibition consists in creating conditions which will force the substances to flow where we want growth to occur.

VIII. The rules and mechanism of inhibition in regeneration

We cut off the base and tip of the main stem of a plant of *Bryophyllum*, remove all the leaves, and suspend the stem in a closed aquarium saturated with water vapor. Only the two buds at the highest apical node will grow out (fig. 23); it does not matter whether the stem is hung upright or inverted. The buds at the more basal nodes are all inhibited from growing by the growth of the two apical buds; for if we isolate any of the lower nodes, their buds also may grow (fig. 24). This is the well known example of an inhibition of one part by another. In the terminology of REINKE, we might call the two apical buds the "dominants." What is the source of their dominance? By way of an answer we intend to show that the following relation exists: *If an element a inhibits the growth in an element b, b very often accelerates or makes possible the growth in a.* When we cut off a single node near the top of the main stem of *Bryophyllum*, remove the two leaves, and suspend it in an aquarium saturated with water vapor, as a rule the two buds will not grow out. If, however, we leave it in connection with one or more of the lower nodes of the stem, it will regenerate, and incidentally inhibit the growth in the lower nodes (figs. 23, 24). The regeneration and growth of the two shoots at the apical node will as a rule be the quicker the more nodes are left in contact with it. Hence the lower part of the stem whose regeneration is inhibited by the apical node, at the same time accelerates the latter's regeneration or makes it possible.

The second example is the following: When we cut off one leaf with a piece of the main stem (as in fig. 2) and suspend it in water, the bud opposite the intact leaf will grow out into a shoot (*S* in fig. 2). We have seen that the growth of this shoot has a share in the inhibition of the growth of the notches of the leaf in this experiment. It can be shown that conversely the leaf accelerates or renders possible the growth of the bud in the stem. As stated, the isolated node near the top will not be able to form shoots if suspended in moist air.

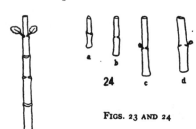

24

FIGS. 23 AND 24

23

If, however, one leaf or even a fraction of one leaf is left in connection with the stem, the bud on the opposite side will grow out (figs. 25–29). In the isolated nodes (figs. 28, 29) cut off near the apex no buds could grow in moist air. When, however, only a piece of a leaf was left in connection with such a stem, as in fig. 25 or 26, the bud could grow out. Here we see again that the presence of the leaf accelerates or renders possible the growth of that shoot in a stem whose formation inhibits or retards the growth of the notches of the leaf.

This gives us a clue to the nature of the dominance and the power of inhibition. The inhibition seems to consist in this, that the dominant part receives something from the inhibited part which accelerates growth or renders it possible in the former.

When we put an isolated node from near the top (whose leaves are removed and which cannot regenerate in moist air) in a very thin layer of water, new shoots grow out (figs. 30–32). This looks as if the "something" which the inhibited part supplies to the dominant part were water. But the writer is suspicious that the water may be only indirectly needed, namely to render the flow of material in the conducting vessels possible. In animals we know that the blood vessels must be filled to render a closed circulation possible. It would seem as if in plants a flow of substance through

conducting vessels should be possible only if a certain minimum
amount of water is contained in the conducting cells or vessels.

The buds of an isolated node nearer the base of the stem may
grow out if suspended in moist air, probably because such a piece
does not dry out so easily.

The following experiment rarely fails. If we suspend a piece of
stem consisting of several nodes and stripped of all leaves in moist
air (fig. 33), the two most apical buds (*b*) will grow out. Their
growth, which is usually slow, is greatly accelerated if we leave
one leaf (or more) on the stem (*b* in fig. 34). In two weeks the
growth of the apical buds (*b*) in fig. 33, which had no leaves, was
very slight, while it was strong in the stem (fig. 34) in which one

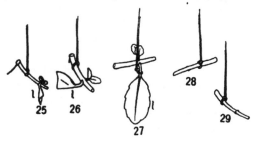

FIGS. 25-29

leaf was left. Here we have the combined accelerating effect of
stem and leaf upon the growth of the apical bud.

Why is it that the apical bud grows out first? Should this be
connected with the anatomy of the conducting vessels, possibly
in the way that the majority of these vessels go directly from the
leaves to the growing point at the apex?

Since the rapid growth of the bud on a stem inhibits or retards
the growth of adventitious roots of the opposite leaf (fig. 2), it
follows that the removal of the bud or the inhibition of its growth
should favor the growth of adventitious roots in the notches of the
leaf. This is indeed the case. If we suppress the growth of the two
buds in an isolated node, we favor the growth of adventitious roots in
the leaves if they are submersed in water (fig. 3). The same happens if we split the node longitudinally (figs. 35, 36). The leaf

(fig. 35) connected with a longitudinal half of a node was submersed in water and formed adventitious roots in nine days. The leaf (fig. 36) attached to a whole node formed no adventitious roots under the same conditions.

IX. Isolation, inhibition, and the flow of material through the plant

These rules give us some basis on which we may try to form a preliminary idea on the nature of the mechanism of inhibition.

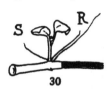

30

As we mentioned already, the rules are comprehensible if we assume a flow of certain (possibly specific) substances (or formed cells) from the places where the dormant buds are ready to grow, or the prevention of such a flow toward these dormant buds.

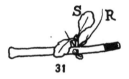

31

We will first show in a few simple examples that this idea leads us easily through the maze of facts in which the terms isolation or inhibition have no more than a metaphorical value.

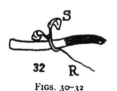

32

FIGS. 30–32

When we isolate a leaf and suspend it in moist air or put it into a Petri dish the bottom of which is covered with water, as a rule only a few of the notches will grow out into shoots. Why do not all grow out? From what was said in the previous section it was natural to expect that the growth of the shoots in some of the notches of a leaf inhibits the growth in the rest of the notches of the same leaf, and that if all the notches could be isolated from each other this inhibiting effect would cease and they would all grow out. This idea was put to a test in a way indicated in fig. 37. Five notches on one side of a leaf were isolated from the leaf and from each other. The rest of the leaf and the isolated notches were put into a Petri dish whose bottom was covered with a layer of water. All five isolated notches grew out into shoots, while only three of the ten or twelve notches left on the leaf grew out. This

experiment, which has often been repeated, succeeds easily if the
piece of leaf left around each notch is not too small. It is noticeable
that the rapidity of growth is greater in shoots which grow out
from a whole leaf than in those growing out from the isolated
notches. Here we see again an application of the rule that if an
organ a inhibits the growth in b, the presence of b accelerates the
growth in a. This is intelligible on the assumption that the leaf
furnishes a flow of liquid containing material for the growth of shoots;
and that the flow of
this material away from
the notches (wherever
this may be) leads to
the inhibition of the
growth in the notches.

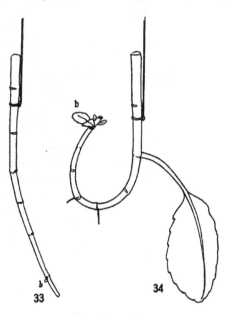

When the piece of
leaf around an isolated
notch is too small, no
growth may occur or
only tiny roots or shoots
will grow out. This
observation again agrees
with the assumption
that a notch of a leaf
will grow into roots and
shoots if certain sub-
stances or formed con-
stituents of the leaf flow
toward a notch or are
prevented from flowing
away.

FIGS. 33 AND 34

We can understand the experiment illustrated in fig. 37 on the
assumption that if in a leaf one or more notches begin to grow out
into roots and shoots, these shoots determine a flow in the rest of
the leaf in a similar way as if a piece of the stem had remained
attached to the leaf; and with the same inhibitory or retarding
effect upon the growth of the other notches of the leaf. If, however, each notch is isolated and given enough water (for example, if

it is put into a Petri dish which has a very thin layer of water), each notch can grow out, since the inhibitants through the establishment of currents in the leaf to growing shoots are lacking.

We have stated that if a leaf is suspended in moist air the growth of the shoots is prevented if a piece of a stem is left attached to the leaf. It seemed of interest to find out if this inhibiting effect would show itself even in a leaf in which a number of lateral incisions were made. This is indeed the case, as figs. 38 and 39 show. In these experiments the incisions were such that the pieces of the leaf had to be kept together by stitches of a thread. The leaves were suspended in moist air. Yet complete inhibition of the growth of

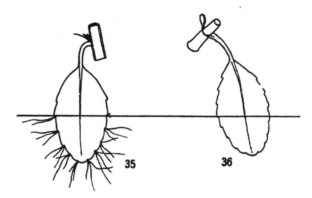

FIGS. 35 AND 36

the notches occurred in the leaf which was connected with a piece of stem (fig. 38). If a flow of substances from the leaf to the piece of stem is the cause of inhibition, such a flow must have taken place along a zigzag path in the leaf. One finds occasionally in such an experiment that in the extreme apical piece of the leaf the inhibiting effect of the stem may cease and that there a growth of roots may occur in the notches.

That the flow of water and of the material it carries in a stem may be deviated and altered by the growth of new shoots is rendered obvious by such observations as are represented in figs. 40 and 41. In this and similar cases thick pieces of the stem of *Bryophyllum*

were cut out from the plant, deprived of their leaves, and put on
moist soil. As is usual, shoots grew out very soon from the top
buds of the stem; very soon after-
ward the piece in front of the top
node began to shrink and wilt, not
directly to the top node, but to
within a few millimeters (fig. 40).
When by chance the new shoot
grows out not from the top node
but from the one next to it, the
whole piece in front of the top
node may wilt (fig. 41).

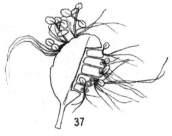

FIG. 37

These observations were made
on stems kept in the laboratory rooms (not in the greenhouse).
When the root of the stem is left intact and in its natural position,
this wilting of the piece of
the stem in front of the
node from which the new
shoots grow out will not
occur. The ascending
flow of liquid or material
in the stem was deflected
in this experiment into the
most apical bud, and there
was not enough root pres-
sure to maintain a flow
through the pieces of the
old stem more apical than
the new shoot.

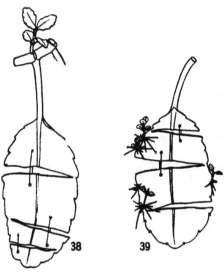

FIGS. 38 AND 39

By way of parentheses
we may here briefly men-
tion that light exercises a
great influence on the
growth of the notches of
an isolated leaf. If such
leaves are suspended in moist air but free from light, as a rule
none of the notches will grow out, while they will grow out

promptly as soon as they are exposed to the light; provided they had not been kept too long in the dark. If the leaves are kept in the dark in a Petri dish whose bottom is covered with water, a few notches may grow out, but they are not nearly as numerous as

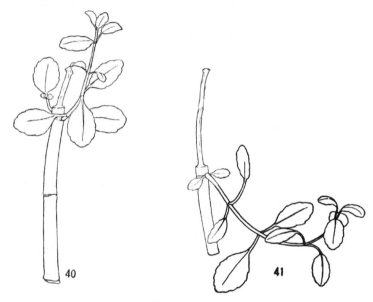

FIGS 40 AND 41

those growing out in the light. Whether we are dealing here with a direct chemical effect of the light or with an indirect effect on the flow of substances in the leaf remains to be seen. It should not be overlooked that as soon as the leaves of *Bryophyllum* turn yellow they become less turgid and easily fall off from the stem.

X. Theoretical remarks

We may now go back to the first experiments mentioned in this paper and try to analyze them on the basis of the old idea that the flow of material in the plant is responsible for phenomena of growth. We start from the assumption that a notch of a leaf can grow out only if there is no flow of liquids (carrying non-formed or possibly formed material) away from the notches. This is not the case in the normal plant when the circulation is normal, and WAKKER and DEVRIES have shown the rôle which the root pressure plays in this case. But the root pressure is not the only factor which influences this flow. The experiments in figs. 1–3 seem to indicate that different factors aside from the root pressure can determine the flow away from the notches of the leaf, provided our assumption is correct that such a flow is the cause of the phenomena of growth and regeneration observed.

If we go back to these first experiments in this paper and try to formulate them in harmony with this idea, we should have to state that in a completely isolated leaf the flow away from the notches ceases. As a consequence, one or more of the notches may grow out, and as soon as this happens the flow in the leaf is directed toward the growing notches. They act as if they exercised a "suction" on the flow of liquids in the leaf, and they may inhibit the growth in other notches of the leaf.

If the leaf is in connection with a piece of stem, the latter exercises this "suction," and the flow of liquids is away from the leaf to the stem; hence the inhibiting effect of the stem upon the growth of the notches of the leaf. This "suction effect" is especially great if the bud opposite this leaf can grow out, as in fig. 2. If both leaves are left attached to the piece of stem (as in fig. 3), the flow from a leaf will be deflected from the buds and may go into the opposite leaf. This might explain why when both leaves remain attached to a piece of stem the growth of the notches of the leaves is favored again, though it is not so rapid as in a completely isolated leaf.

This idea of a deflection of the current away from the leaf toward the opposite side of the stem is in harmony with the fact that the bud opposite a leaf grows out very quickly if its own leaf is removed (fig. 2); while the growth of the axillary bud of the leaf which is not removed is inhibited in this case. If we split the stem longitudinally, this deflection ceases and the leaf ceases to inhibit the growth of its own axillary bud. This idea is supported by the fact that if the leaf attached to a longitudinally split node is partly suspended in water its notches will grow out rather rapidly.

We have assumed that if we have a node with its two leaves attached, the flow will be deflected from the buds; this again is in harmony with the facts that in such a case as a rule neither of the buds grows out.

When a plant is normal, it is almost or possibly absolutely impossible to induce the notches of a leaf which is connected with the plant to grow. The writer has submersed such leaves in water, but in months not a single notch ever formed a root or a shoot. If, however, the flow of substances in a plant is abnormal, either because the roots or the apical parts or both have suffered, a growth of shoots may occur in moist air from the notches of leaves which are in contact with the plant. This fact is mentioned by DeVries and is well known to those who have seen the plants in their natural conditions in Bermuda.

If we now return to the question from which we started, namely, why it is that the notches of the leaves of *Bryophyllum* will not grow out while in connection with the normal plant, the answer should be that the flow of material from the root and from the leaves into the stem and to the apical end of the latter prevents this growth. Through this flow material is carried away from the notches of the leaves. The anatomy of the conducting vessels and tissues, which is inherited, and the dynamical factors determining the flow are the factors concealed in the term "correlation." We understand why it is that if we isolate a part, buds may grow out which without the isolation would not have grown, the reason being that in the mutilated part material can flow to and be retained at places where if the part had remained in the whole it could not have been retained. This assumption agrees with the older ideas of Dutrochet, Sachs, DeVries, and Goebel on regeneration in plants. We understand on this basis why it is that the term isolation of parts or the inhibiting effect of growing parts on others may express some but not all the facts of regeneration. It is not the isolation in itself, but the retention of material in places where there would not have been such a retention under ordinary conditions which apparently determines the growth of dormant buds in an isolated piece; and so it may happen that while this term expresses adequately some results, it fails in others.

SACHS assumed that specific organ-forming substances were needed for growth, and that the consumption of these substances in the growing regions was the cause of the inhibition of growth in the dormant buds of a plant. While the first half of the theory may be correct, the second part is not tenable, since in each stem of a *Bryophyllum* there is enough "formative" material to allow each bud in all the nodes to grow out, while as a matter of fact only the most apical ones will do so. This is intelligible on the assumption that these apical nodes retain the "formative" material in excess of what they need for their own growth.

The ideas expressed in this paper agree in the main with the results and conclusion of the author's older experiments on regeneration in animals. The writer had found that if a piece be cut out from a stem of a *Tubularia* a new polyp may form at each end of the stem, but that the formation of the polyp at the oral end precedes that at the aboral end; and the difference in time may be from one or two days to as many weeks, according to the species or the temperature. He found also that the formation of the oral polyp is the cause of the delay in the formation of the aboral polyp, and that if he prevented the formation of the oral polyp this delay in the formation of the aboral polyp was no longer observed.[7] This is the same rule which we have found for the relation between the growth of the bud of the stem and the formation of adventitious roots and shoots in the opposite leaf of *Bryophyllum*. The growth of this bud causes a delay in the growth of adventitious roots and shoots in the opposite leaf, and this delay is suppressed or diminished if the bud is prevented from growing.

The writer suggested that a flow of substances was the cause of these phenomena of correlation in *Tubularia*. He had found that pigmented cells which come from the entoderm and are carried in the circulation are always collected at the spot where regeneration of the natural growth of the hydroid is to start.

These remarks may suffice to indicate that the rules of inhibition observed in *Bryophyllum* may have a wider application.

[7] LOEB, J., Untersuchungen zur physiologischen Morphologie der Tiere. I and II. Würzburg. 1890 and 1891; Pflüger's Archiv 102:152. 1904.

XI. Summary of results

The phenomena of inhibition of regeneration have been studied in *Bryophyllum calycinum* and it was found that they are governed by the following simple rule:

If an organ *a* inhibits the regeneration or growth in an organ *b*, the organ *b* often accelerates and favors the regeneration in *a*.

This rule is best understood on the assumption that the inhibiting organ receives something from the inhibited organ necessary for regeneration.

It is pointed out that this harmonizes with the older assumption of botanists and of the writer that the flow of material and the block to such a flow after mutilation is responsible for the phenomena of inhibition in regeneration, as well as for the phenomena of correlation.

ROCKEFELLER INSTITUTE FOR MEDICAL RESEARCH
NEW YORK CITY

VASCULAR ANATOMY OF THE MEGASPOROPHYLLS OF CONIFERS

CONTRIBUTIONS FROM THE HULL BOTANICAL LABORATORY 208

HANNAH C. AASE

(WITH 196 FIGURES)

Introductory and historical

The megasporophyll in Coniferales has been the subject of much investigation and discussion, as on the interpretation of this structure depends to a great extent the views held in regard to the relation of living conifers to fossil forms, and the interrelation of living genera. If the scale in *Pinus*, the ligule in *Araucaria*, and the epimatium in *Podocarpus* represent a dorsal outgrowth of the bract, there is added a strong argument in support of the contention that the conifers have sprung from lycopod stock. If the megasporophyll represents a metamorphosed fertile shoot and its subtending bract, there exists a suggestive likeness to the Cordaitales, in which the presence of bracts on the shoot makes its identification as a shoot less difficult. Again, if the scale in the Abietineae represents an axillary shoot and the ligule in *Araucaria* represents a dorsal outgrowth, two other possibilities may be suggested: either the Coniferales have a double origin, or the Cordaitales included not only forms with compound strobili in which the scale is a metamorphosed structure, but also forms with simple strobili in which the scale is a ligular outgrowth of the bract. All these four views are supported by various investigators. Correlative with the views taken as to the origin of the group as a whole are the views as to the interrelation of genera; forms which may be considered as progenitors according to one theory may be the descendants according to another, or there may exist no relation.

The investigators before 1868 were concerned chiefly with gross observations of development and abnormalities. Rather complete summaries of these investigations are given by RADAIS (2) and WORSDELL (11).

VAN TIEGHEM (9) in 1869 was the first to attack the problem from the standpoint of vascular anatomy. He studied forms from all the six large groups. He concluded that the megasporophyll in all Coniferales is a compound structure. The seminiferous scale represents the first and only leaf of an axillary shoot, as the vascular supply to the scale is arranged in an arc. The ovules are borne on the dorsal side of the leaf except in *Araucaria*, where the ovule is reflexed toward the ventral side and hence appears to be located between the bract and scale. In the Podocarpineae and Taxineae the leaf is reduced to such an extent that it is represented practically only by the ovule. In *Podocarpus* the leaf is folded on its dorsal surface to form an anatropous ovule, while in *Pherosphaera*, species of *Dacrydium*, *Phyllocladus*, and the Taxineae, the leaf remains erect and the ovule is orthotropous. The inversion of the ovule in certain forms is probably related to the greater elongation of the sporophyll beneath the ovular insertion in those forms.

STRASBURGER (6, 7) in 1872 and 1879 gave comprehensive descriptions of forms from all the groups. He held that in all cases the ovule-bearing organ is an axillary structure. In *Taxus* and *Torreya* the ovule is borne at the end of a secondary leafy shoot; in *Cephalotaxus* the secondary shoots are reduced to ovules. In the podocarps the secondary shoot is leafless and often reduced to an ovule as in *Phyllocladus*, or provided with a stalk as in *Dacrydium* and *Podocarpus*. In the Araucarineae it appears as if a stalk bearing an inverted ovule were fused to the dorsal side of the bract. In *Cunninghamia*, which he classified with the Araucarineae, there is a fusion of an inflorescence to the bract. In the Abietineae the scale is a flattened axillary structure which is folded inward and hence bears the ovules inverted. The two ovules suggest that the axillary shoot is an inflorescence, a primary and two secondary shoots similar to the two-flowered inflorescence in *Cephalotaxus*. In the Cupressineae and Taxodineae the scale and bract are fully welded together. Where many ovules occur, as in *Cupressus*, he left it undecided whether the ovules represent a reduced branch system, or the large number of ovules is a new feature.

RADAIS (2) in 1894 made a rather intensive study of a number of cones of the Abietineae and Taxodineae. He notes that the bundles to bract and scale are distinct in origin in the Abietineae, *Sciadopitys*, and some of the Taxodineae, as *Cryptomeria*, *Taxodium*, and *Sequoia*, and how this distinctness is on its way to obliteration in species of *Arthrotaxis* and more so in *Cunninghamia*, and is lost in *Araucaria Rulei*.

WORSDELL (10, 11) in 1899 made a comparative study of types from the different tribes. He believes that in the megasporophyll in all conifers there is an axillary structure concerned. Speaking of *Araucaria* he says:

> Holding to the theory of the axillary bud as the explanation of the structure of the appendage of the cone in *Araucaria*, I believe, with CELAKOVSKY, that the ligule represents the seminiferous scale which is itself the vegetatively developed outer integument of a sporangium situated in the anterior position on an axillary bud. This outer integument has become almost completely fused with the subtending bract in *Araucaria*, completely so in *Agathis*.

Concerning the Taxeae and Podocarpeae he says:

> The Taxeae differ from the other groups in the fact that the sporangia occur in a position terminal instead of lateral to the axis on which they are borne. The anatomy points clearly to the fact that no axial foliar appendage of any kind exists upon which the sporangia are inserted, the cylinder of the axis being directly continuous into the base of the sporangium. This latter difference, however, amounts to very little if we regard, with CELAKOVSKY, the seminiferous scale of the other groups as being the morphological equivalent of the outer integument of the Taxeae, which has become, with the exception of Podocarpeae, vegetatively developed. In the Podocarpeae the relationship is precisely the same as in the Taxeae, with the exception of the axillary instead of terminal position of the sporangium. In this order the bundle system belonging to the sporangium (which is in all the other groups the sole representative of the sporophyll according to the view I here adopt) becomes obvious, owing to the fact that the latter gets by the basal intercalary growth on to the upper part of the bract. In the four other groups the bundle system pertaining to the vegetative development of their outer integuments, which, in the form of the widely expanded seminiferous scale, possesses a pronounced vascular tissue.

SEWARD and FORD (3) in 1906, in a somewhat extensive article on living and fossil Araucarineae, offer no interpretation of the megasporophyll in the Abietineae, but in the Araucarineae they consider it as a simple structure which may or may not be homologous with the double structure in other conifers. They regard the Araucarineae as one of the oldest if not the oldest of the conifers. They favor the lycopod origin of the Araucarineae and set them apart under the name of Araucariales.

THOMSON (8) in 1909, in a paper on *Saxegothaea* and *Microcachrys*, admits that the brachyblast theory is inevitable in the Abietineae, Taxodineae, and Cupressineae. Accepting BRAUN'S conception that the scale in the Abietineae represents the first and only two leaves of an abortive shoot, which have fused by their adaxial margins, he says:

> The first inversion is explained and the ovules in the group are borne on the morphological under side. The second inversion is analogous to the single one in *Saxegothaea* and of the nature of a sporangial supply. There are then two great groups of conifers from the standpoint of this study, the simple and the complex scaled series. Both forms have the ovules on the physiologically upper surface, a position rendered almost imperative by the necessities of the seed habit. This position however has been attained in two very different ways.

STILES (5) in 1912 investigated several species of *Podocarpus*. He concludes that the original position of the ovule was erect and axillary as in *Pherosphaera*, but that owing to growth of the scale at the base of the sporophyll it has been carried away from the axis. As a result it has become inverted, and correlated with the inversion is probably the development of an incomplete epimatium.

> Whether this epimatium is an outgrowth of ovular or sporophyll tissue it is at present impossible to say. The evidence of development in *Saxegothaea* and *Microcachrys* suggests the former, while a somewhat older state in *Dacrydium cupressinum* suggests the latter.

In the latter form the ovule is borne on the epimatium, while in *Podocarpus* the epimatium has elongated into a stalk. The development of a strong and independent vascular supply in the epimatium he thinks is the result of a required need of a larger ovule. The epimatium in the podocarps and the scale in the Abietineae are homologous, but both are new structures. Both these complicated structures have been derived from a simple sporophyll.

The Abietineae and Podocarpineae have come from a common primitive stock. The Abietineae are more advanced in the development of the scale, but more primitive in holding on to a perfect cone. "The evidence at present is much in favor of the lycopodean ancestry of the conifers." He has little faith in the brachyblast theory, as it depends for its support mostly on abnormalities and the vascular anatomy of the cone scales. "But abnormalities, especially when they are supposed to be more or less of the nature of reversions, afford by themselves unsatisfactory evidence of phylogeny." Vascular anatomy disproves the double nature of the megasporophyll of the araucarians and podocarps, except in some species in the latter, and there the compound structure is of recent origin.

SINNOTT (4) in 1913 gave a very clear account of the strobilar anatomy in a number of podocarps. He is of the opinion that the podocarps and araucarians, along somewhat parallel lines of development, have been evolved from ancient abietinean stock. The scale in the Abietineae, the ligule in the araucarians, and the epimatium in the podocarps are all homologous and vestiges of an axillary shoot, and a simple sporophyll has arisen either by the fusion of both of its parts or by the abortion of one. Of the podocarps he considers those most primitive in which the epimatium is well developed and has a strong vascular supply, as *Podocarpus;* and those most advanced in which there is a reduced epimatium, as *Dacrydium.* In *Podocarpus dacrydiodes* there is a definite step in the direction of *Saxegothaea, Microcachrys,* and *Pherosphaera.* The resemblance in reproductive structures between certain members of the Podocarpineae and *Cephalotaxus,* the most primitive genus of the Taxineae, suggests that the latter family has arisen from some ancient member of the Podocarpineae.

EAMES (1) in 1913, in a paper on *Agathis,* considered also the megasporophyll situation in other conifer groups and concludes that the megasporophyll is compound in origin in all Coniferales. "Even within themselves the Araucarineae show a complete series from a form with strobilar units of a distinctly double nature to one most simple through reduction." EAMES has traced a similar reduction in the Taxodineae.

Investigation

ABIETINEAE

The ovulate strobilus in all the Abietineae is composed of a comparatively large number of sporophylls. The sporophyll here is obviously composed of two organs. In some forms, as *Keteleeria* and *Pseudotsuga*, each of the three-pronged bracts is reflexed over the scales below, giving the strobilus a bristly appearance; while in others, as *Cedrus Libani*, the bract is a minute flap, and its bundle dies before it reaches the free portion. Between these two extremes are many intermediate forms. That the bract is a modified leaf seems evident at least in some genera. In *Pseudotsuga* and *Larix* there is a gradual transition from ordinary foliage leaves to bracts of a well developed sporophyll; and in abnormal cones of *Picea*, *Larix*, etc., the bracts are like the vegetative leaves. The scale is well developed in all the Abietineae.

The vascular anatomy of the megasporophyll is less variable in the different genera of the Abietineae than is the case in any of the other five groups.

In the lower one-third or more of the strobilus of *Pinus maritima* and *P. Banksiana* there is a general sterilization, beginning with failure of the ovules to produce seed, followed lower down by the abortion of the ovules, and finally at the base of the strobilus the reduction in size and final disappearance of bract or scale or both. In *P. Banksiana* the bract disappears before the scale, but both are finally lost, and between the lowest ovuliferous scales and the bud scales is a region where the strobilus stalk is smooth except for slight elevations (fig. 2). Each of these elevations is supplied with a small vascular strand and suggests a vestige of a megasporophyll. In *P. maritima* only the scale suffers reduction and loss, and the bract, reduced throughout the strobilus, increases in size toward the base (fig. 1).

Correlative with the sterilization and reduction of the appendages in the lower portion of the strobilus are variations in mode of origin of their vascular supplies (figs. 3–29). In the upper half of the strobilus the bract supply originates as a single bundle at the base of the cylinder gap. The scale originates as three or four bundles instead of two as in other Abietineae, one at each side of the

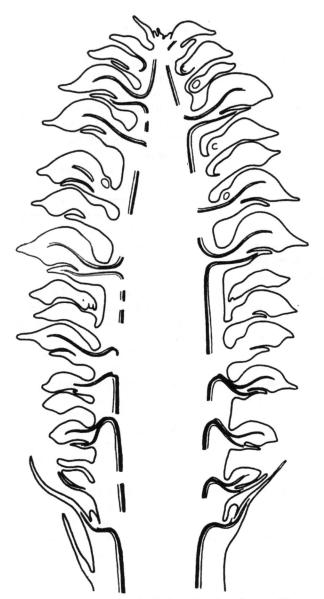

FIG. 1.—*Pinus maritima:* longitudinal section of ovulate strobilus; vascular supplies to bract and scale of each sporophyll, separated in cortex of strobilus by mass of parenchyma in upper portion, approach one another near middle and merge in lower portion of strobilus; in lower portion of strobilus scale decreases in size, while bract becomes comparatively larger.

gap some distance above the bract bundle, and above these again
one at each side of gap, or one at one side only.   The two uppermost

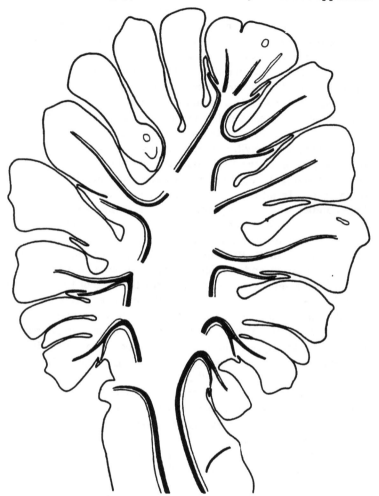

FIG. 2.—*Pinus Banksiana:* longitudinal section of ovulate strobilus, bract and
scale supplies, as in *P. maritima,* separated in upper region of strobilus but become
merged in lower; unlike *P. maritima,* bract becomes very small in lowest sporophylls;
beneath lowest sporophylls are humps, each supplied with a bundle.

bundles soon unite into one, so that four bundles in all with xylem facing result; the lower passes undivided to the end of the bract, the three remaining spread out and give rise in the scale to a row of inverted bundles, seven or more in *P. maritima*, twelve or more in *P. Banksiana*. No branches are seen to bend toward either of the two inverted ovules. Below the middle of the strobilus the uppermost one or two bundles, as the case may be, originate near the lower scale bundles or become united with the latter so that the scale supply starts as two bundles. One of the two bundles soon divides, so that four bundles including that of the bract result; the subsequent course in the appendages is as described before. In the lower sporophyll, the scale supply originates as one or two bundles very near the bract bundle. In the lowest sporophylls the bract and scale supply originates as a single bundle at the base of the gap. In either case there is a subsequent separation into four bundles which supply their respective appendages as before described. In this region of the strobilus the four bundles remain closer together in their course through the cortex than is the case in the upper. In *P. Banksiana* the phloem is continuous around the four xylem strands, which except for a few parenchyma cells would form a solid xylem strand. Near the base of the strobilus, in this species, where the sporophyll supply begins as a single bundle, the xylem creeps around its protoxylem as a pivot and the phloem about the xylem, so that a concentric bundle results. On nearing the appendage the bundle drops into four. In this species the gaps are small, owing to a shortening of the strobilus axis. In both species the sporophyll supply takes a diagonal downward course through the cortex in the lower part of the strobilus, in contrast to the diagonal upward course taken near the tip. This is also likely due to a shortening of the strobilus axis and a consequent crowding of the appendages.

For the sake of comparison, the anatomy of a young vegetative shoot was investigated. In *P. maritima* and *P. Banksiana* (figs. 30–35), whether the bud is to give rise to a spur shoot or a long branch, a single bundle springs from the base of the cylinder gap and supplies the bract, while two other bundles, one at each side of the gap, supply the bud. The bud bundles increase in size and divide, giving rise to a semicircle of bundles with xylem on the concave side, facing the xylem of the bract bundle. The semicircle

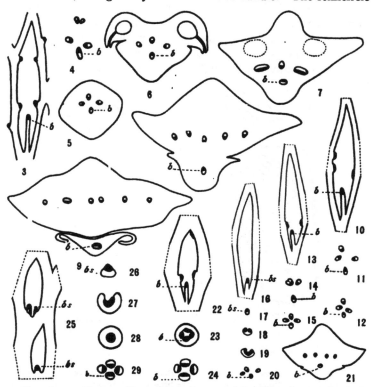

FIGS. 3–21.—*Pinus maritima:* figs. 3–9, course of bundles to bract and scale in upper half of strobilus; fig. 3, origin of bract bundle (*b*) and scale bundles (4 remaining) in strobilus cylinder; fig. 4, two upper bundles uniting; figs. 5–9, transverse sections of sporophyll, bract bundle remaining undivided, scale bundles giving rise to several bundles; figs. 10–12 and figs. 13–15, from lower half of strobilus, scale bundles decreasing in number and originating nearer to bract bundle as base of cone is approached: figs. 16–20, from lowest sporophylls, bract and scale bundles (*bs*) one in origin; fig. 21, transverse section of one of lowest sporophylls; ×16.

FIGS. 22–29.—*Pinus Banksiana:* figs. 22–24, from middle of cone, bundles closing up to form concentric cylinder; figs. 25–29, from lower portion of cone, phloem creeps about xylem so that concentric bundle is formed; ×34.

In all cases four bundles result (figs. 5, 12, 15, 20, 24, 29), the lower supplying the bract, the remaining three the scale.

gradually approaches a circle, and in a slightly older bud closes to form the cylinder of the shoot.   The bud is united with the bract at its base, and the general appearance of a young bud and its subtending bract is suggestive of a young abietinean sporophyll.

In *Keteleeria Fortunei* (figs. 36–57), one bundle originates near the base of the gap in the strobilus cylinder and supplies the bract. It remains undivided throughout its course.   Two bundles, one

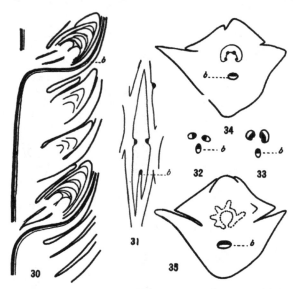

FIGS. 30–35.—*Pinus Banksiana*, vegetative bud: fig. 30, radial longitudinal section of lower portion of young vegetative branch; a vegetative bud in axil of each bract except at base of branch; fig. 31, origin of bract (*b*) and bud bundles; figs. 32–35, transverse sections, young bundles of bud beginning to form cylinder of branch; ×16.

from each side of the gap, supply the scale.   The two bundles soon unite, forming one inverted bundle, that is, its xylem faces the xylem of the bract.   The inverted bundle then breaks, forming a semicircle of bundles which spread and divide, forming in the scale a row of sixteen or more bundles.   Near the ovular insertion a small concentric branch is given off to each of the two inverted ovules.

In the lower sporophylls the bract and scale bundles originate in close proximity at the base of the gap. In the lowest sporophylls the vascular supplies to both organs originate as one bundle, and

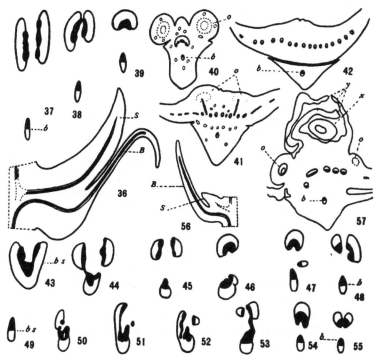

FIGS. 36–57.—*Keteleeria Fortunei:* fig. 36, longitudinal section of sporophyll, ×16; figs. 37–42, transverse sections of sporophyll from upper half of strobilus; figs. 36–39, two scale bundles divide actively, a single strand is given off to each ovule (fig. 41, *o*), ×7; figs. 43–48, from one of lower sporophylls, bract and scale bundles, close together in origin (fig. 47), gradually separate, ×34; figs. 49–55, from one of lowest sporophylls, bract and scale bundles united in origin gradually become distinct, ×34; fig. 56, longitudinal section of young sporophyll, scale (*S*) still comparatively small, ×7; fig. 57, transverse section of one of lowest sporophylls showing outgrowths (*x, y*) between ovules (*o*), ×16.

the two systems separate later for their respective organs. In some of the lowest sporophylls the scale supply begins blindly in the scale.

Between the ovules of the lower sporophylls is found an out-growth which suggests an ovule with a poorly developed nucellus. At each side of and behind this median outgrowth are others which are more bractlike. In some of these xylem cells are present. Whether these abnormal excrescences mean reversions to ancestral features may be difficult to determine. The young strobilus shows no such outgrowths.

CUPRESSINEAE

The general features of the ovulate strobilus in the Cupressineae are the great reduction in the number of the sporophylls, the cyclic arrangement of the sporophylls, complete coalescence of bract and scale, and erect ovules variable in number.

The strobilus of *Cupressus Benthamii* (figs. 58–70) is composed of four decussate pairs of sporophylls. Many erect ovules are packed at the base of each sporophyll. At the megaspore mother cell stage the only evidence of the scale is a slight elevation back of the ovules and the differentiation of the scale supply near the strobilus cylinder. In somewhat later stages the free part of the scale projects outward almost as prominently as that of the bract. Contact and interlocking of epidermal cells takes place between neighboring sporophylls in such a way that the ovules become entirely inclosed.

The vascular anatomy is slightly variable in the individual sporophylls of a strobilus. As in other cyclic forms, the cylinder gap fails to close after the departure of the appendage vascular supply, and hence is continuous with the one above and below. The bract supply may arise as a single strand at one side of the continuous gap, or as two, one from each side of the gap, in which latter case the two unite into one. The scale supply originates as two bundles, one from each side some distance above the bract supply. In other cases two bundles, one from each side of the gap, compose both bract and scale supply. A strand may be given off from one of the bundles to form the bract bundle, or a strand may be given off from each bundle and the two strands unite to form the bract bundle. Whatever may be the origin, the bract bundle remains undivided to the tip of the bract. The scale bundles divide actively; the majority come to lie near the dorsal side of the sporophyll and have inverted orientation of xylem and phloem, but several twist around to each side of the bract bundle and have

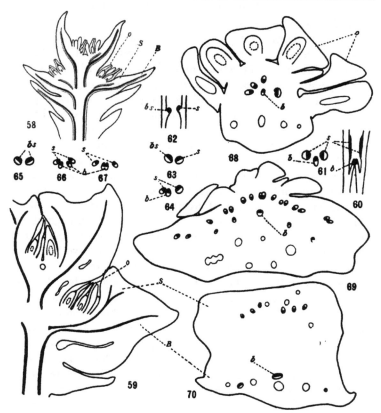

FIGS. 58–70.—*Cupressus Benthamii:* fig. 58, longitudinal section of young strobilus, scale (*S*) evident only by slight elevations on upper side of sporophyll, and beginning of scale bundles; fig. 59, longitudinal section of older strobilus, scale (*S*) evident but incorporated with tissues of bract (*B*), ×7; figs. 60–67, three methods of origin of sporophyll bundles; figs. 60, 61, bract (*b*) and scale bundles (*s*) distinct in cylinder gap; figs. 62–64, bract bundle derived from one of scale bundles (*bs*); figs. 65–67, bract bundle derived from both scale bundles (*bs*); figs. 68–70, transverse sections of sporophyll, scale bundles divide vigorously, some scale bundles persist in bract portion (fig. 70, *B*), ×16; *o*, ovule.

the same orientation of parts as the latter bundle. A fact to be
noted in the further course of the bundles is that some of the lower
scale bundles extend into the bract, though not as far as the bract
bundle.

The ovulate strobilus of *Thuja occidentalis* (figs. 71–89) consists of about five decussate pairs of sporophylls. The two lower
pairs are sterile, the third may have only one ovule to each sporophyll, the fourth two ovules to the sporophyll, and the fifth is
sterile. The tissues of the bract and scale are so closely welded that
it is impossible to distinguish one from the other except at the very
tip. In the upper fertile pair of sporophylls the vascular supplies
to bract and scale are distinct in origin. The supply of the first
arises as one bundle at one side of the continuous gap; that of the
latter as two bundles, one at each side and at a higher level. The
bract bundle does not divide; the scale bundles give rise to numerous bundles, a few of which become inverted and lie near the dorsal
side of the sporophyll; the majority, together with the bract bundle,
form an irregular lower row of inverted bundles. In the two sterile
pairs at the base, the bract and scale supply have their beginning as
one bundle at one side of the long gap. This bundle soon drops
into three, the median one supplying the bract, the two lateral the
scale. The further course is as described before for the upper
fertile sporophyll. In the lower fertile sporophyll is found a combination of the two methods of bundle origin mentioned. One
of the scale bundles rises separately, but the other is combined with
the bract bundle and rises at a lower level; it later separates
from the bract bundle. The sterile tip pair of sporophylls receives
the last two bundles of the strobilus axis. One bundle goes to
each sporophyll, and drops into three bundles which divide further.

The strobilus of *Chamaecyparis Lawsoniana* (figs. 90–104)
consists of about four pairs of sporophylls. In the cone represented in cross-section (figs. 100, 101), one of the lowest sporophylls
has one lateral ovule, each of the sporophylls of the second pair
two lateral ovules, and each of the third pair one median ovule;
the fourth pair is sterile. In the early free-nuclear stage the
scale is apparent only as a slight elevation on the dorsal side of the
bract, which at this stage is almost straight. At a little later stage

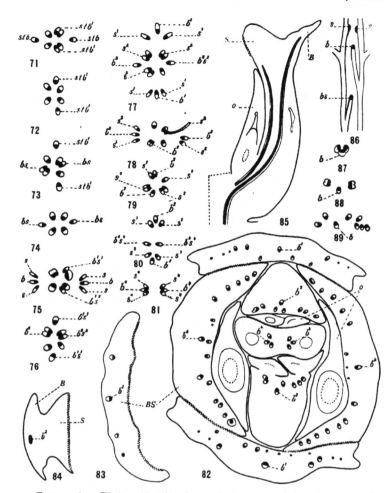

FIGS. 71-89 —*Thuja occidentalis:* figs. 71-81, transverse section of strobilus axis, showing origin of sterile bract bundles (*stb*, *stb¹*) below cone, bract bundles (*b-b⁴*) of sporophylls, and scale bundles (*s-s⁴*); fig. 82, transverse section of same strobilus, many of scale bundles become oriented like bract bundle, two pairs of sporophylls fertile, one pair bearing one ovule per sporophyll, the other two ovules; figs 83, 84, transverse sections near tip of sporophyll, bract (*B*) and scale (*S*) beginning to separate; fig. 85, longitudinal section of sporophyll, bract (*B*) and scale (*S*) closely united; figs. 86-89, sections parallel to cone axis; fig. 86, bract (*b*) and scale (*s*) bundles distinct in origin in upper part of strobilus, united in lower (*bs*); figs. 87-89, bract bundle separating from scale bundles; ×16.

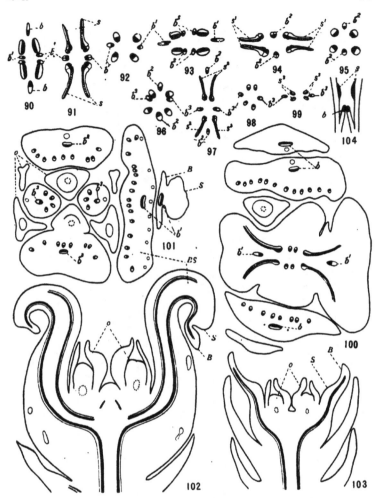

FIGS. 90–104.—*Chamaecyparis Lawsoniana:* figs 90–99, transverse sections of strobilus axis, showing origin of bract (*b–b³*) and scale (*s–s³*) bundles; figs. 100, 101, transverse sections of same strobilus, scale bundles remain close to dorsal side; one of sporophylls of lowest pair bears one lateral ovule, each of second pair two lateral ovules, each of third one median ovule, fourth pair sterile; fig. 102, longitudinal section of strobilus, bract (*B*) and scale (*S*) separate only at recurved tip of sporophyll; fig. 103, longitudinal section of younger strobilus, scale (*S*) evident as slight elevation on dorsal side; fig. 104, origin of bract (*b*) and scale bundles (*s*) in cylinder gap; ×16.

the growth on the dorsal side of the sporophyll is very marked, resulting in the deflexing of the coalesced bract and scale. The bract is free only at the tip of the deflexed portion. The vascular systems to the bract and scale are distinct from the strobilus cylinder. The bract bundle remains single to the tip of the bract. The two scale bundles end in a dorsal row of numerous inverted bundles. The flank bundle may turn partly and lie nearer the ventral side of the appendage.

The strobilus of *Juniperus communis* (figs. 105–113) consists of a whorl of three sporophylls united at the base and surrounding three ovules. In the older stage the sporophylls coalesce also at the tip, forming a fleshy berry-like body. The coalescence of bract and scale is complete.

After the traces to the three bracts beneath the sporophylls have left the cylinder, each of the three remaining bundles divides into three; the median or bract bundle proceeds undivided into the bract; the two lateral or scale bundles divide, giving off some bundles to the dorsal side of sporophyll and some to ventral in such way that, including bract bundle, an oval ring of bundles with xylem facing is formed.

TAXODINEAE

The general features of the ovulate strobilus in the Taxodineae are spiral arrangement of sporophylls, reduction in number of sporophylls in some forms, considerable coalescence of bract and scale, and varying number and orientation of ovules.

In *Cryptomeria japonica* (figs. 114–128) the axis of the young strobilus is very short and the sporophylls are crowded at the broadened summit. The ovules at this stage appear to be inserted on the strobilus axis. The scale begins its development as four to six lobes between the ovules and the dorsal side of the bract. In an older stage the axis has elongated, forming a globose cone. The erect ovules, usually three in number, are definitely inserted on the sporophyll. The lobed scale is united for two-thirds of its length to the bract. In the upper portion of the cone three bundles leave the axis for the sporophyll; one at the base of the gap to supply the bract, and one from each side of gap to supply the scale.

The two scale bundles divide, resulting in the sporophyll in a ring
of bundles with xylem facing inward.   Included in the lower
portion of the ring is the bract bundle.   One and sometimes two
bundles enter each lobe of the scale.   A number of the scale bundles

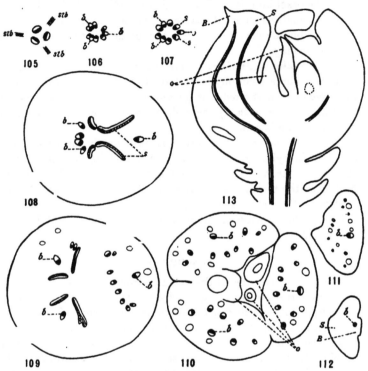

FIGS. 105-113.—*Juniperus communis:* figs. 105-110, course of bundles from base
of strobilus to middle of three sporophylls; sporophylls closely coalesced at base;
some of scale bundles come to lie in same plane as bract bundle (*b*); figs. 111, 112,
transverse sections of tip of sporophyll, bract apparent as small ridge (*B*); fig. 113,
longitudinal section of strobilus; *o*, ovule; *stb*, sterile bracts below strobilus; ×16.

pass into the free portion of the bract at each side of the bract
bundle and have the same orientation of parts as the bract bundle.
The ending of each bundle is accompanied by short irregular
tracheids.   The scale bundle may arise in closer proximity to the

bract bundle; in the lower sporophylls scale and bract supplies
spring from the base of the gap as one large bundle which breaks
into three; these latter supply the sporophyll parts as described
before.

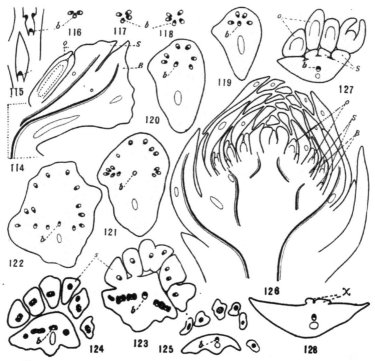

FIGS. 114–128.—*Cryptomeria japonica:* fig. 114, longitudinal section of sporophyll;
fig. 115, origin of bract (*b*) and scale bundles in upper part of strobilus distinct, in
lower part united (*bs*); figs. 116–125, transverse sections, further course of bundles in
sporophyll; bract bundle (*b*) remains undivided, some of scale bundles come to lie
in same plane as bract bundle (figs. 121, 122); lobes of scale separate from one another
and from bract, each lobe receiving usually one bundle; remaining scale bundles enter
free portion of bract (figs. 123, 124); bundles end in large irregular tracheids; ovules
not represented; ×7; fig. 126, longitudinal section of young strobilus, scale (*S*) appears
as small lobes between bract (*B*) and ovules (*o*), ×16; fig. 127, transverse section of
young sporophyll, showing bract (*B*), four lobes of scale (*S*), and ovules (*o*); fig. 128,
transverse section of one of uppermost sterile sporophylls, showing outgrowths similar
to lobes of scale in fig. 127, ×34.

The strobilus of *Taxodium distichum* possesses the same general characters as described for *Cryptomeria japonica*.

The strobilus of *Cunninghamia Davidiana* (figs. 129–141) is composed of a relatively large number of sporophylls. The scale is united to the bract, with the exception of a small edge distal to the insertion of the three inverted and slightly winged ovules. The vascular supply to the sporophyll departs from the cylinder gap as one large bundle; this bundle soon divides more or less definitely into three; a further division takes place until a row of fifteen or more normally oriented bundles results in the expanded part of the sporophyll. In the earlier course of the branching of the bundles weak strands separate from the lateral bundles and swing around 180°. so as to lie on the dorsal side of sporophyll; some of these strands fork, so that an upper row of five or six inverted bundles results; some of these bundles may begin blindly; near the insertion of the ovules one or two strands bend toward the chalaza and end there or a short distance behind the chalaza. Owing to the fact that the bundles in the lower row adhere more or less in the earlier course of division, it is difficult to determine whether the median of the three first bundles passes undivided into the narrow portion of the bract. It is accompanied, however, by branches from the lateral bundles for some distance into the free portion of the bract. At the tip and base of the strobilus are sterile sporophylls. These have a vascular anatomy similar to that of the fertile sporophylls, with the exception of the absence of the upper inverted bundles in the former. Neither scale nor ovules are present, but in place of these appear slight excrescences with different staining reactions.

ARAUCARINEAE

The ovulate strobilus in the Araucarineae is composed of numerous spirally arranged and closely compacted sporophylls. Each sporophyll bears one inverted ovule, which is imbedded in the sporophyll tissues in *Araucaria*, and naked and winged on one side in *Agathis*. Another feature of the Araucarineae of significance in this connection is the branching of the sporophyll bundle in the

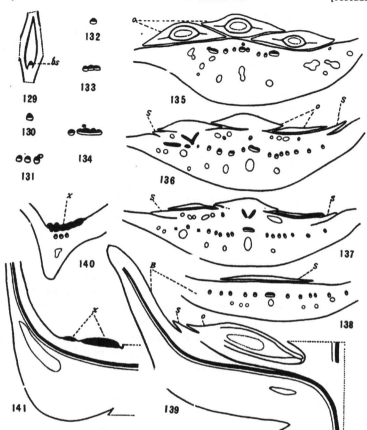

FIGS. 129–141.—*Cunninghamia Davidiana:* fig. 129, vascular supplies to bract
and scale leave cylinder gap as one bundle (*bs*); figs. 130, 131, and 132–134, splitting
up of single bundle; figs. 135–138, transverse sections of sporophyll small strands split
off from lower series or begin blindly (fig. 134, dots) and come to lie on dorsal side of
sporophyll; the majority of these small bundles finally bend toward chalaza of ovules
and end there; the majority of lower bundles enter free portion of bract (fig. 138, *B*)
beyond insertion of scale (*S*); fig. 139, longitudinal section of sporophyll, scale (*S*)
appears as flap beyond insertion of ovules (*o*); fig. 140, transverse section of sterile
sporophyll near tip of strobilus, dark stained elevations (*x*) taking place of scale and
ovules; fig. 141, longitudinal section of sterile sporophyll at base of cone, similar
outgrowths (*x*) as in fig. 140; ×7.

vegetative leaf. This is a characteristic which so far as investigated occurs outside the Araucarineae only in *Podocarpus Nageia*.

The sporophyll of *Araucaria* is characterized by the so-called ligule, which there is good reason to believe is homologous with the scale described in the foregoing groups. In the young strobilus of *Araucaria Rulei* and *Araucaria Balansi* the comparatively small ligule is attached to the bract at its base only, the greater portion being free. In the older strobilus it is free only at the tip. Distal to the line of coalescence of bract and scale the bract remains thick and wide, then becomes narrow and stiff, resembling the bracts at the base of the strobilus.

The vascular supply to the sporophyll of *A. Balansi* (figs. 142–154) arises as a single bundle near the base of the gap in the strobilus cylinder. In the middle of the cortex the single bundle divides into two unequal parts, the smaller of which twists through an angle of 180° so as to lie above the parent bundle with its xylem facing the xylem of the latter. The upper bundle may or may not divide at this stage; the lower bundle divides into two or three; in case of two, one bundle soon divides and a median lower bundle is formed. The lateral bundles divide actively; some of the resulting bundles become inverted and lie on a level with the upper bundle, others normally oriented come to lie on a level with the lower median bundle. Where the sporophyll broadens, branches with normal orientation enter the wings. Behind the ovules the bundles of the upper series begin to converge in groups, and finally end in masses of irregular tracheids. A bundle is sometimes found to continue almost to the free portion of the ligule. Where the bract becomes narrow, the lower bundles also end, with the exception of about three median ones which extend into the slender portion. In the lowest sporophylls, where the ovule and scale are poorly developed or absent, the upper bundles are weak and few in number, or wanting.

In the other forms investigated, *A. Rulei* (figs. 155–161) and *A. excelsa*, the general features are as in the above-described species.

The strobilus of *Agathis australis* (figs. 162–170) is composed of numerous sporophylls which are very closely packed, probably owing to the shortening of the strobilus axis. A ligule, as found in

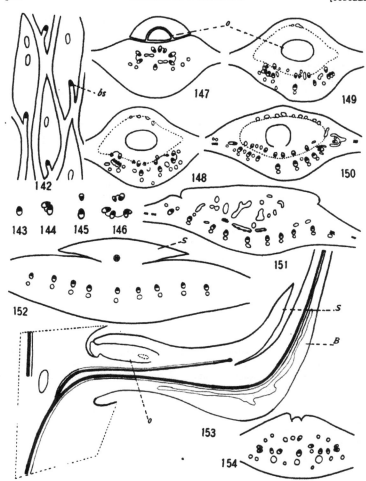

FIGS. 142–154.—*Araucaria Balansi:* fig. 142, vascular supply to sporophyll springs from single strand near base of cylinder gap; figs. 143–146, course of bundles in cortex, figs. 147–152, transverse sections of sporophyll, lines drawn between bundles indicate last division; upper bundles run together in groups and end in large irregular tracheids (fig. 151), one persists almost to free portion of scale (fig. 152); fig. 153, longitudinal section of sporophyll (*S*, scale or ligule; *B*, bract, *o*, ovule); fig. 154, transverse section of sterile sporophyll at base of strobilus, upper bundles few; ×8 5.

Araucaria, is absent, but the general appearance of a slight elevation behind the ovule suggests the possibility of a fused body.
The vascular supply begins as a single stout trace a little to one
side at the base of the short wide cylinder gap.   The trace increases
in size, and in the outer one-third of the cortex a branching begins
which in the lamina of the scale results in about thirteen bundles.
About half-way between the insertion of the sporophyll and the
insertion of the ovule a small strand splits off from one of the median
bundles, swings around 180°, and becomes an inverted bundle

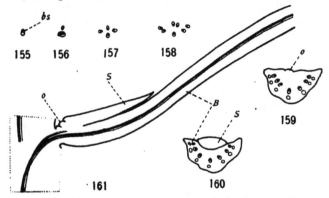

FIGS. 155–161.—*Araucaria Rulei*, young sporophyll: figs. 155–160, course of
bundles from strobilus axis to free portion of bract, upper bundles differentiated in
cortex and base of sporophyll (*B*, bract; *S*, scale; *o*, ovule); fig. 161, longitudinal
section of sporophyll; ×8 5.

opposite its sister bundle.   The inverted strand may fork, and also
inverted strands may spring from several of the median lower bundles
and these strands may divide so that the number of upper bundles
varies from one to several.   In any case, each of the upper
bundles, after giving off a phloem-like strand to each ovule, ends in
large irregular tracheids.

PODOCARPINEAE

In most of the Podocarpineae a definite strobilus is absent, the
fructification consisting in most cases of one or two fertile sporophylls.   In many forms there is a tendency for some part of the
fructification to mature fleshy.

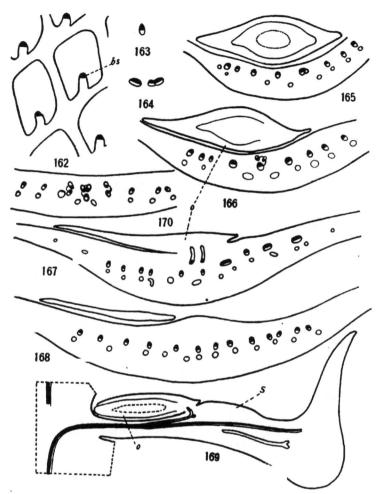

FIGS. 162–170.—*Agathis australis:* fig. 162, vascular supply to sporophyll springs from single strand near base of cylinder gap, figs. 163–168, course of bundles in cortex and sporophyll, a bundle springs from a median bundle, becomes inverted (fig. 166), then breaks into two branches which enter chalaza of ovule (fig. 168), ×8 5; fig. 169, longitudinal section of sporophyll, an elevation appears distal to insertion of ovule, fig. 170, transverse section of sporophyll, inverted bundles have sprung from three of median lower.

In a species of the subgenus *Eupodocarpus* (figs. 171-182) the ovule-bearing branch is a dwarf lateral branch bearing at its tip one and occasionally two fertile sporophylls, and a few small bracts some distance below. The single inverted ovule is imbedded in the tissues of the "epimatium," which there is evidence for believing to be homologous with the ovuliferous scale in other conifers. At maturity the portion of the fruiting branch between the fertile sporophylls and the upper sterile bracts becomes fleshy.

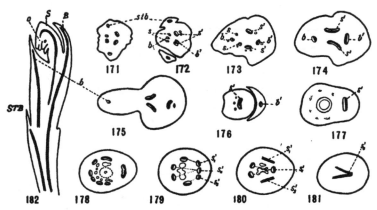

FIGS. 171-182.—*Podocarpus* (sp. of *Eupodocarpus*): figs. 171-181, transverse sections from base of uppermost sterile bracts (fig. 182, *STB*) to tip of strobilus; *slb*, bundles supplying sterile pair of bracts; *b*, bundle supplying sterile bract opposite fertile sporophyll; *s*, bundles inclining toward sterile bract then ending; *b¹*, bundle supplying bract of fertile sporophyll; *s¹*, two bundles uniting into one enter scale of fertile sporophyll where the single bundle breaks into three (fig. 179, $s^1{}_1$, $s^1{}_2$, $s^1{}_3$); distal to chalaza each of three bundles bends to pass down dorsal side of ovule (figs. 180, 181; fig. 179; dotted lines between bundles indicate that one bundle is a continuation of the other); fig. 182, longitudinal section of strobilus; ×7.

The vascular anatomy was investigated only in cases of one fertile sporophyll. The vascular supply of the branch axis, after the traces to the sterile bracts have gone out, consists of three small and three large bundles. The median of the three smaller bundles enters what in the young stages appears like a bract opposite the fertile sporophyll; the two small lateral bundles bend in the same direction as the small median bundle, then end. The median of the three larger bundles supplies the bract of the fertile sporophyll; each of the lateral bundles twists so as to lie with xylem facing the bract bundle, then the two unite into one bundle which proceeds into the scale. The single bundle increases in size, and on nearing the chalaza of the ovule breaks into three; at the base of the ovule each of the lateral branches curves over and passes downward in the tissues of the scale flanking the ovule. The median bundle proceeds slightly farther, then curves over the base of the ovule, and at the same time breaks into two branches which pass downward in the portion of the scale on the dorsal side of the ovule. The recurved branches of the scale supply fork so that

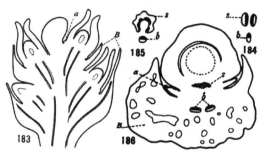

FIGS. 183–186.—*Phyllocladus alpinus:* fig. 183, longitudinal section of young strobilus, aril (*a*) appearing around the base of the ovule; figs. 184, 185, scale (*s*) bundles unite to form a semicircle; fig. 186, transverse section of an older sporophyll, bract bundle (*b*) forks in this instance, scale bundle (*s*) ends at base of ovule; ×61.

in cross-section a ring of bundles with xylem facing outward surrounds the ovule.

In another species of *Podocarpus* investigated the scale bundle divides into two instead of three branches. Each of these two curves at the chalazal end of the ovule and passes downward, one at each side of the ovule. The crests of the two curves are connected by irregular xylem cells, and a strand of similar tissue extends from each crest outward into the protuberance on the scale behind the ovule.

Phyllocladus alpinus (figs. 183–186) consists of a globose strobilus of few sporophylls. The ovule is sessile in the axil of the bract. The vascular supply to the bract springs from the base of the gap.   Two bundles from the sides of the gap unite to form a semicircle which ends at the base of the ovule.

TAXINEAE

Perhaps the most outstanding features of the Taxineae in this connection are the distinctly terminal ovule in some forms and the development of the aril in the later stages.

In *Taxus baccata* (figs. 187–196) the bud of the primary fertile and dwarf branch develops in the axil of a leaf of a long branch of the first season's growth; the second season the ovule matures. Near the tip of the primary fertile and dwarf branch is a secondary dwarf branch bearing a few decussate pairs of bracts and terminally an ovule.   Occasionally two secondary dwarf ovule-bearing branches are produced, one at each side of the terminal vegetative bud of the primary branch.   Beneath the ovuliferous branch or branches the primary shoot is covered with scalelike bracts.   The growth of the terminal vegetative bud of the primary shoot is usually arrested by the active development of the secondary ovuliferous branch, so that the latter appears to be terminal.   Ovules are found maturing on the second, third, and fourth seasons' growth, instead of on the second season's only.   This implies that the ovule in some cases fails to mature the second season, or that the vegetative bud of the primary dwarf branch may resume activity and produce other ovuliferous branches; the latter seems most probable; also the presence of branch scars on the older and more elongated dwarf shoots suggests that the latter conclusion may be drawn.

After the bundles to the upper sterile bracts have either passed into their respective appendages or have ended before entering the latter, the vascular supply of the axis of the ovule-producing branch consists of four bundles.   These bundles unite in pairs, one bundle from each side of the bract bundle of the next lower pair, and not one from each side of the bract of the last pair as is usually the case where the united bundle is to supply some axillary structure.   The xylem creeps around the phloem in each of the two bundles formed.   Sometimes one of the four bundles in the axis ends, whereupon the odd bundle behaves like the fused bundle

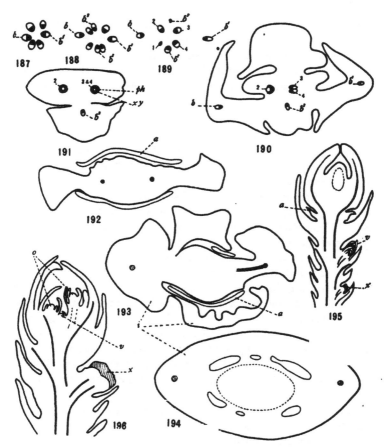

Figs. 187–196.—*Taxus baccata:* figs. 187–194, transverse sections upper part of fertile branch to middle or ovule, *b*, *b¹*, bundles to supply pair of bracts, *b²*, *b³*, bundles to supply uppermost pair of bracts, *b³* ends before entering bract; 1, ends in axis; 2, xylem surrounds phloem; 3, 4, unite into one bundle, xylem surrounds phloem (fig. 191), the two bundles continue as weak strands of xylem cells in base of each wing of ovule, xylem cells are soon replaced by elongated slightly thickened cells without true xylem markings, ×16; fig. 195, longitudinal section of fertile branch, terminal vegetative bud (*v*) is turned aside by rapidly growing ovule bearing branch; *x*, scar left by vegetative or fertile bud or branch, ×7, fig. 196, young fertile branch showing two lateral ovule-bearing branches and a terminal vegetative bud (*v*); *x*, scar as in fig. 195, ×16.

(figs. 183, 186). At the base of the ovule each bundle passes into a mass of short irregular tracheids of large caliber. From each mass arises a small strand consisting of poorly developed xylem soon replaced by elongated thickened cells without xylem markings. Each strand passes upward in the thickened ridge of the ovule.

Discussion

Assuming that the megasporophyll in all the forms is a composite organ, consisting of bract and axillary scale, the investigated forms may be divided into four general groups based on the relation of bract and scale to one another.

In the first group the bract and scale are separate almost to the base of the appendage and both are about equally prominent. Here belong *Keteleeria* (fig. 36), *Pseudotsuga*, species of *Abies*, and species of *Larix*.

In the second group the bract and scale are separate as in the first group, but the bract, at least in the later stages, is much less prominent than the scale, and in certain instances appears distinctly to be on its way to obliteration. In this group belong species of *Abies* (2), species of *Larix*, *Tsuga*, *Picea*, *Pinus* (figs. 1, 2), and species of *Podocarpus* and *Dacrydium* (4); *Cedrus Libani* (2) and the lower sporophylls of *Pinus Banksiana* (fig. 2) show the bract in process of extinction. In species of *Podocarpus* the scale has folded toward its dorsal side, thus forming the second integument or epimatium of the inverted ovule (fig. 182).

In the third group the bract and scale are considerably to completely welded, but the fused structure shows some evidence of its double nature. Within this group are *Sciadopitys* (2), *Sequoia* (2), *Cunninghamia* (fig. 139), species of *Arthrotaxis* (1), *Cryptomeria* (fig. 114), *Taxodium*, *Thuja* (fig. 85), *Cupressus* (fig. 59), *Chamaecyparis* (fig. 102), *Juniperus* (fig. 113), *Araucaria* (fig. 153), and *Podocarpus dacrydioides* (4). In young strobili of *Araucaria Rulei* (fig. 161) and *Cryptomeria japonica* (fig. 126) the bract and scale are distinct almost to the base, and the fused portion becomes comparatively large in the subsequent development of the organ. In *Cupressus Benthamii* (fig. 58), *Thuja occidentalis*, *Chamaecyparis Lawsoniana* (fig. 103), and *Juniperus communis* the scale in the young strobilus is one with the bract and becomes evident in the later stages by the comparatively rapid growth of the tissues on the dorsal side of the composite organ.

In the fourth group there is no external evidence of more than one organ. Into this group fall *Arthrotaxis selaginoides* (1), *Agathis* (fig. 169), *Saxegothaea* (5, 8), *Phyllocladus* (fig. 183), *Taxus* (fig. 191), *Torreya* (6, 7), and *Cephalotaxus* (6, 7). It should not be surprising to find forms in which the welding has taken place beyond the recognition of more than a single structure when one considers to what extent this process has taken place in *Chamaecyparis, Juniperus, Thuja, Cunninghamia,* and *Podocarpus dacrydioides* (4). The low cushion behind the ovule in *Agathis australis* suggests the complete fusion of a scale to a large bract; a similar fusion is nearing its completion in *Cunninghamia Davidiana.*

On the basis of vascular anatomy the investigated sporophylls fall into two general groups.

In the first group the bract and scale supply arises as separate bundles in the cylinder gap. In this group belong in general those forms in which the two sporophyll parts are separate and fairly well developed, as the seed-producing sporophylls of *Pinus Keteleeria, Picea, Larix, Tsuga, Pseudotsuga,* and *Abies* (2). To this group belong also many in which the two sporophyll members present considerable to complete fusion, as *Araucaria Bidwilli* (1, 3, 10), *Chamaecyparis Lawsoniana, Juniperus communis,* and the upper sporophylls of *Thuja occidentalis, Cupressus Benthamii,* and *Cryptomeria japonica.* This group includes also some in which the sporophyll is evident only as a single organ, namely *Phyllocladus* and *Cephalotaxus* (6, 7).

In *Podocarpus* and *Dacrydium,* where the strobilus consists of one or two sporophylls, and in *Juniperus communis,* the sporophylls receive the final bundles of the axis. There is in these instances no cylinder gap, and the bract and scale supplies, at least in the forms investigated by the writer, result from the division of one of the final bundles in the axis. The early division of the bundle in the tip of the axis perhaps justifies the placing of these forms in this group.

In the second group based on vascular anatomy the bract and scale vascular supplies are more or less intimately united into one bundle which springs from the base of the cylinder gap.

Of sporophylls with parts separate there belong here the lower sporophylls of *Pinus Banksiana*, *P. maritima*, and *Keteleeria Fortunei* (the ovules in these sporophylls are to greater or less extent abortive), and *Cedrus Libani* (2). Of the sporophylls with parts considerably united there fall into this group those of *Cunninghamia*, *Arthrotaxis laxifolia* (1), most species of *Araucaria* (1, 3, 6, 7, 9, 10), and the lower sporophylls of *Cryptomeria japonica*, *Cupressus Benthamii*, and *Thuja occidentalis*. Most of the apparently simple sporophylls are included in this group, as those of *Agathis*, *Saxegothaea* (5, 8), and *Arthrotaxis selaginoides* (1).

The degree of welding of the bract and scale vascular supplies varies considerably. In *Arthrotaxis selaginoides*, *Agathis*, *Araucaria*, and *Saxegothaea* the two remain united into one bundle for greater or less distance in the cortex; in most of the others the single bundle divides early; but in many cases, where the two sporophyll parts have fused extensively, branches of the scale supply swing about to lie on the ventral side of the appendage at each side of the bract bundle. This fact is well illustrated in *Thuja occidentalis*, *Juniperus communis*, *Cupressus Benthamii*, and *Cryptomeria japonica*. In *Cupressus Benthamii* and *Cryptomeria japonica* the scale bundles at either side of the bract bundle even accompany the bract bundle into the free portion of the bract.

Bundle distribution is generally directly related to the size of the organ supplied, hence the bundles extend into the most expanded region of the sporophyll, whether that particular region represents bract or scale.

In *Cunninghamia*, *Araucaria*, and *Agathis*, in which absence of sporophylls with separate bract and scale supplies makes comparison impossible, it is difficult to determine with certainty what is bract and what is scale supply. The matter is further complicated by the presence in the last two genera of a branching bundle in the vegetative leaf, a condition which probably implies a branching bract bundle in the bract of the sporophyll as well. And, further, *Cupressus Benthamii* and *Cryptomeria* have clearly shown that the final destination of a bundle does not always determine to which sporophyll part it belongs. It is to be suspected that the bundle system of *Agathis* and *Araucaria* represents a complex of bract and scale bundles. Judging by the course of events in other conifers with single-veined vegetative leaves, it may be suggested that the large median lower bundle in *Cunninghamia Davidiana* is the bract bundle proper, which is accompanied for some distance into the free portion of the bract by a few scale bundles.

Taxus presents some features which perhaps ought to be mentioned. The single ovule is produced terminally on a secondary dwarf branch clothed with a few pairs of decussate bracts. The primary dwarf branch may occasionally become a long branch by the resumption of growth by its terminal bud. In all of the many ovules examined the ovule is flattened transversely to the upper-most pair of bracts. The four final bundles of the branch of the axis which fuse in pairs before entering the two wings of the ovule fuse in pairs across the next lower pair of bracts, and not across the uppermost pair of bracts, a behavior which is contrary to what should be expected if the fused bundle were destined to supply an axillary structure. The dying out of bundles near the tip of the axis and the consequent failure to supply the uppermost bracts or enter into the formation of the ovuliferous supplies, as the case may be, suggest that a general reduction and loss of parts is taking place. The terminal position of the ovule, the flattening of the ovule transversely to the uppermost bracts, and the fusion in pairs of the final bundles of the axis in the definite way to form the two bundles of the wings of the ovule suggest a structure which might result from a process well under way in *Juniperus communis*, namely the fusion of sporophylls to form a single structure. This in *Taxus* would imply the reduction of the ovules to one, the complete fusion of two sporophylls to the integument of the ovule, and finally the reduction of the vascular supply of each sporophyll to the single weak bundle present in the wing of the ovule. In view of the modifications that are apparently taking place in other conifers such a course of events may be possible, but further investigation is necessary.

The ovule of *Phyllocladus glauca* presents a slightly simpler situation than that of *Taxus*. The two bundles beginning from the sides of the gap and ending at the base of the ovule are probably the only vestige of the scale. The aril may be no more than an outgrowth such as appears in connection with the ripening of ovules in the podocarps. The outer integument of *Torreya* may be a more complex organ.

If any probable conclusions can be made concerning the ovulate structure of the Taxineae, they must in any event be preceded by a more thorough investigation of the different forms.

In spite of striking modifications, the origin of the megasporophyll is homologous throughout the conifers. As to the identity of the organ in question there seems nothing new to be added. The gradation from foliage leaves to bracts of sporophylls is so definite in many forms, as in *Larix* and *Pseudotsuga*, that the homology of one with the other need not be questioned. The scale is the organ in doubt. Its axillary position and the origin of the vascular supply when separate from that of the bract justifies the theory that it is some modification of a fertile branch. Through shortening of the axis and rather delayed development of the shoot in general, together with a relatively earlier development of the ovules, the semicircle of bundles at the base of an ordinary shoot failed to form the cylinder, but instead flattened out into an arc and in some cases even into a straight line.

Summary

1. In the evolution of the ovulate strobilus in members of the Coniferales, two general tendencies are apparent: (1) the reduction in number of sporophylls in the strobilus; (2) the modification of a compound sporophyll into an apparently simple sporophyll; the latter appears in diverse disguises, but in general implies loss of one of the sporophyll members or welding of the two.

2. Strobilus reduction has reached its highest expression in members of the Cupressineae, Taxineae, and Podocarpineae; one type of strobilus reduction is represented by the general sterilization and reduction of parts in the lower sporophylls of *Pinus*.

3. Simplification of a compound sporophyll has been attained to fullest extent in *Arthrotaxis selaginoides*, *Agathis*, and *Saxegothaea*, and possibly others; an extensive reduction of bract occurs in *Cedrus Libani* and the lower sporophylls of *Pinus maritima;* the scale in *Phyllocladus* is probably reduced so as to be represented only by a distinct ovular supply; the welding of the two organs is complete in *Juniperus communis* and *Chamaecyparis Lawsoniana.*

4. Fusion of bract and scale vascular supplies does not directly parallel fusion of bract and scale.

5. Separate origin of bract and scale vascular supplies occurs most generally in the Podocarpineae and Abietineae; fusion of bract and scale supplies has reached its highest expression in the Araucarineae; both types of bundle origin are represented in the same strobilus in *Cryptomeria japonica*, *Cupressus Benthamii*, and the lower sporophylls of *Pinus*.

6. The bract bundle in plants with uninerved vegetative leaves divides only slightly if at all; the extent of the scale bundle system is directly related to the size of the organ supplied.

7. The scale bundles in the Abietineae and *Chamaecyparis Lawsoniana* form in the expanded portion of the organ a straight row or arc; in members of the Taxodineae and Cupressineae scale bundles swing around so as to lie at each side of the bract bundle.

8. In *Cryptomeria japonica* and *Cupressus Benthamii* and perhaps *Cunninghamia Davidiana* scale bundles accompany the bract bundle into the free portion of the bract.

9. A branching bundle in the vegetative leaf in *Araucaria* and *Agathis* probably implies a branching bundle in the bract of the sporophyll; the vascular system in the megasporophyll is probably a complex of bract and scale bundles.

10. In species of *Podocarpus* the scale bundles continue in the portion of the scale folded toward the dorsal side, forming the epimatium of the ovule.

The author is indebted to Professor J. M. COULTER and Dr. C. J. CHAMBERLAIN for valuable suggestions, and to the latter also for excellent material.

UNIVERSITY OF CHICAGO

LITERATURE CITED

1. EAMES, A. J., The morphology of *Agathis australis*. Ann. Botany 27:1–38. *figs. 1–92.* 1913.
2. RADAIS, M. L., Anatomie comparée du fruit des conifères. Ann. Sci. Nat. Bot. VII. 19:165–368. *pls. 1–15.* 1894.
3. SEWARD, A. C., and FORD, SIBILLE O., The Araucariae, recent and extinct. Phil. Trans. Roy. Soc. London B 198:305–411. *pls. 23, 24. figs. 28.* 1906.
4. SINNOTT, E. W., The morphology of the reproductive structures in the Podocarpineae. Ann. Botany 27:39–82. *pls. 6, 7. figs. 8.* 1913.
5. STILES, W., The Podocarpineae. Ann. Botany 26:443–515. *pls. 46, 47. figs. 4–8.* 1912.
6. STRASBURGER, E., Die Coniferen und Gnetaceen. 1872.
7. ———, Die Angiospermen und die Gymnospermen. 1879.
8. THOMSON, R. B., The megasporophyll of *Saxegothaea* and *Microcachrys*. BOT. GAZ. 47:345–354. *pls. 22–25.* 1909.
9. VAN TIEGHEM, PH., Anatomie comparée de la fleur femelle et du fruit des Cycadées, des Conifères, et des Gnetacées. Ann. Sci. Nat. Bot. V. 10:269–304. *pls. 13–16.* 1869.
10. WORSDELL, W. C., Observations on the vascular system of the female "flowers" in Coniferae. Ann. Botany 13:127–147. *pl. 27.* 1899.
11. ———, The structure of the female "flower" in Coniferae. Ann. Botany 14:39–82. 1900.

THE INDEX OF FOLIAR TRANSPIRING POWER AS AN INDICATOR OF PERMANENT WILTING IN PLANTS

A. L. BAKKE

(WITH ONE FIGURE)

The index of transpiring power as put forward by LIVINGSTON[1] is taken as a measure of the power of leaves to supply moisture to the surrounding air. The reciprocal of this index is a measure of the power of leaves to retain water against the drying influence of the surroundings. The index of transpiring power for any leaf as found by means of cobalt chloride paper represents the power of the leaf to give off water, this being measured in terms of the similar power possessed, at the same temperature, by a saturated blotting paper surface blanketed by a millimeter of air. The index for an entire leaf is obtained by averaging the indices obtained for the two leaf surfaces. Indices less than unity represent leaf surfaces with lower power of giving off water than is possessed by the standard evaporating surface. Those greater than unity can give off water faster than does the standard surface at the same temperature. The work of BRIGGS and SHANTZ,[2] that of CALDWELL,[3] and that of SHIVE and LIVINGSTON[4] have recently brought forth the concept of permanent wilting in plants. This is that stage of progres-

---

[1] LIVINGSTON, B. E., The resistance offered by leaves to transpirational water loss. Plant World 16:1-35. 1913.

[2] BRIGGS, L. J., and SHANTZ, H. L., The wilting coefficient and its indirect determination. BOT. GAZ. 53:20-37. 1912.

———, The relative wilting coefficient for different plants. BOT. GAZ. 53:229-235. 1912.

———, The wilting coefficient for different plants and its indirect determination. U.S. Dept. Agric., Bur. Pl. Ind. Bull. 230. 1912.

[3] CALDWELL, J. S., The relation of environmental conditions to the phenomenon of permanent wilting in plants. Physiol. Res. 1:1-56. 1913.

[4] SHIVE, J. W., and LIVINGSTON, B. E., The relation of atmospheric evaporating power to soil moisture content at permanent wilting in plants. Plant World 17:81-121. 1914.

sive wilting at which the plant cannot recover turgidity if inclosed for 24 hours (without watering of the soil) in a chamber of saturated air.   Interest in this stage of wilting centers mainly about the determination of the residual moisture content of the soil at the time when permanent wilting is attained.   It is unnecessary here to review the important contributions already made in this connection, but it may be safely stated that the concept of permanent wilting promises to be of great importance in soil moisture studies for ecological and agricultural interpretation.   Besides the authors above mentioned, ALWAY[5] has studied the relation of plants to soil moisture, with results bearing upon this general question.

It is somewhat difficult to determine just when the stage of permanent wilting is attained; an observer can never be quite sure whether a plant is permanently wilted or not until after the 24-hour exposure in the moist chamber, and there is always considerable danger that the wilting may be carried too far.   This difficulty suggested to the writer the possibility of employing the cobalt chloride paper test of foliar transpiration power as an indicator of attainment of permanent wilting.

In the summer of 1913, at the Desert Laboratory of the Carnegie Institution, some preliminary studies were carried out upon the relation of the index of foliar transpiring power to the stage of wilting in plants.   It was the writer's privilege to carry out these tests upon plants actually employed in the work of SHIVE and LIVINGSTON, then in progress.   The results of some of these tests have already been published.[6]   They clearly indicate a direct relation between the value of the index and the extent of wilting.

During the summer of 1914 it was again the writer's privilege to carry out studies in this connection at the Desert Laboratory, and he wishes here to express his appreciation of the kindness of Dr. D. T. MACDOUGAL, director of the Department of Botanical Research of the Carnegie Institution, in placing at his disposal the facilities of the Laboratory.   At the suggestion of

[5] ALWAY, A. J., Studies on the relation of the non-available water of the soil to the hygroscopic coefficient.   Nebraska Agric. Exp. Sta. Research Bull. 3. 1913.

[6] BAKKE, A. L., Studies on the transpiring power of plants as indicated by the method of standardized hygrometric paper.   Jour. Ecol. 2:145-173. 1914 (see pp. 166-168).

Professor B. E. LIVINGSTON, an attempt was made to obtain further information concerning the march of the wilting process by means of cobalt chloride paper and the index of foliar transpiring power.

Various writers have shown that the transpiration rate falls suddenly with wilting, but rises again thereafter to a considerable extent, subsequently to fall once more as desiccation finally occurs. It appeared that these alterations in the transpiration rate must be due to internal conditions (within the plant), and that a similar march should be shown by the value of the index of transpiring power. To test this supposition, sunflower plants (*Helianthus annuus* L.) were lifted from the open soil in the forenoon, the adhering soil was shaken from the root system, and the plants were allowed to wilt in the laboratory. The temperature of the room was about 32° C. and varied but little during the experiments. Determinations of the index of transpiring power for both the upper and lower leaf surfaces were made at hourly intervals, the first determination just preceding the lifting of the plant.

An example of the sort of results obtained is given in table I. Five leaves of different ages were employed, the same leaves being tested at each time of observation. Although there were considerable differences in the indices for different leaves at any one time, according to their various ages or positions upon the plant, all five indices have been averaged to give a single index for each time. The first observation was made just before the plant was lifted from the soil. No wilting was apparent at this time. The last observation was made after drying of the leaves became pronounced. It required 10–15 minutes to complete each observation, the tests being begun at the time indicated in the table.

From the data given in table I it is at once clear that the indices first decrease rapidly, then increase, and then decrease again. This march for the upper and lower surfaces is shown graphically in fig. 1, where abscissas denote time and ordinates denote index values which are shown upon the graphs.

Wilting began almost immediately after the removal of the plant from the soil. The indices are seen to decrease very rapidly during the first hour. This first rapid decrease is followed by a continued but very gradual decrease during the four following hours. Then a marked increase in the index values is evident, lasting for two hours, after which another decrease is clearly shown.

TABLE I

MARCH OF THE INDEX OF TRANSPIRING POWER FOR UPPER, LOWER, AND ENTIRE LEAF SURFACE OF UPROOTED PLANT OF *Helianthus annuus* L. DURING PROGRESSIVE WILTING AND DRYING OF THE LEAVES

TIME OF OBSERVATION	INDEX OF FOLIAR TRANSPIRING POWER		
	Lower surface	Upper surface	Entire surface
8:30....	0.9468	0.9020	0 9243
9:30....	0.1175	0.1289	0 1232
10:30....	0.1054	0.0865	0.0959
11:30 ..	0.0632	0.0699	0.0665
12:30....	0.0657	0.0616	0 0636
13:30........	0.0571	0.0578	0.0574
14:30........	0.1029	0.1013	0.1021
15:30 ...	0.1723	0.2029	0.1876
16:30. ..	0.0899	0.1216	0.1058

To interpret these alterations in the index values, it may safely be supposed that the great initial fall in the graphs represents the prompt increase in incipient drying[7] within the leaves, which must have followed the uprooting of the plant and the consequent breaking of its connection with the soil moisture films.[8] Temporary wilting had already set in before the end of the first hour.

After wilting began, the increase in incipient drying apparently continued, but at a low rate, as indicated by the gradual downward slope of the graphs for the hours 9:30 to 13:30. During this time it may be supposed that the continuous water columns[9] of the plant remained mainly intact, so that the leaves were still slowly drawing water from the stem. With increasing tension on these

[7] LIVINGSTON, B. E., and BROWN, W. H., Relation of the daily march of transpiration to variations in the water content of foliage leaves. BOT. GAZ. **53**:305-330. 1912.

[8] BROWN, W. H., The relation of evaporation to the water content of the soil at the time of wilting. Plant World **15**:121-134. 1912; CALDWELL, J. S., *loc. cit.*

[9] DIXON, H. H., Transpiration and the ascent of sap. Prog. Rei Bot **3**:1-66. 1909, RENNER, O., Experimentelle Beiträge zur Kenntnis der Wasserbewegung. Flora **103**:171-247. 1911.

water columns the incipient drying of the leaves should be increased, which should in turn lower the transpiring power.

The relatively high transpiring power indicated for the hours 14:30 and 15:30 appears to have followed the breaking of the water columns, by which the entire water mass of the plant had previously been united. Such a breaking should remove the tensile strain from the leaf cells, decreasing the saturation deficit of their exposed cell walls and increasing their transpiring power. From the time when this rupture occurred, the removal of water from the leaves should be accompanied by air entrance and no further tensile strength could be developed. If this is the correct interpretation of the observation here set forth, the rupture of the water columns extended over the period from 13:30 to 15:30, during which time the transpiring power was markedly increasing.

After hour 15:30, it is to be supposed that practically no more water entered the leaves from the stem, and the leaves gradually dried out, exhibiting the decreasing transpiring power which should accompany desiccation.

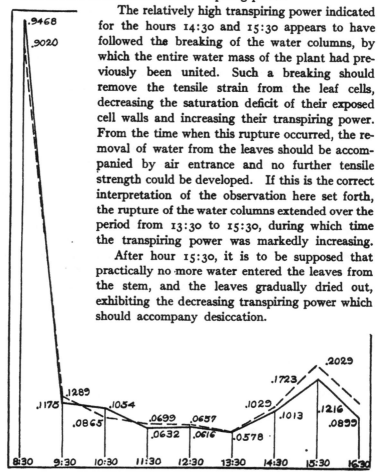

FIG. 1.—Graphs showing march of index of foliar transpiring power for wilting sunflower plant: lower leaf surface full line; upper leaf surface broken line.

From the nature of permanent wilting, and from the theoretical considerations above stated, it appears highly probable that this stage of wilting represents simply the *most intense wilting possible without serious rupture of the water columns of the plant.* At any rate, this rupture should represent a rather definite critical point in the march of the power of the plant to extract water from the soil, probably the same critical point about which the concept of permanent wilting has been developed. If this suggestion be correct, then the graphs of fig. 1 show that this critical point occurred in the particular case under discussion about hour 13:30, or five hours after the roots actually ceased to absorb moisture from the soil. In other words, it appears that the plant in question probably attained the stage of permanent wilting about five hours after it was uprooted and taken into the laboratory.

To test the validity of this supposed relation between the minimum in foliar transpiring power and the attainment of permanent wilting, it will of course be necessary to employ the hygrometric paper tests upon plants in which permanent wilting is subsequently established by the method of BRIGGS and SHANTZ—the 24-hour exposure to a saturated atmosphere, the plants being still rooted in the soil. The press of other experimentation prevented the writer from undertaking this comparison, and the foregoing suggestions have been made with the hope that such a study as that just mentioned may be accomplished at no distant time. Enough has been done to render it highly probable that the index of foliar transpiring power determined by means of cobalt chloride paper may furnish a somewhat precise criterion for the determination of that state of plants heretofore vaguely defined as permanent wilting.

IOWA STATE COLLEGE
AMES, IOWA

BRIEFER ARTICLES

A TROPISM CHAMBER

(WITH THREE FIGURES)

A few years ago the writer planned some classroom experiments to illustrate the geotropism of roots. The students wasted much time and many seedlings because of the wilting of the roots while they were transferring them to the observation chambers and fastening them to their supports. To overcome this difficulty some special tropism chambers were designed. These have given excellent satisfaction, and have proved to be so well adapted to the study of tropic responses that their use has suggested a number of new classroom experiments, which would have been impracticable without this apparatus.

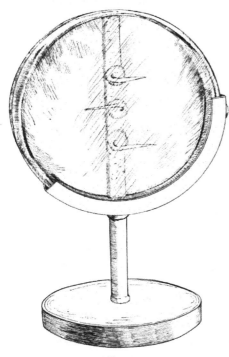

FIG. 1

The first makeshift contrivance consisted simply of a Petri dish, 4–5 inches in diameter, with a small piece of cork cemented to the bottom, near one side. The dish was partly lined with wet filter paper. The seedling was pinned to the cork and the dish set on edge on top of an ordinary tumbler. The curvature of the top of the tumbler made a convenient support for the cover of the Petri dish. The Petri dish could be rotated and the radicle of the seedling quickly brought to any desired angle with the vertical.

The makeshift apparatus served the purpose fairly well, but the tumbler support was not always convenient. A special support was then constructed. As will be seen from the drawings, the support consists of a circular trough (of brass) mounted on a pedestal. There are two sockets for attaching the trough to the pedestal, which screws into the socket. These sockets are 90° apart. The purpose of the extra socket is shown in fig. 2, where the tropism chamber is shown attached to a regular laboratory support. This arrangement is for use with the

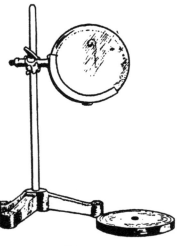

horizontal microscope.[1] The chamber may be transferred from the base to the laboratory support without altering the position of the seedling.

Fig. 1 shows the chamber provided with a strip of cork so as to make room for several seedlings. The drawing is from an experiment to illustrate traumatropism.

Fig. 3 shows the improved apparatus. The support is provided with a scale marked in degrees. This makes it possible to place the seedling at any desired angle with the vertical. Four marks etched or scratched on the cover of the Petri dish 90° apart serve as pointers.

FIG. 2

Into a hole, bored through the bottom of the Petri dish, a cork is inserted. The seedling is pinned to the inner face of the cork. The cork can be moved from the outside and it is possible, therefore, to make the final adjustments of the position of the seedling (for example, bringing the radicle parallel with the faces of the chamber) without opening the Petri dish.

[1] The tube from any compound microscope, clamped to a laboratory support, will serve as a horizontal microscope. A hair stuck to the diaphram in the eyepiece will serve as a cross-hair.

The bottom of the Petri dish is ruled into 5 mm. squares. The squares not only aid in determining the angles made by the bending root,

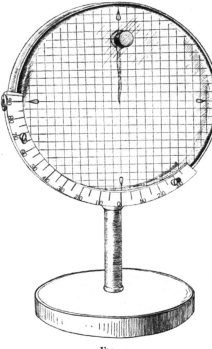

but also serve as markers to locate the regions of growth and curvature. A sheet of coordinate paper pasted over the bottom of the dish will serve nearly as well as the cross-ruling. Thin paper, such as is used for making blue-print tracings, is best. The paper can be made more transparent by wetting it with a mixture of alcohol, glycerine, and water.

Seedlings may be grown for a number of days in the chambers. It is possible, therefore, to place the seedlings in the chambers some time previous to the time set for the experiments. If the filter paper lining the chamber is placed symmetrically, so as to avoid hydrotropic curvatures, the radicles will grow straight

FIG. 3

downward. The chamber can now be rotated and the seedling quickly brought to any desired angle with the vertical, and without exposing it to the dry air of the laboratory. The chambers are excellent for experiments on perception and latent periods, and for periodic responses.—W. T. BOVIE, *Laboratory of Plant Physiology, Harvard University.*

CURRENT LITERATURE

The "graft hybrids" of Bronvaux.—Since the work of WINKLER in producing so-called "graft hybrids" among several species of *Solanum*, the older chimeras, *Laburnum Adami* and the "Crataegomespili of Bronvaux" have attracted renewed attention. The graft hybrids of Bronvaux originated from the callus formed at the junction between a *Crataegus monogyna* stock and a scion of *Mespilus germanica* These aberrant branches were of two types, and have been described under the names *Crataegomespilus Asnieresii* and *Crataegomespilus Dardari*. The first of these resembles more closely *Crataegus monogyna*, while the *Crataegomespilus Dardari* has a much closer resemblance to *Mespilus*. A thorough study of these two chimeras has been made by MEYER,[1] in order to find means of identifying the tissues which have been derived from the two parent species In most respects the tissues of *Crataegus* and *Mespilus* are so similar that few distinguishing features can be found Numerous chemical tests failed to discover any method for certainly differentiating the tissues of the two species. A study of the chromosomes, however, gave one good criterion, for, although the number of chromosomes in the two species is the same (32 in the diploid nucleus), the chromosomes of *Mespilus germanica* are considerably longer and thinner than those of *Crataegus monogyna*. The capacity to produce anthocyan, which is present in the epidermis of the fruits of *Crataegus*, is lacking in the fruits of *Mespilus*, while the reverse relation with respect to anthocyan is seen in the fact that *Mespilus* flowers turn reddish in aging, while those of *Crataegus* remain white. The fruits of the two chimeras show no anthocyan, and the flowers of both turn reddish with age.

The only clear distinction that could be found in the internal anatomy of the two species was visible only in longitudinal sections. The libriform vessels in *Mespilus* possess spiral thickenings which are absent in those of *Crataegus*. In both chimeras the libriform vessels lack spiral thickenings. The form of the epidermal cells is also strikingly different in the two species, those of *Mespilus germanica* being oblong, while those of *Crataegus* are nearly spherical. The cuticle of *Crataegus* is level, while that of *Mespilus* follows the contour of the rounded ends of the underlying cells. The epidermis of both chimeras agrees with that of *Mespilus*. The fruits of *Mespilus* are five-loculed and those of *Crataegus* one-loculed. Both of the chimeras have one-loculed fruits.

---

[1] MEYER, J., Die Crataegomespili von Bronvaux. Zeitschr. Ind. Abstamm. Vererb. 13:193–233. 1915.

The view of BAUR is sustained throughout, that these two "graft hybrids" consist of a core of *Crataegus* tissue overlaid by a mantle of *Mespilus*. In *Crataegomespilus Asnieresii* the mantle is the single epidermal layer, while in *C. Dardari* the first subepidermal layer is also of *Mespilus* tissue. By periclinal divisions this subepidermal layer may come to be represented by a number of cell layers, *Mespilus* chromosomes having been identified as deep as the eighth cell layer in one case. As the initiation of lateral branching results from periclinal divisions in the second subepidermal layer, the author argues that no chimera would be able to maintain itself as a chimera in which the mantle should consist of more than two layers. Several "reversions" to one or the other component species, and changes from one of the chimeras to the other, are described and easily explained, and one sectorial branch is figured and described.—G. H. SHULL.

Albinism in maize.—The important studies of EMERSON[2] on the inheritance of albinism and partial albinism in maize have been continued by one of his students.[3] Two different forms of albinism are found, one in which the seedlings are pure white, the other in which they are yellowish white, the latter turning slightly greenish as they grow older and sometimes developing enough chlorophyll to reach maturity. Both of these sorts of albinism prove to be simple Mendelian recessives to the normal green stains. The pure white seedlings could not be used in breeding, but the yellowish white supplied several mature plants which were selfed and which gave progenies consisting entirely of yellowish white seedlings. When plants heterozygous for the pure white were crossed with plants heterozygous for yellowish white, all of the offspring were green, showing that the normal green plants possess two determiners, the absence of one of which gives rise to pure white seedlings, while the absence of the other gives yellowish white seedlings. In confirmation of this interpretation, the second generation from these crosses between the heterozygous plants consisted of four different kinds of families. (*a*) all green; (*b*) green and pure white in the ratio 3:1; (*c*) green and yellowish white in the ratio 3:1; and (*d*) green, yellowish white, and pure white in the ratio 9:3:4. These results demonstrate the existence of the same genotypic situation in maize that NILSSON-EHLE[4] assumed to be present in rye in which pure white and yellowish white albinos were also found.

A continuation of the work on yellowish green (chlorina) plants described by EMERSON confirmed that investigator's conclusions that the yellowish

---

[2] EMERSON, R. A., The inheritance of certain forms of chlorophyll reduction in corn leaves. Rep. Nebr. Agric. Exp. Sta. **25**:89–105. 1912.

[3] MILES, F. C., A genetic and cytological study of certain types of albinism in maize. Jour. Genet. **4**:193–214. *pl. 1.* 1915.

[4] NILSSON-EHLE, H., Einige Beobachtungen über erbliche Variationen der Chlorophylleigenschaft bei den Getreidearten. Zeitschr. Ind. Abstamm. Vererb. **9**:289–300. 1913.

. green is a simple Mendelian recessive to the normal, and a similar result was obtained in crosses between the striped forms known as *Zea japonica*, in crosses with the normal green strains, the striped being recessive to normal, though in the latter cross there is some confusion when certain differences in aleurone colors are also involved in the same cross. The exact nature of this relation between aleurone colors and leaf colors was not worked out. Crosses between striped plants and chlorina, and between striped and yellowish white yielded in each case normal green plants, owing to the bringing together of complementary factors.

MILES also made a study of the chloroplasts in the several strains with which he worked. He could find no plastids in the pure white seedlings, and only a few small plastids in the yellowish white, which became more numerous and larger as the plants grew older, until they resembled, in the better developed individuals, the normal condition. In the case of striped-leafed plants, the arrangement of the plastids showed a sharp distinction between the cells of the green portions of the plants and those of the white stripes.—G. H. SHULL.

The physiology of pollen.—In his work on the physiology of pollen, TOKUGAWA[5] dealt with three main points of interest: factors determining germination, factors determining the direction of the pollen tube, and factors determining the rate and extent of the growth of the pollen tube. The investigator adds evidence against the views of MOLISCH and of BURCH that specific substances on the stigma generally determine whether the pollen will germinate or not. Already many cases have been found in which the physical conditions are the important ones in determining germination. JOST showed for various species in several families that restricted water supply is the main requirement for the germination of the pollen. He secured this condition by germination of the pollen in a saturated atmosphere or on leaf epidermis or parchment paper. TOKUGAWA adds many more to the list He has evidently failed, however, to notice the work of MARTIN,[6] which shows the important newly discovered fact that conditions giving a free water supply to the stigma may lead to sterility. This holds for alfalfa and certain clovers in the central Mississippi Valley when pollination occurs at moist or wet periods.

The author confirms the statement that sugars and proteins are important chemotropic substances for the pollen tube. In certain plants (*Narcissus Tazetti* and *Prunus mume*) sugars are effective, and in other (*Camellia japonica*) proteins. He concludes that chemotropism determines the entrance of the tube into the stigma canals and the micropyle, and that the tube is directed in a physical manner in the rest of its course.

The conditions affecting growth of the pollen tube are considered in relation to their significance in determining self-sterility and failure to

[5] TOKUGAWA, Y., Zur Physiologie des Pollens. Jour. Coll. Sci. Tokyo **35**:1–35. *figs. 2.* 1914.

[6] BOT. GAZ. **56**:112–126. 1913.

hybridize.   In this connection it is found that tubes penetrate rather deeply into
agar.   The work with several species of lilies shows that the tubes penetrate
most rapidly in stigmas of the same species, and progressively slower as the
species becomes more distant.   When the tube penetration was slow in these
crosses, it became continuously slower as the depth of penetration increased,
and finally ceased short of the embryo sac.   The writer thinks that he has
shown that this behavior is due to shortage of nutrient or stimulative materials
and not due to toxic materials.   His evidence, however, is not at all against
the gradual formation of antibodies, a suggestion made by JOST.   Self-sterility
and failure to hybridize, so far as it is due to lack of tube penetration, deserves
a thorough physiological study in the light of our modern knowledge of antibodies and of several other phases of physiology.   The later evidence (perhaps not entirely conclusive) that the character of self-sterility mendelizes does
not subtract in the least from the need of such an exhaustive physiological
study.—WILLIAM. CROCKER.

   A new type of embryo sac.—The evolution of the sporophyte and the
gradual reduction of the gametophyte are well known to every botanist.   In
the angiosperms, where the reduction of the gametophyte generation is most
extreme, intensive research has revealed several types of embryo sac.   Doubtless the most common type and the one long believed to be practically the only
type is the familiar 8-nucleate sac, two of whose nuclei fuse to form the endosperm nucleus.   This sac is formed by one of a row of 4 megaspores.   Soon the
Lilium type, looking like the preceding but formed from 4 megaspores, was
discovered.   Sacs with 16 nuclei, some formed from four megaspores and
some from a single megaspore, were added to the list.   *Cypripedium* has a
4-nucleate sac formed from two megaspores.   Since the discovery of the
Cypripedium type, other 4-nucleate sacs have been found, some formed from
one megaspore and some from two.
   A new type of embryo sac has been found in *Plumbagella*, one of the
Plumbaginaceae.[7]   The development starts as in *Lilium*, there being 4
megaspores, not separated by walls, but there are no further nuclear divisions.
One nucleus becomes the nucleus of the egg, two fuse to form an endosperm
nucleus, and the remaining nucleus, which is at the antipodal end of the sac,
disintegrates   At the time of fertilization, there are only two nuclei in the
sac.   The most important feature is that a megaspore functions directly as the
egg.   The fusion of two megaspore nuclei to form an endosperm nucleus
is also new.   Without question, this is the most reduced female gametophyte
ever described, and in the nature of the case the reduction can go no farther.
DAHLGREN recognizes that the reduction is as extreme as in animals, and he
compares this sac with the egg and three polar bodies.

---

   [7] DAHLGREN, K. V. OSSIAN, Der Embryosack von *Plumbagella*, ein neuer Typus.
Arkiv fur Botanik **14**:1–10. *figs 5.* 1915

Ten years ago, in a paper on alternation of generations in animals,[8] the reviewer predicted that gametophytes more reduced than those known at that time might still be found. Miss PACE soon described the *Cypripedium* type and afterward various investigators found digressions from the "normal" type in *Peperomia, Penaea, Oenothera, Clintonia, Lawia, Podostemon, Dicraea, Helosis*, and *Plumbagella*.

It is interesting to note that the most reduced sac has not been found in the most advanced family. It has been a prevalent custom for investigators to focus their attention upon forms which at the moment seemed to have particular phylogenetic significance, forgetting not only that phylogenetic schemes might be awry, but also that facts of great importance in interpreting phylogeny might be found in plants not in the supposed line of ascent.—C. J. CHAMBERLAIN.

Phytopathology in the tropics.—Dr. JOHANNA WESTERDIJK presented a paper at the twenty-fifth anniversary celebration of the Missouri Botanical Garden[9] dealing with the general facts concerning plant diseases in the tropics, based upon recent observations she had made in the Dutch colonies of the East Indies. The combination of high temperature and moisture would seem to be peculiarly favorable to fungi, and therefore to diseases of economic plants induced by fungi; but in fact there are only a few such diseases of real importance. Not only among cultivated plants, but also among the native plants are attacks of fungi rare. Dr. WESTERDIJK has concluded that the tropical temperature is too high for many fungi, a conclusion confirmed also by experimental work. The condition of the tropical host is also unfavorable for invasion by fungi because of the high water content and small air content of the tissues concerned. Among the disease-producing fungi in the tropics, mention is made of the root fungi (certain Hymenomycetes) which attack practically all cultural woody plants. The fungi (certain Ascomycetes and Fungi Imperfecti) inducing die-back diseases of orchards and forests, so common in temperate regions, are represented in the tropics only by a *Corticium* (a Hymenomycete); and there are no representatives of the powdery mildews (Erysiphaceae). The relation of fungus attacks to temperature is well shown in the behavior of *Phytophthora infestans* in Java, where potatoes are cultivated in mountain districts between 1500 and 6000 feet altitude. In the lower areas the infected regions are rare, but in ascending to the higher areas of lower temperature, the more destructive does *Phytophthora* become. Among the rusts there is only one representative of importance in tropical agriculture, namely the coffee leaf disease (*Hemileia vastatrix*). Leaf spot diseases are much less frequent than in Europe or in the United States; while bacterial

[8] BOT. GAZ. 39:137–144. 1905.

[9] WESTERDIJK, JOHANNA, Phytopathology in the tropics. Annals Mo. Bot. Gard. 2:307–313. 1915.

diseases are scarcely to be found. The general conclusion is that the small part which fungi play in the plant diseases of the tropics is not due to the absence of fungi, but the tropical conditions that influence the hosts in their relation to parasites.—J. M. C.

Soil acidity.—DAIKUHARA[10] finds that more than 75 per cent of the soils of Japan and Korea are acid, while the Chinese, South Oceanic, and European soils show little or no acid. This is due to the difference in geological formations, climatic conditions, and fertilization methods. Soils of acid rock origin show the most prevalent acidity, those of basic origin less, and those of laval ash are generally free from acids. Mesozoic formations are most commonly acid; tertiary, paleozoic, and diluvial next; and alluvial formations least. The condition in Japan and Korea is related to the common occurrence of acid soils in the United States. The author finds that more than half of the cases of acidity are due to aluminium and iron compounds of acid reaction that are adsorbed by the colloids of the soil and set free upon the addition of such fertilizer salts as KCl, K_2SO_4, KNO_3, and $NaCl$. In these soils fertilizing with neutral salts alone often proves very detrimental, but fertilizing with neutral salts plus lime is highly beneficial. More than 50 per cent of the cases of soil acidity in Japan and Korea are due to this colloidal phenomenon. The author cites from mineralogies a number of iron and aluminium compounds that are acid in reaction, as phosphates, double salts of silicates, etc. It was already known that negative colloids of the soil often render lime-poor soils acid by adsorbing the basic ion of neutral salts and setting free the acid. The author speaks of his finding as a newly discovered source of soil acidity. He has developed a test for soil acidity that shows advantages over the litmus, Baumann and Gully, or the Loew tests. He has also evolved a method for the quantitative determination of soil acidity. The acid soils generally bear little lime, and the lime factor is unfavorable, owing to the excess of magnesium.

We are now coming to recognize that many acid-forming processes take a part in the dissolution of lime from the soil and the final rendering of it acid. The two absorption processes described above are only two of the several known.[11]—WILLIAM CROCKER.

Field physiology of cotton.—BALLS and HOLTON[12] have published an article on analysis of agricultural yield which exemplifies the application of

[10] DAIKUHARA, G., Über saure Mineralböden. Bull. Imp. Cent. Agr. Exp. Sta. Japan 2:1–40. 1914.

[11] Readers will be interested in the following citations from American literature, in addition to the literature in the foregoing paper: HARRIS, J. E., Soil acidity and methods for its detection. Science N.S. 40:491–493. 1914. TROUG, E. A., A new method for the determination of soil acidity. Science N.S. 40:246–248. 1914. BARKER, J. F., and COLLISON, R. C., Ground limestone for acid soils. Bull. 400. Geneva Exp. Sta. New York. 1915.

[12] BALLS, W. L., and HOLTON, F. S., Analysis of agricultural yield. Part I. The spacing experiments with Egyptian cotton, 1912. Phil. Trans. Roy. Soc. London B 206:103–180. 1915.

very accurate scientific methods to the problem of crop yield. The work, with its excellent analytical methods, deserves the careful attention of students of field crop production. When such methods are in general use in this phase of agronomy, the results and conclusions gained will carry with them much more weight and dependence. One is especially impressed with daily determinations of growth rate and flower and boll opening, which make possible the evaluation of accidental and temporary factors. Aside from the important contribution to method in analysis of agricultural yield, which after all is its greatest value, the article also contributes some clear-cut conclusions upon spacing as effecting production in the Nile valley, as follows: "(a) the experiment shows that the yield of a cotton crop is primarily dependent on the number of flowers which it forms, (b) the normal extension of the root system of an isolated cotton plant can utilize more than 2 m.^{2} of soil surface in soil which is more than 2 m. deep; (c) the plants in the field crop have only 0.18 m.^{2} allowed them or less; most of the phenomena of field crop physiology in the fruiting seasons are traceable to the interference of one root system with another; (d) the yield per unit area of the conventional spacing of the Egyptian Fellah is the maximum obtainable under the limitations of field cultivation (two plants per hole, each hole 0.34 m.^{2}); (e) the sources of error in field experiments with cotton can be traced to (1) soil variation, especially below one meter depth; (2) insufficient frequency of observation, whereby accidental episodes cannot be distinguished from normal sequences; (3) fluctuations of single plants, heterogeneity of commercial varieties, and normal physiological variations from day to day."—WILLIAM CROCKER.

Geotropism of the grass node.—It is well known that lodged grass stems recover their vertical position by growth on the lower flank of the older mature nodes. Gravity, acting transversely on the stem rather than longitudinally, incites growth in these otherwise mature regions of the stem. ELFVING showed that these nodes are incited to growth when the stems are rotated with transverse exposure on the clinostat, thus giving a diffuse all-sided action of gravity; but in this case growth is equal on all flanks and no bending results. RISS[13] has attempted to analyze more fully the mode of action of gravity in this behavior. She finds that when the gravity stimulus is applied intermittently but equally (intermittent clinostat) on two opposite flanks, the growth is greater than when it acts equally (continuous clinostat) on all flanks. By means of a compound centrifuge clinostat,[14] she has applied a centrifugal stimulus of one gravity transversely (intermittently and continuously as above), at the same time the organ held its vertical position in relation to the pull of gravity. While the transverse stimulus thus applied incites growth, its effect is far less than in the absence of the longitudinal pull of gravity.

[13] RISS, M. M., Über den Geotropism der Grasknoten. Zeitschr. Bot. 7:145-170. 1915.

[14] See BOT. GAZ. 58:89. 1914.

She finds no conclusive evidence for or against the view that the tropic and growth stimuli of gravity in this organ are distinct. Her work makes it evident that the lack of growth in the older nodes with the stem vertical is due both to the lack of the stimulating transverse action of gravity, and to the presence of the inhibiting longitudinal action. Thus we see these responses to stimuli becoming more and more complex. One wonders whether a study of changes, acid and otherwise, induced in the tissues of the nodes by these various exposures might not simplify the matter. Such, I believe, is the possibility of real progress in this field.—WILLIAM CROCKER.

Phylogenetic significance of endosperm.—Nuclear endosperm and cellular endosperm, and also endosperm beginning its development with a free nuclear period and later passing into a cellular condition, have been known since the days of HOFMEISTER; and since that time various modifications and peculiarities have been described, some of them characterizing genera or families or even orders; while others seem to be confined to species. Whether the character of the endosperm has any phylogenetic significance or not, is a question which has often been discussed and often answered, both in the affirmative and in the negative.

The most recent discussion[15] is also the most comprehensive. It is a study of the literature rather than a laboratory investigation. For all the orders of Dicotyledons and Monocotyledons, the literature dealing with the endosperm has been assembled and discussed and charts have been made, so that it is possible to see at a glance just what the endosperm and haustorium conditions are in any order. In this bird's-eye view, the names of the principal investigators are given and full citations appear in an extensive bibliography. After describing the endosperm and haustorium situation in each order, often treating the families separately, sometimes the genera, and occasionally the species, the author adds a long summary dealing with orders. Both in the introduction and in the conclusion it is very plainly stated that the endosperm character is only one factor among many, but nevertheless endosperm and haustoria characters have great phylogenetic significance.—CHARLES J. CHAMBERLAIN.

Temperature and photo-perceptions.—In studying the influence of temperature upon phototropism in the coleoptile of *Avena sativa*, Miss DEVRIES[16] has determined the influence of temperature upon the rate of photo-perception and photo-reaction and the influence of previous heating upon the rate of these processes. She finds that van't Hoff's law of rate of chemical reaction

[15] JACOBSSON-STIASNY, EMMA, Versuch einer phylogenetischen Verwertung der Endosperm- und Haustorialbildung bei den Angiospermen. Sitzungsb. Kaiserl. Akad. Wiss. Wien **123**:1–137. 1914.

[16] DEVRIES, M. S, Der Einfluss der Temperatur auf den Phototropismus. Extrait du Rec. Trav Bot. Néerland **11**:195–291. *figs.* 7. 1914.

applies to photo-perception, with a coefficient of about 2 6 for 10° C. rise in temperature up to 30° C. It also applies to photo-reaction with a considerably smaller coefficient. Above 30° C. an injurious time effect of high temperature sets in, and finally at 40° C. the power of perception and response is soon lost. These results stand in contradiction to those of NYBERGH, who claimed that temperature ranging from −3° C. to 47° C. have little influence upon the photo-perception rate, indicating that the process is a strictly photo-chemical reaction such as occurs on the photographic plate. DeVRIES' work, on the other hand, lines these two processes up, so far as they are influenced in rate by temperature, with chemical reactions in homogeneous solutions in general, and with photosynthesis, respiration, and geo-perception.[17] From 0° to 25° C. the perception speed was independent of the time of previous warming. Long previous warming at 27.5–30° C. hastened perception rate, and such previous heating at 32 5° C. or higher temperatures slowed the perception rate. One hour's heating at 39° C. lowered the perception speed at 20° C. more than fourfold. This effect entirely disappeared, however, after four hours' storage at 20° C., and is therefore considered rather a matter of hysteresis than of the accumulation of poisonous materials.—WILLIAM CROCKER.

Invasion of a prairie grove.—In the high prairie just outside of Lincoln, Nebraska, a grove was started about forty years ago by running furrows at intervals of 4–6 feet through the prairie and dropping the tree seeds into the furrows. At present about 20 acres are thus forested with *Fraxinus pennsylvanica*, *Juglans nigra*, *Ulmus fulva*, *Acer saccharinum*, and *A. Negundo*. No culture has been attempted at any time during the history of the grove. nor has there been any damage by fire or grazing, hence the forested area affords an exceptionally good demonstration of the fact that trees grow freely once they are planted in this prairie soil, although they almost never invade the grasslands, and it also provides an unusually good opportunity of studying the changes in the undergrowth vegetation resulting from the changed conditions due to the tree growth. POOL[18] in investigating the character of the invasions has found that not only has the prairie sod gone, but nearly every one of the original prairie species has entirely disappeared, being replaced by some 90 invading species, of which 85 per cent are mesophytic and 60 per cent are distinctly woodland. Lists of these species prove how completely the area has been transformed from prairie to forest in a very short period. Doubtless changes in soil moisture, evaporation intensity, and light as the trees developed led to the changes in the undergrowth. POOL has these and other factors under investigation and doubtless his results will form a valuable contribution to the understanding of the problems of the relations existing between the grasslands and the forests.—GEO. D. FULLER.

[17] BOT. GAZ. **50**:233–234. 1910, **51**:239. 1911.

[18] POOL, R. J., The invasion of a planted prairie grove. Proc. Soc. Amer. For **10**:1–8. 1915.

Embryology of the Rosaceae.—A literary study of the embryology of the Rosaceae brings JACOBSSON-STIASNY[19] to some conclusions in regard to the interrelationships of the subdivisions of this large family. The characters which were traced through the various forms are number and character of integuments, number of archesporial cells and megaspores, number of tapetal cells, presence or absence of an obturator, character of young and mature endosperm, fate of the nucellus, character of the suspensor, orientation of the ovule, and form of the embryo. After tabulating and comparing all these features, the author concludes that the Rosaceae do not represent a single developmental line, but that the Spiroideae are the primitive stock which has produced two principal lines, one consisting of the Pomoideae and Prunoideae, and the other of the Rosoideae.

While everyone recognizes that in an investigation of such scope it would be impossible to make a laboratory study of all the forms, still those who are familiar with such investigations realize that descriptions are not always reliable, and that observations upon embryology, made before technique had reached its present efficiency, may be quite misleading. However, such an assembling of the literature and the graphic presentation of the results will be useful to workers in various fields.—CHARLES J. CHAMBERLAIN.

Growth forms.—On the basis of RAUNKIAER's classification of growth forms, TAYLOR[20] has made an interesting analysis of his own *Flora of the vicinity of New York*, both as a whole and as to certain of its constituent elements. He notes the high percentage of water plants and of perennials possessing rootstocks and bulbs; and calls attention to the much greater abundance of woody forms among the southern types in the flora than among the northern ones. The "biological spectrum" of growth forms in the region covered is compared with spectra of several other areas, arctic, temperate, and tropical, and certain conclusions as to climate are suggested. Such a method of studying climate as this of RAUNKIAER's, by rigid comparative analyses of the floras of widely separated regions, seems to be open to the general objection that it underestimates the importance of the historical factor. Temperate areas of the Northern Hemisphere, for example, contain a much higher proportion of herbaceous species than do regions with a corresponding climate south of the equator, presumably owing to the fact that the herbaceous type has originated for the most part in the north temperate zone. Such careful analyses of growth habit as the present one, however, are of much value for drawing critical comparisons between floras, and for other purposes. It is to be hoped that their number will continue to increase.—E. W. SINNOTT.

[19] JACOBSSON-STIASNY, EMMA, Versuch einer embryologisch-phylogenetischen Bearbeitung der Rosaceae. Sitzungsb. Kaiserl. Akad. Wiss. Wien 123:1–38. 1914.

[20] TAYLOR, N., The growth forms of the flora of New York and vicinity. Amer. Jour. Bot. 2:23–31. 1915.

Modification of germ plasm.—In connection with the twenty-fifth anniversary celebration of the Missouri Botanical Garden, MACDOUGAL[21] has published an account of the work done at the Desert Laboratory of the Carnegie Institute in experimental modification of germ plasm. It is directed against the old conception of "an inviolable germ plasm," in the interest of the conception that germ plasm is a responsive structure, with its chemistry and physics. At the Desert Laboratory over 200 species of seed plants, selected for their suitability and promise of response, have been included in the cultures, subjected to various conditions (mountain top, desert, and seashore). Less than 80 of these have survived, and about 20 continued in all three habitats. The notable feature in the behavior of these plants in unaccustomed habitats is the variation in sexual reproduction and seed formation. A second method of experimentation has been to use "inciting agents" applied directly to the reproductive bodies. In this kind of experimental work the forms chiefly used at the Desert Laboratory were certain cacti, *Penstemon Wrightii*, and an undescribed species of *Scrophularia*, the results reported in the paper being chiefly obtained from the last form. The paper concludes with a condensed statement of the conditions involved in any experimental work upon the germ plasm.—J. M. C.

Bacterial diseases of plants.—SMITH,[22] in a paper presented at the twenty-fifth anniversary celebration of the Missouri Botanical Garden, has summarized our present knowledge in reference to the bacterial diseases of plants, all of which has come within a generation. It is an interesting historical fact that our first knowledge of such diseases was the announcement by BURRILL in 1878 that pear blight is a bacterial disease. SMITH states that he is now ready to venture the sweeping statement that eventually a bacterial disease will be found in every family of plants, from the lowest to the highest. At present such diseases are known to occur in 140 genera, representing more than 50 families. The paper includes a list of the families of seed plants, those in which bacterial diseases have been found being indicated. There is also discussion of the following interesting topics: period of greatest susceptibility; what governs infection; how the infection occurs; time between infection and appearance of the disease; recovery from disease; agents of transmission; extra-vegetal habitat of the parasites; action of the parasites on the host; the reaction of the host; prevalence and geographic distribution; and methods of control.—J. M. C.

Stem anatomy of Isoetes.—LANG has begun an investigation of *Isoetes*, whose problems, as he remarks, remain difficult and fascinating. The first

---

[21] MACDOUGAL, D. T., The experimental modification of germ plasm. Annals Mo. Bot. Gard. 2:253–274. *figs. 4.* 1915.

[22] SMITH, ERWIN F., A conspectus of bacterial diseases of plants. Annals Mo. Bot. Gard. 2:377–401. 1915.

paper[23] deals with the morphological nature of the so-called stem, which he calls "stock," the reason being that one of the questions is whether this tuberous axis represents only a short stem. After a presentation of the anatomical details, LANG concludes that the stock consists of an upwardly growing shoot region, and a downwardly growing region giving rise to the roots; the latter region he calls the rhizophore. He suggests that the origin of the rhizophore may hold some relation to the deep-seated secondary meristem at the base of the shoot, but that, once initiated, "the growing region of the rhizophore behaves like the primary axis which is congenitally sunken and inclosed." This region, it seems, is very suggestive of the basal root-bearing region of *Lepidodendron* and its allies, and confirms WILLIAMSON's suggestion that *Isoetes* may be the nearest living representative of that paleozoic stock.— J M. C.

Phylogeny of the Ascomycetes.—ATKINSON,[24] in a paper presented at the twenty-fifth anniversary celebration of the Missouri Botanical Garden, discusses at length the vexed question of the origin of the Ascomycetes and their interrelationships. In general his thesis is that the Ascomycetes have been derived from the Phycomycetes, rather than from the red algae, the possible transitions being suggested most strikingly by such a form as *Dipodascus*. The details of the argument are too numerous to cite here, but it is well sustained, and more convincing than any argument hitherto favoring the algal origin of the group. A chart presents in graphic form the conclusions as to interrelationships, the Protoascomycetes arising from the Phycomycetes, and in turn giving rise to the Euascomycetes through the *Dipodascus* "stock." The divergent lines of the Euascomycetes are represented as emerging from two primitive overlapping stocks (*Gymnoascus* and *Monascus*). The paper is a very important contribution to our knowledge of a perplexing group.— J. M. C.

Cecidology.—One of the latest American papers on cecidia is by FELT,[25] in which the author describes a very large number of species of midges, many of which cause galls, while others are more or less closely associated with galls. Although this paper is primarily entomological, it contains many descriptions of galls which are of value to the botanist.

---

[23] LANG, WILLIAM H , Studies in the morphology of *Isoetes*. I. The general morphology of the stock of *Isoetes lacustris*. Mem. and Proc. Manchester Lit. and Phil. Soc. **59**: no. 3. pp. 28. 1915.

[24] ATKINSON, GEO. F., Phylogeny and relationships in the Ascomycetes. Annals Mo. Bot. Gard. **2**:315-376. *figs. 10.* 1915.

[25] FELT, E. P., A study of gall midges II. Itonidinae. Rep. N.Y. State Entomol. 1913 pp. 79-211.

In the recent European literature we find a very interesting article by CHRISTY,[26] in which the author describes a gall previously unrecorded for the British Isles. It reaches its full size of 2–10 inches in length by the middle or end of May. It is a malformation of the pistillate flower and has the appearance of a bunch of moss. The same or a very similar gall due to *Eriophyes triadiatus* Napela is well known on the Continent, but thus far the author has failed to find the mites in the British forms. It appears to be somewhat similar to our American *Acarus aenigma* Walsh.—MEL T. COOK.

Anthocyan pigments in plants.—In an examination of the recent work upon the occurrence and chemical nature of the red, purple, and blue plant pigments known as anthocyans, and yellow pigments designated flavones or flavonols, EVEREST[27] has summarized the present state of our knowledge in a very concise manner. He shows that it has been established that (1) the anthocyans always occur as glucosides, and that some seven of these pigments have now been isolated; (2) the same pigment may be capable of showing a blue, purple, or red color, according as it exists as alkali salt, free pigment, or oxonium salt of some acid; all anthocyans do not, however, form blue alkali salts; (3) the anthocyans may be obtained from flavonols by reduction followed by spontaneous dehydration; and (4) glucosides of flavonols can pass, by reduction, to glucoside anthocyans without intermediate hydrolysis.—GEO. D FULLER.

Morphology of Gnetum.—THOMPSON[28] has published a preliminary note on the embryo sac conditions in *Gnetum*, several species of which he has investigated. There are no vegetative cells in the male gametophyte, which is the expected contrast with *Ephedra*. Only free nuclei occur in the embryo sac before the pollen tube enters, although cells are formed before fertilization takes place, and one or more eggs are definitely organized. Perhaps the most significant observation is that before fertilization the female gametophyte becomes divided into a large number of multinucleate compartments, all the nuclei in each compartment later uniting to form a fusion nucleus, the endosperm being formed by the division of the fusion nuclei in the lower compartments. This situation is certainly very suggestive of a historical relation to the polar fusion in the embryo sac of angiosperms.—J. M. C.

Origin of stipules.—The much debated question of the origin of stipules has received fresh light from the anatomical studies of SINNOTT and

---

[26] CHRISTY, MILLER, Witches brooms on British willows. Jour. Botany 53: 97–102. 1915.

[27] EVEREST, ARTHUR E., Recent chemical investigations of the anthocyan pigments and their bearing upon the production of these pigments in plants. Jour. Genetics 45:361–367. 191.

[28] THOMPSON, W. P., Preliminary note on the morphology of *Gnetum*. Amer. Jour. Bot. 2:161. 1915.

BAILEY.[39] The former had already concluded that in the primitive angiosperms the vascular supply of the leaves arose as three strands. It is now found that such leaves are usually provided with stipules, the vascular supply of which is connected with the lateral leaf traces; where there is a single leaf trace at the node, stipules are absent, but if the leaf is supplied by many strands it has a sheathing base. It is further observed that leaves with an entire margin generally have no stipules, even though three strands supply the leaf. Stipules are regarded as an integral part of a leaf, and are homologous with sheaths, ligules, and similar modifications of the base of the petiole.—M. A. CHRYSLER.

Scientific phytopathology.—APPEL presented a paper at the twenty-fifth anniversary celebration of the Missouri Botanical Garden which has just been published,[30] dealing with the scientific aspects of plant pathology. This point of view is rapidly developing in this country, but still needs to be emphasized. As APPEL states, until recent times there were no places where scientific phytopathology was taught. The thesis of the paper is illustrated by the biological work in connection with the smut problem, the culture work in establishing polymorphic life histories and, in many identifications, the experimental work on the air content of host tissues, the work in physiological chemistry, and finally the histological study of the host tissues involved.—J. M. C.

A new genus of Erysiphaceae.—ITO[31] has described a new genus (*Typhulocaeta*) of Erysiphaceae from Japan, parasitic on *Quercus glandulifera*. The asci are several in the globose perithecium, and 8-spored; while the appendages are simple and clavate. The conidia have not been observed. The genus is most closely related to *Erysiphe*, but differs in its appendages.—J. M. C.

---

[39] SINNOTT, E. W., and BAILEY, I. W., Investigations on the phylogeny of the angiosperms 3. Nodal anatomy and the morphology of stipules Am. Jour. Bot. 1:441–453. *pl. 44.* 1914.

[30] APPEL, O., The relations between scientific botany and phytopathology. Annals Mo. Bot. Gard. 2:275–285. 1915.

[31] ITO, SEIYA, On *Typhulocaeta*, a new genus of Erysiphaceae. Bot. Mag. Tokyo 29:15–22. *pl. 1.* 1915.

Put a piece of metal in acid, and

The University of
CHICAGO, ILLINO

VOLUME LX   NUMBER 5

THE

Botanical   Gazette

NOVEMBER 1915

OENOTHERA GIGAS NANELLA, A MENDELIAN MUTANT

Hugo DeVries

In a recent book GATES has studied the significance of the experiments made with species of *Oenothera* as proofs for the general theory of mutation, and has given an exhaustive and critical review of the facts in this rapidly increasing field of research.[1] He has laid special stress upon the results of crosses, which show the great diversity of these phenomena when studied in some wild plants, as contrasted with the now prevailing doctrine of Mendelism; for among the mutants of *Oenothera* instances of Mendelism are rare. The first known example is that afforded by *O. brevistylis*, which follows the law of MENDEL as a recessive in all its crosses with the parent species, with other mutants, and with other species of the same group.[2]   But, unfortunately, the production of this form by means of mutation from *O. Lamarckiana* is so rare that it has not, as yet, been repeated under experimental control. Another instance is *O. rubricalyx*, discovered and studied by GATES (*op. cit.*, p. 103), which behaves as a dominant in its crosses with *O. Lamarckiana*.

In this article I hope to show that the dwarf character, which in so many instances complies with the formulae of MENDEL, but which behaves in a different way in crosses of the derivatives of *O. Lamarckiana*, may, at least in one instance in this group, follow that law as exactly as in any other pure Mendelian case.   This

[1] GATES, R. R., The mutation factor in evolution.   London. 1915.

[2] DeVries, Hugo, Die Mutations-Theorie.   1:223; 2:151-179, 429.

instance, therefore, affords a means for the experimental study of the origin of such a form by mutation. The main result of this study is the proof of the occurrence of mutant Mendelian hybrids besides the pure dwarfs.

In my book on the mutation theory I have pointed out that the origin of *O. brevistylis* from *O. Lamarckiana* may have been induced by the mutation of a single sexual cell. If this combined in fertilization with a normal gamete, a hybrid would be produced which would not be distinct from the parent species in its external features. This hybrid would then, in its self-fertilized seeds, follow the law of MENDEL and produce, besides constant *Lamarckiana* plants, partly hybrids of the same type and partly specimens of the type of *O. brevistylis*. From this origin and the subsequent free intercrossing in the field, the yearly appearance of *O. brevistylis* would receive a sufficient explanation (DEVRIES, *op. cit.*, p. 506).

If the process of mutation into this type were more often repeated, it should be possible to discover the original hybrids. They would, it is true, not be discernible from their normal sisters by external marks, but would yield, after artificial self-fertilization, about 25 per cent of *brevistylis*. And since mutants are produced ordinarily in a proportion of 1–2 per cent or less, the difference would be large enough to be noticed. Until now, however, such cases have not been observed.

I have, therefore, been looking for another example in which a Mendelian behavior of the mutants might be associated with a normal coefficient of mutation from the parent species. Such cases would betray themselves by exceptionally high coefficients in single parent plants. Instances of such individual deviations are very rare, partly on·account of the necessarily limited number of mother plants from which the seeds of our cultures are taken. But SCHOUTEN[3] has observed that *Oenothera gigas*, which ordinarily produces 1–2 per cent dwarfs, may be seen to throw them off in as large a number as 15 per cent. The same phenomenon has been described by GATES (*op. cit.*, p. 137), who counted 9 per cent and 11 per cent of dwarfs among the offspring of two self-fertilized plants of *O. gigas*.

3 SCHOUTEN, A. R., Mutabiliteit en Variabiliteit. 1908.

From time to time I have noticed the same deviating percentages in my own cultures. Thus, for instance, I fertilized in 1910 a specimen of *O. gigas* by its own pollen, and among 50 seedlings of its offspring 10 were dwarfs, pointing to a percentage of about 20 per cent.[4] Similar facts have since occurred more than once in my cultures.

SCHOUTEN and GATES have interpreted these figures as indicating a Mendelian proportion of dwarfs, and on this assumption the parent plant would have been a mutant hybrid in the same sense as explained above for *O. brevistylis*. Mutant hybrids would then occur in a race which produces dwarf mutants also, and the latter would then, of course, have to be considered as the products of the combination in fertilization of two sexual cells, both of which had mutated into *nanella*. The production of dwarfs from *O. gigas* would then follow the same process which is to be assumed for the origin of *O. gigas* itself from *O. Lamarckiana;* and the copulation of two similarly mutated cells would then more easily be accessible to experimental investigation.

In order to verify the exactness of this conception I have followed up the progeny of such a presumed mutant hybrid, and on the other hand have made crosses between *O. gigas* and *O. gigas nanella*. In both cases the truth of the assumption was easily ascertained.

Mutations of single gametes may be discovered by different means in other instances also, the production of potential *nanella* gametes by *O. Lamarckiana* being the most likely to be betrayed in this way.[5] I have observed such cases in crosses between *O. Lamarckiana* and *O. rubrinervis*. From these ordinarily two types arise in the first generation, one of which resembles the mother and the other the father. In my book on *Gruppenweise Artbildung* I have called them "*Lamarckiana*" and "*subrobusta*." Both types are usually constant after self-fertilization. But, from time to time, individuals appear which in their progeny produce an unexpected number of dwarfs. The following cases may be adduced.

[4] DE VRIES, HUGO, Gruppenweise Artbildung. p. 340. 1913.

[5] Besides the production of gametes for *gigas* by *O. Lamarckiana*, as shown by the occurrence of specimens of *semigigas* in self-fertilized strains of the parent species, or by the production of the Hero-type in crosses of *O. Lamarckiana* with allied species.

The *rubrinervis* strain for these experiments had arisen as a mutant from *O. Lamarckiana* in 1895, and its second generation was cultivated in 1905. No dwarfs were produced in the first generation after the crosses, and in the second only from single individuals, the remainder giving either no dwarfs at all or only about 1 per cent, by ordinary mutation.

TABLE I

EXCEPTIONAL PRODUCTION OF DWARFS BY SINGLE PLANTS OF *Oenothera subrobusta*

Cross	Cross	1 Gen.	2 Gen.	Number of individuals	Percentage of nanella
Lamarckiana×rubrinervis ...	1905	1913	1914	140	9
rubrinervis×Lamarckiana. .	1905	1907	1913	70	11
rubrinervis×Lamarckiana. ...	1907	1913	1914	70	16

If we compare these figures with the results of the crosses between *O. rubrinervis* and *O. nanella* itself, as described in my *Gruppenweise Artbildung* (p. 215), we find a complete analogy, since these crosses give no dwarfs in the first generation, and in the second about 10–14 per cent from the self-fertilized specimens of *O. subrobusta*. It is evident, therefore, that the exceptionally high yield of dwarfs in these crosses of *O. Lamarckiana* and *O. rubrinervis* must be the product of latent mutations which occurred in some of the sexual cells of one of the parents. And since *O. Lamarckiana* is known to produce ordinarily 1–2 per cent dwarfs, while *O. rubrinervis* does not show signs of such a mutability, we may confidently assume that our figures indicate latent mutations of sexual cells of *O. Lamarckiana*.

BARTLETT[6] recently described a similar instance of an unexpectedly high mutability, and proposed for it the same explanation, on the assumption of a latent mutation of a sexual cell in a previous generation. This case is of the greatest interest since it relates to a pure species and not to the discovery of mutated gametes by means of crosses as in the experiments just described. The mutating species was *O. Reynoldsii* Bartlett, one of the forms of the old *O. biennis*. It produced in 1913 three marked types, one repeating the parental form, and the two others being dwarfs and called

[6] BARTLETT, H. H., Mutation en masse. Amer. Nat. 1915.

mut. *semialta* and mut. *debilis*.   The latter is, on the average, about half as high as the former.   This curious segregation repeated itself in the next generation in 1914, not from all the individuals, but from only one of the two whose offspring have been tried in this respect.

Similar proofs of latent mutations of sexual cells may evidently be expected to occur in other strains also and will have to be looked for in all cases of an unexpectedly high degree of mutability.

I will now return to my experiments on the production of dwarfs by *O. gigas*.   In order to obtain specimens of *O. gigas* yielding a high percentage of dwarfs from their seeds, I sowed in 1911 seeds of my pure strain, cultivated the plants as biennials, and fertilized them in 1912 by their own pollen, in bags.   They were vigorous plants of the fourth generation (*Gruppenweise Artbildung*, p.175), and yielded a large harvest of seed, which was sown in 1913, and served as a criterion, since no essential differences were to be seen on the plants themselves.   Moreover, I used the seeds of some good biennial specimens of the previous or third generation.   The ancestors of all these plants had been fertilized by myself in bags down from the mutant in 1896 which started the race.   The harvest of 1912 and 1910, sown in 1913, gave the result as shown in table II.

TABLE II

A.   PERCENTAGES OF DWARFS AMONG OFFSPRING OF *O. gigas*

Generation	Number of seed-bearer	Total of seedlings	Dwarfs	Percentage of dwarfs
4th generation....	1	174	0	0.0
"	2	176	1	0.6
"	3	191	34	17.8
∴	4	154	1	0.6
"	5	166	1	0.6
3rd generation... ...	6	164	0	0.0
"	7	43	1	2.3
"	8	52	0	0.0
"	9	132	2	1.5
"	10	130	0	0.0

From a second strain, derived from the same mutant and described in my *Gruppenweise Artbildung* (p. 175), I had in 1911–1912 nine biennial specimens which yielded a sufficient harvest.

Tried in the same way, they gave the percentages of dwarfs shown in table III.

TABLE III

B. DWARFS OF *O. gigas*

Generation	Number of seed-bearer	Total of seedlings	Dwarfs	Percentage of dwarfs
6th generation... ..	1	132	0	0.0
"	2	165	0	0.0
"	3	155	1	0.6
.. ..	4	161	0	0.0
....	5	159	25	15.7
....	6	76	0	0.0
....	7	151	0	0.0
"	8	130	. 0	0.0
....	9	124	19	15.0

All in all, 19 specimens were studied. Among them three gave a percentage of 15–15.7–17.8, but the others gave only 1–2 per cent or no dwarfs at all. The dwarfs produced by this latter group were evidently due to ordinary mutability, but the figures for the former group differed too widely from these to be looked at in the same way. I consider them to be due to Mendelian segregation, and assume that the fact that they fall short of the expected 25 per cent is due to the difficulties of cultivation and to a less viability of the dwarfs as compared with the normal specimens.[7] I chose no. 3 of the first group (17.8 per cent dwarfs) for continuing the experiment.

If the segregation in this second generation followed the law of MENDEL, then among the plants of normal stature one-third must be constant in their progeny and the remainder must split up according to the same law. I succeeded in having a dozen of plants flower and ripen their seeds as annuals, fertilized them purely, and sowed the harvest in the spring of 1914. The result is given in table IV.

Three of the individuals yielded no more dwarfs than in ordinary mutation, and the seven others showed figures which approach the Mendelian law as nearly as might be expected. If we combine these figures with the 17.8 per cent of dwarfs of the former generation, we find for this about 18 per cent dwarfs, 57 per cent hybrids

[7] See GATES, *op. cit.*, p. 89.

of high stature, and 25 per cent normal high specimens. This may be considered as sufficient proof that the splitting took place after the law of MENDEL.

TABLE IV

C. DWARFS AMONG THE OFFSPRING OF *O. gigas* (A, no. 3)

Number of seed-bearer	Total of seedlings	Dwarfs	Percentage of dwarfs
1...........	242	1	0.5
2...........	276	0	0.0
3...........	177	1	0.5
4...........	237	39	16.0
5...........	238	52	22.0
6...........	236	50	21.0
7...........	196	42	21.0
8...........	81	25	31.0
9...........	269	59	22.0
10...........	265	57	21.0

The dwarfs were counted in June and July, and the degree of development at this time corresponded with the photographs given in my *Gruppenweise Artbildung*, p. 316, *figs. 115* and *116*. At this period they are clearly distinct from the normal specimens and so there was no difficulty in counting them. In some specimens of *O. gigas* mut. *nanella* the number of chromosomes has been determined and was found to be the same as in *O. gigas* itself (28), as was to be expected. Partly on account of this fact, partly in consequence of the nearer relationship, the fecundations did not experience the difficulties which are connected with crosses between *O. gigas* and *O. Lamarckiana* mut. *nanella*. They succeeded fairly well and yielded, as we have seen, relatively large numbers of seeds.

The Mendelian behavior of the production of dwarfs by means of mutation from *O. gigas*, moreover, may be proved in another way. If the mutant hybrids of this form are fertilized by the pollen of *O. gigas nanella*, the expectation will, of course, be the production of 50 per cent of tall specimens and 50 per cent of dwarfs. But, on account of the smaller viability of the latter, we should have to be content with somewhat smaller numbers. In 1913, therefore, I crossed some specimens of apparently normal *O. gigas* with the pollen of a constant race of *O. gigas nanella*, my culture being the third generation derived from a mutant of 1910 (*Gruppenweise Artbildung*, pp. 315–316). I was fortunate in choosing, among

some normal plants, two mutant hybrids, and will give the constitution of their progeny, together with that of two normal individuals
of *O. gigas*, in table V. The numbers of seedlings have been very
small in this case, owing to the small degree of fertility of the pollen
of *O. gigas nanella*.

TABLE V

DWARFS IN THE FIRST GENERATION OF *O. gigas*×*O. gigas nanella*

Number of seed-bearer	Number of seedlings	Dwarfs	Percentage of dwarfs
1.............	38	11	30
2.............	65	28	43
3.............	28	1	} 3
4.............	59	2	

The first two seed-bearers had evidently about one-half of their
egg cells mutated into *nanella*, which by the fertilization with the
pollen of dwarfs must, all of them, become *nanella* specimens. The
two last-named plants, although externally not differing from the
others, had only very few mutated sexual cells, and therefore produced only about 3 per cent of dwarfs.

TABLE VI

DWARFS IN THE SECOND GENERATION OF *O. gigas*×*O. gigas* MUT. *nanella*

Seed-bearer	Total of seedlings	Dwarfs	Percentage of dwarfs
A. O. gigas nanella×O. gigas			
No. 1............................	291	45	15
No. 2...........	69	12	17
B. O. gigas×O. gigas nanella			
No. 1............................	60	16	27
No. 2............................	310	73	24
No. 3............................	304	62	20
No. 4............................	74	14	19
No. 5............................	283	46	16
No. 6............................	210	30	14
C. O. gigas mut. hybrid×O. gigas nanella	326	52	16

The experiment showed at the same time that hybrids between
O. gigas and *O. gigas nanella* have the features and the stature of the
former type, and thereby justified the assumption made above in
the explanation of the behavior of mutant hybrids.

I made the reciprocal cross in the same year, fertilizing some
dwarfs of my race by the pollen of normal plants of *O. gigas*. The

fecundation was a difficult one and I got only 38 seedlings, all of which developed into tall plants of the stature and character of *O. gigas* (1914).

In order to study the segregation of dwarfs in the next generation I fecundated a number of specimens of the three described groups of artificial hybrids and sowed their seed in 1915. On the basis of Mendel's law the expectation is, for all of them, 25 per cent dwarfs, or somewhat smaller numbers on account of the lesser viability of these dwarfs. The sowings of 1915, counted in May and June, gave the results shown in table VI.

These figures give sufficient proof that the crosses between *O. gigas* and its dwarfs follow the law of MENDEL.

Summary

1. *Oenothera gigas* produces dwarfs (about 1–2 per cent) and mutant hybrids of normal stature, which after self-fertilization give 15–18 per cent, theoretically 25 per cent, of dwarfs.

2. These mutant hybrids split up, after self-fertilization, according to the law of MENDEL, yielding about 18 per cent dwarfs, 25 per cent normal specimens of tall stature, and 57 per cent hybrids of the same type. The latter gave about 21 per cent of dwarfs among their progeny.

3. The mutant hybrids, fertilized by *O. gigas nanella*, yield 30–43 per cent, theoretically 50 per cent, of dwarfs.

4. In artificial crosses with *O. gigas* the dwarfs follow the law of MENDEL.

5. The production of dwarfs from *O. gigas* by means of mutation, therefore, is to be considered as requiring the copulation of two gametes, both of which are potentially mutated into dwarfs. The mutant hybrids must then be the result of the fertilization of a mutated gamete by a normal one. They are correspondingly less rare than the dwarfs themselves.

6. In combination with the fact that the dwarfs of *O. Lamarckiana* do not follow the law of MENDEL, either in their origin by mutation or in artificial crosses with the parent species, these conclusions reveal a new differential character between *O. gigas* and its parent species.

THE INFLUENCE OF THE TANNIN CONTENT OF THE HOST PLANT ON ENDOTHIA PARASITICA AND RELATED SPECIES[1]

MELVILLE THURSTON COOK AND GUY WEST WILSON

The differences between species, genera, and families by which certain groups of plants are resistant to certain parasites while others are more or less susceptible are very generally recognized by all botanists. *Actinomyces scabies* attacks potatoes, turnips, beets, carrots, and parsnips, representing four different families. *Pseudomonas tumefaciens* attacks a much larger range of host plants. Many familiar parasitic fungi and bacteria are restricted to certain families, to a few genera, or even to a single genus; others have similar restrictions as regards species; and still others are restricted to the races within the species. The same law will apply to the insects which cause the peculiar physiological or pathological structures known as galls. *Cecidomyia pilulae* occurs on a very large number of oaks, *Amphibolips confluentus* on four species, while the very common *Andricus seminator* occurs on but one.

Individual plants which are more or less immune to destructive parasitic organisms have attracted the attention of the plant pathologists and plant breeders and have been the starting point for long series of selection experiments resulting in resistant varieties which are of great value to the practical agriculturists. Many theoretical explanations of the resistance or susceptibility of related varieties or species have been offered by botanists who should know better, but very few facts of real value have been collected.

MARSHALL-WARD's studies indicated that the histological characters were of but little if any importance in aiding the plant to resist its parasitic enemies. He finally says: "Infection and resistance to infection depend on the power of the fungus protoplasm to overcome the resistance of the cells of the host by means of enzymes or toxins; and reciprocally, on that of the protoplasm of

---

[1] A more complete discussion of the experiments given in this paper will appear in some of the publications of the New Jersey Agricultural Experiment Station, New Brunswick, New Jersey.

the cells of the host to form auto-bodies which destroy such enzymes or toxins, to excrete chemotactic substances which expel or attract the fungus protoplasm."[2]

The little work that has been done indicates that these problems involve both plant pathology and plant physiology, and that the plant pathologist must give more and more attention to the fundamental problems in plant physiology.

Some years ago the senior author and Dr. J. J. TAUBENHAUS conducted a series of experiments on the relation of parasitic fungi to the cell contents of the host plants.[3] The primary object of this work was to determine to what extent tannin might be a factor in enabling the host plant to resist parasitic fungi. Although the fungi which attack fruits were used for most of these experiments, some attention was given to *Endothia parasitica*. This organism gave a good growth of mycelium and scant spore formation when grown on an agar medium containing 0.6 per cent tannin, but a less toleration to tannin when grown on a liquid medium. It was also evident that its toleration of tannin was somewhat dependent on the character of the food supply.

A little later, CLINTON[4] made similar experiments with *Endothia*, using commercial tannin (M. C. E. Brand, U.S.P.) in amounts varying from 0.2 per cent to 14 per cent. All cultures grew in media containing as high as 4 per cent tannin; about one-half of the cultures of *E. gyrosa* grew on media containing 8 per cent tannin, but all failed to grow on cultures containing more than 12 per cent. *E. parasitica* was a little more tolerant to tannin than the other species used. CLINTON used a potato agar, and therefore his results are not comparable to those obtained by COOK and TAUBENHAUS, who used synthetic media.

Although the results of all of these studies indicate that tannin is in a measure toxic to fungi, the report of the chemist of the

[2] MARSHALL-WARD, H., Recent researches on parasitic fungi. Ann. Botany 19:1–54. 1905.

[3] COOK, MEL. T., and TAUBENHAUS, J. J., Relation of parasitic fungi to the cell contents of the host plant. 1. Toxicity of tannin. Delaware Agric. Exp. Sta. Bull. 91. pp. 77. *figs. 43.* 1911.

[4] CLINTON, G. P., Chestnut bark disease, *Endothia gyrosa* var *parasitica* (Murr) Clint. Ann. Report Conn. State Agric. Exp. Sta. 1912. pp. 359–453. *pls. 21–28.* 1913.

Pennsylvania chestnut tree blight disease commission indicated that the tannin content of the diseased bark was higher than that of the healthy bark. This report appears to contradict the idea that tannin in the bark is toxic to the fungus, otherwise the fungus would be destroyed by the tannin. However, this point was also investigated by KERR, who says "the increment of tannin is only apparent and does not really occur. We have found all decayed wood and bark give higher tannin contents, no matter what causes the decay. It simply means that other constituents have decomposed and disappeared, while the tannin remains practically stable."

It is very evident to the writers that commercial tannin is a very uncertain substance, as packages of tannin from the same manufacturers and supposed to be the same were found to give different results when used in cultures. It was also evident that ordinary methods of determining tannin are unsatisfactory. Therefore, KERR, a well known technological chemist of Lynchburg, Virginia, who has devoted considerable attention to the study of tannin, was asked to cooperate in this work. He furnished us with three extracts ("1–X," "2–X," "3–X," and "A") which are described in connection with the experiments. In addition to these extracts we also used commercial tannin (MERCK) for comparison.

Source of cultures

Cultures of various American species of *Endothia*, as well as foreign strains of some of the species, were obtained from various laboratories. In addition, some strains of *E. parasitica* were isolated in our own laboratory. We have indicated these by the names and the serial numbers used by the laboratories when they came to us, except in the case of *E. parasitica*. In this instance we have uniformly referred to this fungus as *E. parasitica*, without regard to whether it was considered a species or a sub-species at the source of supply.

The use of the specific names of *E. gyrosa* and *E. radicalis* varies in different laboratories, according to SHEAR and STEVENS.[5] In the light of this paper it appears that the fungi from CLINTON

5 SHEAR, C. L., and STEVENS, F. E., Cultural characters of the chestnut blight fungus and its near relatives. U.S. Dept. Agric. Bur. Pl. Ind. Circ. 131. pp. 18. 1913.

labeled *E. gyrosa* and that from STEVENS called *E. radicalis* are in reality identical. Our culture work also leads us to the same conclusion. Therefore, in order to avoid confusion, we will use the name *E. radicalis*, but will in such case indicate the origin of our original culture.

A careful study of our cultures indicates that we have only three distinct species: *Endothia parasitica* (American and Chinese races), *E. radicalis* (*E. gyrosa* and *E. virginiana*), and *E. radicalis missis-sippiensis*.

Several months were devoted to preliminary work to determine the most desirable medium, best methods for mixing the tannin into the medium, and for perfecting the technique of the work. The formula for the most satisfactory medium, the one which was used in all our work, is as follows: water 1000 cc., glucose 20 grains, peptone 10 grains, potassium phosphate (monobasic) 0.25 grain, magnesium sulphate 0.25 grain. A given series of cultures was always made from the same lot of medium, treated with the same extract, inoculated in the same manner, and kept under exactly the same conditions.

The difficulties arising from the use of tannin in a medium containing proteid were not overcome. The first difficulty encountered is the fact that commercial tannin (MERCK) is an unstable and variable substance. According to FISCHER (Ber. Deutsch. Chem. Gesells. 36:3252. 1913), tannin is an anhydrous glucoside of gallic acid. This relationship makes it easily convertible by hydrolysis into gallic acid and related substances. *It is therefore entirely possible that no sterile culture medium can be prepared which contains all the tannin unchanged.*

The usual statement that tannin in contact with proteid forms an insoluble precipitate has not been borne out by our work. Indeed, comparatively large quantities of tannin may be added to the agar formula which we used without changing perceptibly either the tannin or the proteid, so far at least as we were able to determine.

The experiments were conducted with two lots of MERCK's tannin. The first of these was already in stock at the time the work was undertaken. By using a 10 per cent solution of this tannin, as much as 2 per cent of tannic acid could be added to the agar without changing the composition of either. However, to accomplish this it was necessary to allow the tannin solution to run slowly from a pipette into the melted agar while the latter was constantly agitated. If the tannin was added too rapidly, or the agitation of the agar was insufficient, more or less coagulation resulted. With a 20 per cent aqueous solution of tannic acid, less than half this amount (0.8 per cent) could be added without coagulation. Moreover, even a very small amount of tannic acid in its solid form would cause coagulation in the agar. Another lot of MERCK's tannin was of such character that only about half as much tannic acid could be added to the agar without change.

When first placed in the agar, the tannic acid caused a milky appearance, which disappeared on sterilization. Where high percentages of tannin were used, the agar upon sterilization showed a tendency to become viscid (about 0.8–2 per cent) or even liquid (about 2–2.5 per cent). The transition between viscosity and liquefaction is gradual in such a series as we used, where each member differed from the next by 0.2 per cent of tannic acid. In no case was the distinct curd which various investigators have described to be observed. In agar with 3 per cent of tannic acid the entire mass of medium becomes a clear liquid with a thin film of solid matter on one side of the test tube if set for a slant, or in the bottom if set upright. This solid material gives the same reaction both to Millon's proteid test and to the ammonium molybdate test for tannic acid as does the solid agar of the lower members of the series. Similar results were obtained by testing the liquid portion of the medium. Evidently, the explanation of this liquefaction is to be found in some other direction than the chemical interaction of tannin and proteid.

If the agar medium used is titrated to various degrees of acidity and a series of such tubes sterilized, it is found that the agar ranges from solid through viscid to liquid; that is, the same phenomenon can be induced by acidulating the medium as by the addition of the tannic acid. In each case a more careful test of the nature of the proteid substance in the liquid from the acidulated agar shows that proteid digestion has progressed so far that the power of solidification has been lost.

These considerations naturally raise the question of the acidity of the culture medium containing tannic acid. Our tests showed that a 3 per cent aqueous solution of tannic acid is about +65 Fuller's scale. This is considerably higher than is indicated by CLINTON (Report Conn. Agric. Exp. Sta. 1912. p. 432). However, as CLINTON used a vegetable (potato agar) medium, while we used a synthetic medium, the results are in no wise comparable, as no account is taken by CLINTON of the effects on the tannic acid of the various organic constituents of the medium to which he added it.

Gallic acid was similarly tested, but failed to show any coagulating effect either on the agar medium or on its constituents.

The various materials furnished by KERR behave in much the same way toward agar as does commercial tannin. His purer tannin extracts, however, do not liquefy the agar at as low percentages as does commercial tannin. Those extracts from which the coloring matter had not been removed had a more pronounced action on the culture medium than even the second lot of MERCK'S tannin. The original acidity of the agar and the quantity and nature of the impurities which may be present in the tannin appear to modify to a great extent the chemical activities upon the admixture of the two substances in completing the culture medium.

To test tannin fermentation

To test the ability of the fungi to live in pure solutions of tannin and related substances, two experiments were made with varying strength of aqueous solutions of these materials. In the first series, the spores were sown on the medium, and after germination cubes about 5 mm. in size were transferred to the tannin solution. In these tests *E. parasitica* and the European strain of *E. radicalis* were used.

The first test included MERCK'S tannin and KERR'S extracts "A" and "1-X," of 2.05 per cent and 5 per cent solution. The strength of the solution appeared to have less effect on the fungus than did the nature of the material. The two species of fungi showed no more difference in quantity and vigor of growth than would be expected for two strains of the same species.

On tannin neither species grew even as much as might have been expected from the nutriment stored in the agar block.

On extract "A" a fair growth was made, using up the food material stored in the original agar block and forming masses of mycelium about 2 cm. across and producing abundant pycnospores.

On extract "I–X" an abundant growth was secured, entirely filling the liquid in the flask with a thick growth of mycelium which rose above the liquid and produced abundant pycnospores.

The second series was made up in strength of 0.2, 0.4, 0.6, 0.8, and 1 per cent tannic and gallic acid. These were sown with spores of the same fungi used before. In no case was growth made, although check sowing on agar showed the spores to be viable.

Endothia on tannin

GROWTH OF ENDOTHIA ON COMMERCIAL TANNIN (MERCK)

The cultures used for this series were made to contain 0.1 per cent, 0.2 per cent, 0.4 per cent, and by intervals of 0.2 per cent up to 2.4 per cent of tannin. In this series of experiments the agar remained firm, except in one or two of the cultures containing the highest percentages of tannin, in which there was a slight tendency to a semifluid condition. No cultures showed sufficient proteid digestion to allow the formation of a liquid. The following strains of *Endothia* were used: *E. parasitica* (STEVENS no. 1158), a Chinese strain from the Bureau of Plant Industry,[6] and a strain designated "P.P." of our own isolation from Prospect Park (Brooklyn) material; *E. radicalis* (METCALF no. A, STEVENS no. 2391, a European strain secured from ANDERSON, and CLINTON's *E. gyrosa* no. 7677); and *E. radicalis mississippiensis* (STEVENS nos. 1196 and 3443).

Endothia parasitica (American strain) gave a good growth of aerial mycelium, varying in amount in direct ratio to the percentage of tannin used. Poor growth on check until the end of the second week. Yellow color in mycelium at end of first week and discoloration of the agar during the second week. Cultures originally containing 0.2 gm. tannin (2 per cent) were sent to KERR,

[6] This culture was made from material sent directly from China to the Bureau of Plant Industry.

who was unable to detect any positive trace of tannin, but found a trace of gallic acid not exceeding 0.002 gm. This appears to to indicate that the fungus can use the tannin as a food. Pycnospores appeared on the check at the end of the sixth week, on 0.1 per cent tannin at the end of the seventh, on 1 4 per cent tannin at the end of the eighth, and ultimately on the 4 per cent tannin.

E. parasitica (Chinese strain) grew about the same as the American strain except that the growth was 5 days earlier and a lighter gray color. Pycnospores appeared during the fifth week on 2.4 per cent tannin and ultimately on the entire series.

It is rather remarkable that the Chinese strain is more tolerant to tannin than the American strain, and it raises the question whether this resistance is due to origin, age, or modification of the fungus since its first introduction into America, or some other cause.

E. radicalis (European) was more resistant than the American strain. Pycnospores were produced on 1.2 per cent tannin after 8 weeks' growth.

E. radicalis (American) gave an abundant growth of aerial mycelium, varying in color from yellow in the lower percentage of tannin to ashen in the higher. The presence of 0.2 per cent tannin was stimulating, but 0.8 per cent retarded the growth. No pycnospores were produced.

E. radicalis (CLINTON's E. gyrosa no. 7677) grew well, but no pycnospores were found. The tannin had an inhibiting influence.

E. radicalis mississippiensis grew well, but apparently was not able to use the tannin as food. No pycnospores were produced. The two cultures (nos. 2443 and 1196) used were very resistant to tannin, the former making the more vigorous growth.

From this series of cultures it appears that tannin (MERCK) affects various species of the genus Endothia quite differently. E. parasitica may for a time be retarded in its growth, but subsequently it feeds on the tannin, using the entire supply in the cultures tested. At the other extreme is E. radicalis mississippiensis, which appears to be entirely unaffected by tannin, and does not feed upon this substance. The cultures labeled E. radicalis (and CLINTON's E. gyrosa no. 7677) are inhibited by the action of tannin. This is true of the American strains to a greater extent than of the European.

GROWTH ON KERR'S EXTRACTS

Extract "1–X" is described by KERR as the water soluble tannin of the chestnut bark. It is insoluble in alcohol and in similar solvents. This occurs in quantities of 3–5 per cent in the bark. The sample used was between 95 and 100 per cent pure. The quantity available would not allow as extensive a series of cultures as were used for commercial tannin products. Accordingly, the percentages used were 1, 1.2, 1.6, 2, and 2.4 per cent. The agar remained firm in all cases. Inoculations were made with *E. parasitica* (STEVENS no. 1158), *E. radicalis*, and *E. gyrosa* (CLINTON's no. 7674).

E. parasitica (American) gave an abundant aerial growth during the third week and of pycnospores during the fifth week. The great growth of pycnospores was on 2 per cent, which is about the normal amount in the bark. This extract is stimulating in normal and subnormal amounts.

E. radicalis (CLINTON's *E. gyrosa* no. 7674) gave a growth similar to *E. parasitica*. At the end of the first week the most normal growth was on 1.2 per cent and the maximum on 2 per cent. Pycnospores appeared in 10 days and were present on all cultures at the end of the fifth week. Finally, it may be said that small quantities of this extract are stimulating, and that higher percentages produce vigorous growth of aerial mycelium and reduced number of pycnospores. It is rather surprising that this extract, which is so near pure tannin, is not as toxic as the commercial tannin.

Extract "2–X" is in all its reactions similar to that designated "1–X," except that it is soluble both in water and in alcohol. Its effect on agar is quite different, however, showing a tendency to digest proteids as do acids, and so render the medium viscous. A series of cultures was prepared containing 1, 1.2, 2, and 2.4 per cent of the extract, and inoculated with *E. parasitica* (STEVENS no. 1158) and *E. radicalis* (CLINTON's *E. gyrosa* no. 7674). The cultures of the two upper members of the series were quite noticeably viscous.

E. parasitica (American) gave growth on 1–1.6 per cent in one week, on 2 per cent in two weeks, and on 2.4 per cent in four weeks. Pycnospores were produced rather abundantly on 1.6–2.4 per cent at the end of two months.

Endothia radicalis (CLINTON'S *E. gyrosa* no. 7674) gave a good growth on 1.6 per cent in one week, and on 2 per cent three days later. In two weeks the growth was greater than with any other extract used, the maximum on 1 per cent and 1.2 per cent. Pycnidia began to appear on 1.2 per cent in about three weeks and finally on 1 per cent and 1.6 per cent, the maximum on the last percentage.

It is very evident that extract "2–X" is not so favorable for the growth of the fungus as extract "1–X." Both these results are surprising when compared with those obtained with commercial tannin.

Extract "3–X" is the coloring matter of the bark. While this is estimated as tannin in bark analysis, its real nature is unknown. It precipitates gelatin and combines with hide, but does not give the same distinct reactions with metallic salts as do other tannins. The sample used was between 85 and 90 per cent pure. As the quantity available was very small, it was used only in the proportion of 1 and 2 per cent. Both *E. radicalis* (*E. gyrosa* CLINTON no. 7674) and *E. parasitica* (STEVENS no. 1158) were grown on these media.

E. parasitica (American) gave a slight growth on 1 per cent of the extract in one week. This growth increased slowly, but finally became abundant. Germination was retarded for about 10 days on 2 per cent of the extract, and the growth was never so good as on 1 per cent. Pycnospores were found on cultures containing as much as 1 per cent of the extract during the first week, but were never found on the medium containing 2 per cent.

E. radicalis (CLINTON'S *E. gyrosa* no. 7674) gave a growth very similar to that of *E. parasitica*, but at the end of two months there were few pycnospores on the 1 per cent extract.

This extract was very toxic; the growth was always unhealthy and the production of pycnospores greatly checked. These results are surprising in that this extract, which is primarily coloring matter of the bark, which under ordinary methods of analysis is estimated as "tannin," is more toxic than commercial tannin. KERR in commenting on these results, says:

> The action of "3–X" is also surprising, as it is what we term the coloring principle of the bark, the exact nature not having been determined by anyone that I know of. Its action brings out a rather interesting point, and that is that chestnut trees of northern growth, say on a line north of the southern boundary of Pennsylvania, contain very materially less coloring matter than the growth south of it, and, as we all know, the wood in the latitude referred to seems to have been more susceptible to the disease than that further south [letter December 26, 1913].

GROWTH ON COMBINATION OF KERR'S EXTRACTS

Since "1–X," which is a tannin extract, was stimulating, and "3–X," which is primarily coloring materials giving tannin reactions, was toxic, it was decided to combine the two into one extract. The material was made up into a series of cultures containing 0.2, 0.6, 0.8, 1.2, 1.6, 1.8, 2, 2.2, 2.4, 2.6, and 2.8 per cent. Sowings were made with *E. radicalis* (CLINTON's *E. gyrosa* no. 7674), *E. radicalis mississippiensis* (STEVENS no. 2424), and both American and Chinese strains of *E. parasitica*.

There was a tendency for the agar containing as much as 1.2 per cent of the extract to become less solid, but even with 2.8 per cent there was no approach to a real liquid condition.

E. parasitica (American) made a fair growth on 0.6 and 0.8 per cent in one week, with a slight growth throughout the series except on 0.2 and 2.2 per cent. The growth increased, but was relatively the same on the different cultures throughout the entire period. During the third week pycnidia appeared on 0.6 per cent, and by the end of the fifth week had developed on all cultures up to 2.2 per cent. In cultures containing more than 2.2 per cent the pycnospores decreased. The growth was always subnormal, but not so pronounced as in some other cultures.

E. parasitica (Chinese) also failed to grow on 0.2 and 2.3 per cent, but made some growth on 0.6, 0.8, and 1.2 per cent during the first week. At the end of the second week the cultures containing 0.6 per cent showed a subnormal growth, and the 2.8 per cent a very slight growth. The growth generally was less than that

of the American strain.   Pycnospores were not found on either the
check or the cultures containing tannin.

E. radicalis (CLINTON'S E. gyrosa no. 7674) made a fair growth
on 0.2, 0.6, and 0.8 per cent, and a slight growth on the higher
percentages.   The growth generally increased on all cultures, but
was always subnormal; no pycnospores were found.

E. radicalis mississippiensis grew on only the 0.8 per cent extract
during the first week.   In 10 days there was good growth on both
0.6 and 0.8 per cent, fair growth on 1 and 1.6 per cent and slight
growth on 2 and 2.8 per cent.   The growth increased on all cultures
during two months' observation, and pycnospores were produced
in abundance on all cultures up to 2 per cent.

In general it may be said that E. parasitica and E. radicalis
thrive fairly well on cultures containing these extracts, but not
so well as on the checks.   The American strain of E. parasitica
is more resistant that the Chinese strain, and E. radicalis mississippiensis is the most resistant of any species used.

KERR'S extract "A" is a compound of various forms of tannin
and of other more or less related substances.   It represents about
9 per cent of the dry weight of the bark.   Its composition is as
follows: tannin (containing the forms represented as "1-X,"
"2-X," and "3-X") 60 per cent, fermentable sugars 10 per cent,
gallic acid 7 per cent, pentoses and pentosans 8 per cent, water 5
per cent, undetermined 10 per cent.   This extract produces a rather
advanced proteid digestion, causing the agar to become semifluid, and in the higher percentages used a considerable amount of
fluid was present.   The series included 1, 1.2, 1.6, 2, and 2.4 per
cent of the extract.   Sowings were made with E. radicalis (CLIN-
TON'S E. gyrosa no. 7674) and E. parasitica (STEVENS no. 1158).

E. parasitica (American) made a slight growth on all cultures
during the first week.   In 10 days the growth was good on all
cultures up to 2 per cent.   After that time the growth was slow.
At the end of the first month pycnospores began to appear on
cultures containing 1 per cent, but did not appear on others.   This
organism made its best growth on this medium.

E. radicalis (CLINTON'S E. gyrosa no. 7674) made a slight growth
on cultures up to 1 per cent.   The growth was poor throughout

the entire time. A very few pycnospores were produced on the 1 per cent, but not on the others.

This was the most toxic extract used in the entire series of experiments.

In order to test the possibility of preparing from commercial sources a compound similar in its effects to KERR'S extract "A," a tannin compound was made as follows: tannin (MERCK) 60 gm., dextrose 10 gm., gallic acid 7 gm., arabinose 8 gm., total 85 gm. This material was added to the agar in the same manner as was the tannin. The series of cultures prepared were at intervals of 0.2– 3.0 per cent. The effect on the proteids was the same for this substance as for tannin. The higher percentages digested almost all proteids so that the medium was a clear liquid. Sowings were made with *E. radicalis mississippiensis* (STEVENS no. 2424), *E. radicalis* (CLINTON'S *E. gyrosa* no. 7674), *E. parasitica* (American, CLINTON'S no. 7675, and the Chinese strain).

It seems scarcely necessary to give a detailed statement of these results. In all cases, except the American strain of *E. parasitica*, pycnidial formation was greatly retarded. Growth was very similar to that obtained with MERCK'S tannin.

Summary

1. Results obtained with commercial tannin are not always comparable to each other or to those obtained from specially pre-pared extracts, because of variations in chemical composition and the presence of tannin-like substances other than tannic acid. Commercial tannins of the same brand differ in their behavior in culture media, as indicated by the growth of the various species of *Endothia* used in these experiments.

2. Commercial tannin and tannin in the plant are not the same. No extract will be the same as the substances in the plant.

3. Tannin is an anhydrous glucoside of gallic acid and is easily converted by hydrolysis into gallic acid and related substances. It is very doubtful if any culture medium can be prepared contain-ing as much pure, unchanged tannin as was put into it. Therefore, we cannot know the exact percentage of tannin in a culture medium,

but we can put known amounts into those with which we are working.

4. The quantity and form of tannin compounds present in the substratum each exerts an influence on the growth of the fungus; but when the fungus attacks a plant, we have no way of knowing the form of the tannin with which it comes in contact. However, it is quite evident that the tannin of the plant is associated with coloring materials and other substances, some of which are toxic. Furthermore, the fungus may be selective during either a part of or during its entire existence, and send its mycelium into certain tissues containing little or no tannin. *E. parasitica* is especially destructive because it works in the cambium cells, but later in life it pushes through the outer tannin-bearing cells of the bark for the production of its spores.

5. The character of the food supply influences the vigor of the fungus, and therefore its power to resist the toxicity of the tannin and other materials with which it comes in contact. The amount and character both of the food supply and of the tannin and other materials no doubt vary with the seasons and the growing periods of the host plant.

6. In almost every instance, without regard to the form of tannin used or the fungus grown, a high percentage (0.8 per cent or more) of tannin caused a retardation of germination, frequently followed by an abnormal stimulation to growth of aerial mycelium.

7. Species of *Endothia* show a marked response to the presence of tannin and related substances in the culture medium.

8. The species of *Endothia*, and to a certain extent strains of the same species, show a considerable variation in their response to tannin and other substances. (*a*) *E. radicalis mississippiensis* was unaffected by the tannin, but did not use it for food and did not produce pycnospores in cultures containing tannin. (*b*) *E. parasitica* was slightly retarded in its germination and early growth, but later was able to use as much as 2 per cent tannin as food. It was the only species studied that was able to utilize any considerable amount of tannin for food. *The American strains were more resistant to tannin and associated toxic materials than the Chinese strains.*

(c) *E. radicalis* (including *E. gyrosa*) was very susceptible to the influences of tannin.

9. Tannin (MERCK) affects the various species of *Endothia* very differently. *E. radicalis* (including *E. gyrosa*) is inhibited; *E. parasitica* is at first retarded and later is able to feed on the tannin; *E. radicalis mississippiensis* is practically unaffected and does not feed on tannin.

10. Analyses of chestnut bark made by KERR show a corresponding tannin reduction in diseased areas, which confirms the culture experiments and makes it possible to state that *E. parasitica* is able to use the tannic acid for food.

11. *E. parasitica* appeared to have its power of pycnidia production stimulated by commercial tannin and the true tannin extracts of chestnut bark, but reduced or inhibited by those extracts of chestnut bark which are composed almost entirely of coloring substances, but which are present in tannin extracts and estimated as tannin.

12. Specially prepared extracts of pure tannin were either stimulating or only slightly toxic when combined with coloring materials and other substances associated with tannins and responding in the same or in a similar manner to tannins. (a) KERR's "1–X" extract has a stimulating effect on *E. radicalis* (*E. gyrosa*) and *E. parasitica*. (b) KERR's "2–X" extract has a tendency to retard *E. radicalis* (*E. gyrosa*), *E. parasitica* (both American and Chinese strains), and *E. radicalis mississippiensis*. (c) KERR's "3–X" extract was extremely toxic to *E. radicalis* (*E. gyrosa*) and *E. parasitica*. (d) In *E. radicalis* (*E. gyrosa*) conidia production was at its maximum at 1–1.2 per cent of the extracts designated "1–X," "2–X," and "3–X," while but few if any pycnospores were produced on the other substances used. (e) The European strain of *E. radicalis* showed similar results, but the American strain showed a tendency to remain sterile. (f) A combination of "1–X" and "3–X" is somewhat toxic, but the toxicity of "3–X" appears to be largely overcome by the stimulating influence of "1–X." *E. radicalis* (*E. gyrosa*) and *E. parasitica* were slightly retarded, and *E. radicalis mississippiensis* was very slightly retarded. The Chinese strain of *E. parasitica* was less resistant than the American

strain. (g) KERR's "A" was most toxic of all compounds on *E. radicalis* (*E. gyrosa*) and *E. parasitica*. (h) Tannin compounds gave results similar to MERCK's tannin instead of "A."

13. A supernormal growth of aerial mycelium was usually accompanied by a corresponding reduction in pycnidia formation.

14. Harmful effects of tannin were also frequently shown by the absence of natural pigment from the mycelial pellicle and by the ashen color of the aerial hyphae. One or both of these might be present in the same series of cultures.

15. While tannic acid is no doubt toxic to many parasitic fungi, there are also other substances associated with tannin which are toxic. Some of these substances respond to the ordinary tannin tests and have probably been mistaken for tannin. The factors which enable plants to resist the attacks of parasitic organisms present an extremely complicated problem. The solution of this problem lies in the study of the chemistry and physiology of the cell.

16. Throughout the summary the terms "tannin" and "tannic acid" have been used in the generally accepted sense, but experiments with KERR's extracts "1-X," "2-X," and "3-X" indicate that the toxic property is in the coloring material, which in analytical work is usually estimated as tannin.

The authors are indebted to Mr. GEORGE A. KERR of Lynchburg, Virginia, Dr. HAVEN METCALF, DR. C. L. SHEAR, and Dr. NEIL STEVENS of the U.S. Bureau of Plant Industry, Dr. GEORGE P. CLINTON of the Connecticut Agricultural Experiment Station, and Dr. H. W. ANDERSON of Cornell University for extracts, cultures, and valuable assistance in this work.

RUTGERS COLLEGE AND
NEW JERSEY AGRICULTURAL EXPERIMENT STATION
NEW BRUNSWICK, N.J.

STUDIES OF DIOSPYROS KAKI. I

CONTRIBUTIONS FROM THE HULL BOTANICAL LABORATORY 209

KONO YASUI

(WITH PLATES XII AND XIII AND ELEVEN FIGURES)

Although the Ebenaceae are recognized as a primitive family of the Sympetalae and possess several remarkable features, the only morphological investigation of them has been that of Miss HAGUE[1] on *Diospyros virginiana*. My purpose in undertaking the investigation of *D. Kaki* was to answer the following questions: (1) Are the Ebenaceae really primitive among the Sympetalae? (2) Are there cytological differences among the so-called garden varieties of *D. Kaki*? (3) How are the garden varieties related to one another? (4) How have they been obtained?

Material and method

D. Kaki is in very common cultivation in Japan and is represented by many so-called garden varieties. The fruits of these varieties differ in shape, size, and flavor, and also in their seeds. Each variety has also its characteristic flowers, leaves, etc., to which horticulturists paid little attention in classification, as compared with the differences in fruits and seeds. For example, the horticultural classification recognizes two principal divisions, dependent upon flavor of the fruit, as follows: (1) astringent kaki (shibu-kaki), and (2) sweet kaki (ama-gaki).

Each of these two divisions is divided into four or five groups. based upon the shape of the fruit, as follows. The astringent group contains (*a*) long, (*b*) round, (*c*) square, and (*d*) flat types; while the sweet group contains (*a*) gosho, (*b*) long, (*c*) round, (*d*) square, and (*e*) flat types.

In her report on *D. virginiana*, Miss HAGUE makes the following statement: "So far as the trees from which material was collected

[1] HAGUE, S. M., A morphological study of *Diospyros virginiana*. BOT. GAZ. 52:34-44. *pls. 1-3.* 1911.

were observed, they were dioecious, and bore only perfect flowers. One possible exception has been found recently." In *D. Kaki*, on the other hand, there are many such exceptions, as follows.

1. In some varieties only the pistillate flowers occur in good condition, while all the staminate and some of the pistillate flowers are imperfectly developed.

2. In other varieties (gosho, egosho, yamagaki, jenjimaru) both kinds of flowers are always well developed.

3. A tree in Mr. YABE's garden in Tokyo (no. 1 of my material) has the same type of fruit as in some of the varieties in group 1, but has pistillate, staminate, and perfect flowers, and produces also two types of fruit from the two kinds of flowers.

4. No tree has been discovered by the writer on which only staminate flowers occur.

5. Some varieties have pistillate flowers only, when the trees are young, but as they become older or weaker (on account of poor nutrition or transplanting) they begin to produce staminate flowers also (hyakume-gaki).

6. Some trees show an alternation of fruiting and non-fruiting years, which means that during one year they produce many pistillate flowers and many fruits, while during another year they produce a few pistillate flowers and abundant staminate flowers. Under the latter conditions they bear very little fruit, or no fruit at all.

It may be added, in reference to the fourth condition referred to, that in Japan the staminate trees are being continually destroyed unconsciously, because the better varieties are produced always by grafting, and therefore many seedlings of staminate plants are destroyed, and pistillate plants take their place. The foregoing data gave me the suggestion that *D. Kaki* is naturally a monoecious plant (fig. 1), and that it is in process of losing this character, producing staminate flowers under cultivation.

I have selected five different varieties from one hundred or more for this study, as follows.

No. 1.—I am not certain of the name of this variety, but it resembles "yemon," which is an astringent, flat type. The difference is that yemon has no staminate flowers at the Okitsu station, but this form has three kinds of flowers upon the same individual.

No. 2.—Tenryubo, which is a sweet, round type. It has no staminate flowers at the Okitsu station. The fruit is rather small and has many seeds.

FIG. 1.—Branch of Tsuru-no-ko (one of the varieties of *D. Kaki*), showing the vigorous upper branchlets with pistillate flowers and the lower small branchlets with staminate flowers; ×⅓.

No. 3.—Jenjimaru is a sweet round type; but in this case there are always staminate flowers. The fruit is small and with many seeds, as in no. 2.

No. 4.—Tanenashi (meaning seedless) is an astringent square type. It has no staminate flowers and is generally seedless.

No. 5.—Fuyū is a sweet, gosho type. Usually it has no staminate flowers, but occasionally a very few may be produced. Occasionally the anthers of the perfect flowers contain a number of pollen grains, which germinate well on the stigma and in nutrient fluids.

I have also examined many other specimens, but the main results of the investigation were obtained from these five. The material of no. 1 was collected in Tokyo, but the others were obtained from the experiment station at Okitsu, in Central Japan. The material was fixed with Flemming's weaker solution, and chromo-acetic acid mixture was also used, giving better results than the former medium. The sections were stained with Flemming's triple stain and Haidenhain's iron-hematoxylin.

Development of staminate flower

The cluster of staminate flowers appears on young shoots in early spring. It consists usually of three flowers (fig. 1), but sometimes includes 7 or more. The very young shoot bends downward, and 2-4 of the lower leaves have nectaries on the under surface, near the base. The succession of members is strictly centripetal. The calyx is valvate, but the four petals are spirally arranged (fig. 3). The 16 stamens occur in four cycles (fig. 3), and appear in pairs from the earliest stage (fig. 2). The mature stamen has a very short filament, and the connective extends above the anther. The pistils complete their development very early. A few flowers were found which were 5-merous instead of 4-merous.

Development of perfect flower

The perfect flower always occurs at the top of the cluster. It has the same arrangement of members as the staminate flower when young (fig. 4), and the pistils of the perfect flower are smaller than those of the pistillate flowers. Occasionally they have less than 8 ovules, which is the ordinary number in the pistillate flowers, and sometimes they have no ovules. Abnormalities occur, not only in the number of ovules, but sometimes in the irregular forms of the pistils themselves. The fruit of the perfect flower is always smaller than that of the pistillate flower, and usually it has no seeds.

The foregoing facts suggested that the perfect flower is an abnormal one, produced by favorable conditions of nutrition; and that therefore it does not indicate a general tendency.

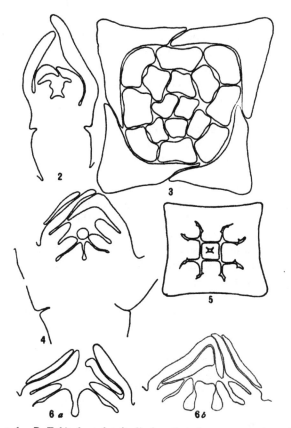

FIGS. 2–6.—*D. Kaki:* fig. 2, longitudinal section of a young staminate flower of no. 1; fig. 3, transverse section of a young staminate flower of no. 1, with calyx removed; fig. 4, longitudinal section of a perfect flower of no. 1, with calyx removed; fig. 5, transverse section of the lower part of a pistillate flower of no. 4, showing the 8 anthers and a pistil; fig. 6, flower of no. 4: *a*, radial section of a young pistillate flower through the two large opposite protrusions, *b*, longitudinal section through the two initial papillae on the large protrusion.

Microspore formation

There is no marked difference between the primary sporogenous cells and the primary wall cells before the latter begin to divide. Two periclinal divisions give rise to three outer wall layers and a tapetal layer. Two or three successive divisions separate the primary sporogenous cells from the mother cells. At the same time the uninucleate tapetal cells become very distinct between the sporogenous tissue and the wall layers, the latter of which become more and more flattened.

During the comparatively long resting period, the spore mother cell increases in size, its protoplasm becomes denser (but not so dense as that of the tapetal cells), and the nucleus becomes large (pl. fig. 1). The nucleus contains rather a small amount of chromatin granules associated with the fine linin network, and also a remarkably large nucleolus and several small ones (pl. figs. 2, 3).

In some cases I observed the larger chromatin granules appearing in pairs, but this did not seem to be the usual situation. The reticulum is denser at the outer part of the nuclear cavity than within. During presynapsis certain connections between the chromatin granules disappear, so that the reticulum gradually becomes simpler in structure. At the maximum of synapsis the chromatin substance appears like a mass of granules (pl. figs. 4 and 5).

There is no morphological connection between the chromatin reticulum and the nucleolus, although they seem to occur in very intimate relationship. A great many nucleolus-like bodies were observed close to the chromatin thread, near the large nucleus, but in other cases there was no such relationship (pl. fig. 4). In the presynapsis stage the double nature of the chromatin thread is not clear, but it becomes gradually evident with the loosening of the thread (pl. figs. 6, 7).

Before 1895, when MOORE called attention to the synapsis stage as an important period in the history of the nucleus, this stage was ignored. Since that time it has attracted chief attention, and is regarded in general as an important event in the history

of the nucleus.   Some investigators, however, do not regard it in
this light.   For example, LAWSON[2] makes the following statement.

> During their development, however, there is a great accumulation of sap
> within the nuclear cavity, which causes a great osmotic pressure.   The pressure,
> acting from within, causes the nuclear cavity to expand.   This expansion con-
> tinues until the nuclear cavity grows to twice or even three times the original
> size.   As the growth proceeds the membrane is gradually withdrawn from the
> chromatin mass within.   The result of this withdrawal of the nuclear membrane
> is the formation of a large clear area of nuclear sap containing the mass of
> chromatin which has been left at one side.   No evidence whatever was found
> to show that any contraction of the chromatin had taken place.

In my material of *D. Kaki* I could not recognize such a remarkable enlargement after the chromatin reticulum began to separate
from the nuclear membrane.   The great enlargement of the nucleus
occurs in the resting stage, so that there is no conspicuous difference in the size of the nucleus at the synapsis stage and at the end
of the resting stage, as LAWSON indicates.

The synapsis stage is a normal occurrence, and during it there
is some rearrangement, but not fusion of chromatin bodies, which
results in the spireme stage.   At the beginning of the spireme
stage the thread is distributed in the outer cavity of the nucleus,
where thickening and shortening occur, resulting in the diakinesis
stage (pl. figs. 6, 7, 8).   There is no difference in the size of the
pairing chromosomes, which are usually parallel, but sometimes
X or V-shaped.   At the same time there appear in the cytoplasm
numerous fibers which penetrate the nuclear cavity and become
attached to the chromosomes by one end, in connection with
which the nuclear membrane begins to disappear.   By the elongation of the fibers the chromosomes, which have been distributed on
the surface of the nucleus, are drawn into the center of the nuclear
cavity, and there are arranged in an equatorial plate by the gather-
ing of the free ends of the fibers at the poles (pl. figs. 9, 10, 11).
The nucleolus disappears entirely during metaphase.

The number of chromosomes is difficult to count by reason of
their small size.   Usually there are 28 in pairs, but sometimes

[2] LAWSON, A. A., The phase of the nucleus known as synapsis. Trans. Roy. Soc.
Edinburgh 47:591–604. *pls. 2.* 1911.

27 in no. 1 (pl. fig. 11).   In *D. virginiana* Miss HAGUE counted 30 or more.

The longitudinal division of the chromosomes in preparation for the homotypic division occurs in metaphase, and at telophase the chromosomes become arranged in the spireme and spread over the outer cavity of the daughter nucleus.   The number of chromatin granules is not the same as the x number of chromosomes; and there appear two or three nucleoli in the daughter nucleus (pl. fig. 12, *a*, *b*).   The homotypic division of the microspore mother cell occurs in the usual way.   The two axes of the spindle fibers of this division are not in any definite relation to one another (pl. figs. 13, 14).   In wall formation the microspores are in tetrahedral arrangement (pl. figs. 15, 16), and the mother cell wall, which becomes mucilaginous and separates the four young microspores (pl. fig. 17), disappears gradually with the formation of the thick wall of the pollen grain.

At the same time the uninucleate tapetal cells enlarge and become multinucleate.   These divisions of the tapetal nuclei are mitotic at first, but later they become amitotic, after which they begin to disorganize when the homotypic division of the mother cell has been completed (pl. figs. 20, 21, 22).

The mature pollen grain is spherical, with remarkable canals upon the surface of the wall (pl. fig. 18).   The cytoplasm occurs as a thin layer surrounding a large central vacuole.   In general only a single nucleus was observed in the pollen grain, pl. fig. 19 representing the only specimen in which two nuclei were observed.

The foregoing account of microspore formation was obtained from material of no. 1.   I have investigated no. 3 also, but could not determine the number of chromosomes.   The pollen grain is a little larger than in the case of no. 1.   The anther of no. 3 contained many withered pollen grains.

Development of pistillate flower

The pistillate inflorescence usually appears upon an upper branch, consisting of one flower and two bracts that suggest the position of flowers that have disappeared.   The bracts usually fall after blooming, but in some varieties they persist even after the ripening of the fruit.

The staminate flower begins to develop as early as the pistillate, but the pistils develop more rapidly than the stamens, resulting in the sterilization of stamens and in some cases in their disappearance.

The mature pistillate flower generally has two sets of stamens (nos. 1 and 2), but in some cases has more (nos. 1, 2, and some of 4) or less (no. 1 and some of no. 2).

I did not discover the following condition described by Miss HAGUE: "In the pistillate flower it is a common occurrence to find the number increased by the branching of one or more of the stamens."

In the development of the pistil there appear first four large protrusions, and then four smaller ones between them (figs. 5 and 6, *a, b*). The larger ones give rise to two small papillae, which are the initials of ovules (figs. 6, *d*, and 7). The papillae grow at first toward the wall of the ovary, then curve downward toward the center, and finally upward (figs. 8 and 9). Simultaneously the small protrusions develop toward the center of the ovary, resulting, along with the development of larger protrusions, in 8 loculi for the 8 ovules.

The ovule has two integuments (fig. 9), which extend beyond the nucellus like a beak at the time the embryo sac is fully developed (fig. 10).

Megaspore formation

In all the varieties examined the archesporial cell is solitary and hypodermal. There is no division resulting in a parietal cell, so that the archesporial cell is the megaspore mother cell. The heterotypic division of this nucleus occurs in the same way as in the case of the nucleus of the microspore mother cell. After nuclear division, the cytoplasm organizes into two parts, the inner one being the larger. Sometimes a wall between these two parts does not appear before another division of nuclei occurs.

Wall formation in connection with the megaspores is variable in direction. Sometimes the two walls are perpendicular to one

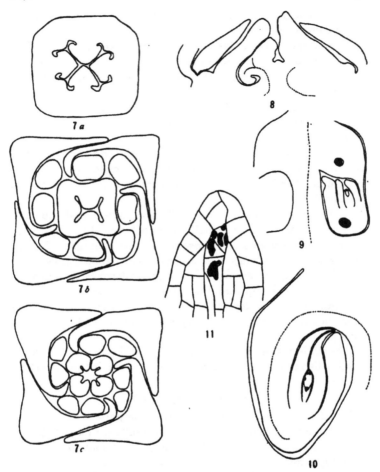

FIGS. 7–11.—*D. Kaki:* fig. 7, cross-section of a pistillate flower of no. 4, at a little older stage than in figs. 5 and 6: *a*, lower part, showing 4 large protrusions which have papillae, and small protrusions among them; *b*, portion of the upper part, showing 4 petals, 8 anthers, and upper part of ovary; *c*, a still higher part, showing 4 petals, 8 anthers, and 4 young stigmas; fig. 8, longitudinal section of a young pistil of no. 4; fig. 9, longitudinal section of the central part of an ovary of no. 4, showing the longitudinal section of an ovule, fig. 10, longitudinal section of an ovule of no. 4; fig 11, longitudinal section of the upper part of a nucellus of no. 4, showing the disorganizing embryo sac mother cell.

another, but at other times more or less parallel, resulting in different arrangements of megaspores (pl. figs. 23–26). The micropylar megaspores degenerate in connection with the maturation of the large innermost one in forming the embryo sac (pl. fig. 27). In my material I observed all of the megaspores disorganizing (fig. 11), and it is evident that this disorganization of the megaspore which ordinarily gives rise to the embryo sac is one reason for seedless fruit. In such a country as Japan, where many different kinds of persimmons are cultivated together, the difficulty of pollination is out of the question.

Summary

1. *Diospyros Kaki* is not a dioecious plant, but a monoecious one whose staminate flowers are disappearing under cultivation.

2. The monoecious habit might have been derived from a condition of perfect flowers; therefore this habit is not a primitive character in this species.

3. Perfect flowers do not indicate the primitive character of the variety in which they occur; they appear among other varieties only through restoration of lost parts.

4. The primitive character of Ebenaceae among Sympetalae is indicated by the spiral arrangement of petals; the stamen situation, although the number of stamens is not definite; and the two integuments.

5. Megaspore formation is also of a primitive character, and suggests, along with other characters of the family, that it may have some relation to the Myrtiflorae.

6. There is no parietal tissue in the microsporangium, which indicates that Ebenaceae come from some higher family of Archichlamydeae, because in the lower families parietal tissue usually occurs in the megasporangium.

7. Embryo sac formation occurs in the usual way, and in general furnishes no evidence for the evolution of dicotyledons. It is true in general, of course, that the gametophytes of angiosperms are of less value for evidence concerning evolution than the sporophytes.

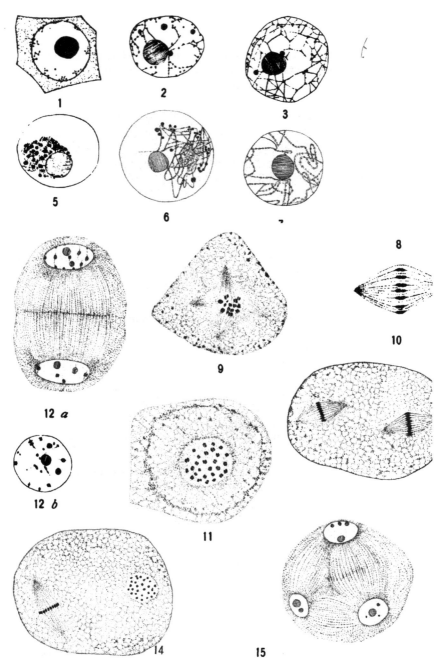

YASUI on DIOSPYROS

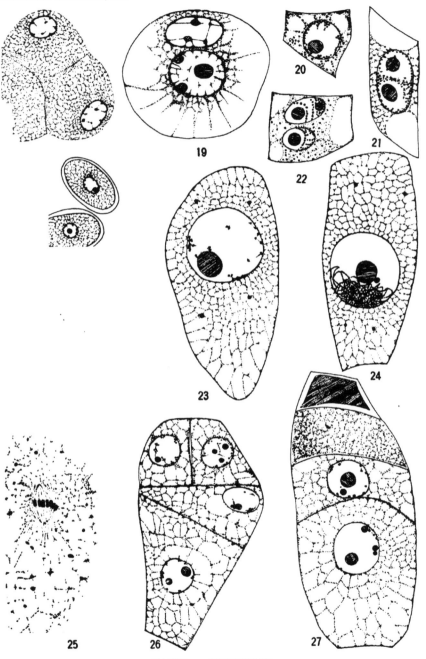

YASUI on DIOSPYROS

8. The $2x$ number of chromosomes is 56 or 54.

I desire to express my thanks to Professor JOHN M. COULTER, under whose direction the investigation was undertaken; to Professor C. J. CHAMBERLAIN, who gave much advice; and to Mr. Y. KUMAGAI and Professor Y. YABE, who assisted in securing material.

UNIVERSITY OF CHICAGO

NEW AND NOTEWORTHY PLANTS FROM
SOUTHWESTERN COLORADO

During the summer of 1913 and the spring and summer of 1914 it was the writer's privilege to make a collection of the plants of western Montrose County, from the Uncompahgre Plateau on the east to the La Sal Mountains of eastern Utah on the west. Of all parts of the state, the southwestern portion is the least known botanically and may therefore be expected to furnish novelties to the systematist for many years to come. This is especially true of the spring flora of the arid or so-called "Upper Sonoran" life-zone. The reason for this is obvious when one considers that most of the collections have been made upon the summer flora, and that by the middle of June most of the vegetation of the arid district has withered and even the seeds have scattered. Though the collections of 1913 and 1914 were made upon the montane as well as the desert flora, most of the interesting forms were discovered at the lower altitudes. The undescribed species, those that had not previously been reported from the state or had been omitted from recent manuals of the district, as well as some perplexing forms, are treated in this paper. A complete set of these plants, containing the types of the species and varieties characterized, is deposited in the Rocky Mountain Herbarium.

CALOCHORTUS FLEXUOSUS Wats.—This lovely and interesting mariposa lily, which has heretofore been known only from farther south and west and has not generally been credited to Colorado, was collected in bloom near Naturita on May 12 and May 26, on a red clay hillside over which it was growing in great profusion. The stems were stiff and flexuous and showed no inclination to become prostrate or twining. The flowers have been described as "purple" or "purplish," but as a matter of fact they are at first white and later become a very delicate pink; it is only in withering that they become "purplish." Nos. 289 and 357.

SALIX.—Nos. 288 and 289, collected in flower near Naturita on May 9, 1915. The stamineal filaments are united nearly to the summit in these specimens. If it were not for this character, I would refer this willow to *S. cordata* Muhl., but because of knowledge of material from one plant only and the lack of mature leaves, I prefer to leave it undetermined for the present.

ERIOGONUM MICROTHECUM Nutt.—Collected near Naturita, September·26, 1914, on a dry rocky hillside; no.·613.

POLYGONUM AMPHIBIUM L.—A form of this species with long fringed sheaths is common in ponds on the Uncompahgre Divide in Tabeguache Basin; no. 173.

Aquilegia pallens, n. sp.—Perennial: stems slender, erect, glandular, pubescent, and sparsely villous throughout; radical and lower cauline leaves biternate; petioles slender, only somewhat glandular, moderately villous, much exceeding the rest of the leaf; leaflets rather thick, broadly cuneate to suborbicular, base often quite truncate, margins inclined to be revolute, under surface slightly glaucous, more viscid and hairy than upper one; upper cauline leaves once or twice ternate, only the few small floral bracts entire: flowers about 3 cm. in diameter and 5 cm. long, mostly erect; sepals 15 mm. long, white or pale blue, lanceolate with narrow claw; petals white, blades about 8 mm. long and 6 mm. broad, truncate, spurs straight, about 3 cm. long; stamens exceeding the petals and exceeded by the mature styles; ovaries pubescent: mature fruit unknown.

Collected only in the canyon of La Sal Creek, Utah, within a few miles of the Colorado boundary, at an elevation between 6500 and 7000 ft. The plants were growing in a moist area at the foot of the sandstone walls of the canyon. The slender stems, the thick leaves, the viscid and villous herbage to which grains of sand were clinging, and the small white or pale blue flowers gave this plant of the Upper Sonoran life zone quite a different aspect from *A. coerulea* of alpine and subalpine stations to which it is evidently related. No. 443, June 16, 1914.

Cleomella montrosae, n. sp.—Glabrous annual, 8–15 cm. high, from a slender taproot: stem diffusely branched from base: leaves trifoliate; petiole 2–3 mm. long; leaflets linear-oblong, 10–14 mm. long, thickish, distinctly petiolulate, more or less folded and with prominent midrib; stipules present, of two or three filiform setae: flowers small, yellow, crowded in corymbiform racemes terminating the branchlets; pedicels 2–3 mm. long, subtended by simple, linear, foliar bracts; calyx segments greenish yellow, 1 mm. long, lanceolate, weakly bristle-tipped; petals elliptical, thickish, three times as long as sepals; stamens exserted; ovary slightly obovate, with two ovules in each cell: capsule ovate, 3 mm. long, 2–3 times as long as the slender stipe and 4–5 times as long as the persistent style; seeds 1 or 2, dark brown, broadly obovate, 2 mm. long.

Probably most nearly related to *C. oocarpa* Gray, from which it is easily distinguished by the possession of a very short stipe. Collected on dry, loose, gypsiferous hillsides near Montrose, alt. 5800 ft., in fruit, October 1912. Flowers produced in May and June. No. 222.

Lupinus crassus, n. sp.—Perennial, caudex branched, forming a dense mat from which arise many stems: stems rather stout, erect, unbranched, sparsely sericeous, 3–4 dm. high: leaves five or six on a stem, green; petioles silky-pubescent, those of lower leaves 8–10 cm. long, of upper ones generally longer than the leaflets; leaflets 5–10 (usually 7 or 8), thickish, somewhat folded, cuneately obovate to oblanceolate, obtuse, silky-pubescent below, upper surface glabrate, 3–4 cm. long, 10–12 mm. wide: inflorescence erect, many-flowered; pedicels 4–5 mm. long, pubescent with short, spreading, white hairs; bracts purplish, lanceolate, 10–12 mm. long, exceeding the buds at the apex of the inflorescence and falling by the time the flowers are fully expanded: calyx not spurred, densely pubescent, lower lip entire, 7–8 mm. long, slightly exceeding the bifid upper lip; corolla white, fading brownish, 12 mm. long, distance from apices of banner and wings about 6 mm., keel included, triangular, acute, scarcely falcate, ciliate except at apex; ovary pubescent, ovules 7 or 8.

This splendid lupine was collected near Naturita, April 21, 1914. At this time it was not in full bloom. Its habit of growth is peculiar in that it forms a dense mat rather than a clump. It seems to prefer soils that are somewhat gypsiferous in character and is often found growing on a loose hillside. No. 239.

Lupinus fulvomaculatus, n. sp.—Perennial: stems nearly glabrous, several to many from a single root, branching upward and

forming a rounded clump 5–7 dm. high: leaves numerous, petioles
of stem leaves about equaling the longest leaflets; leaflets 6–8,
rather unequal, oblanceolate, not folded, green and glabrate above,
sparsely pubescent and somewhat glaucous beneath, 4–6 cm. long,
8–10 mm. broad, obtuse: racemes numerous, terminating the
branchlets, 2–3 dm. long; bracts deciduous; pedicels about 2 mm.
long, softly pubescent as is also the calyx: flowers very small (7–8
mm. long and about 6 mm. deep), crowded and rarely somewhat
verticillate, light blue with a conspicuous rich brown spot on banner;
calyx neither spurred nor gibbous, strongly bilabiate, with two
linear-lanceolate bractlets attached near sinus; upper lip 3 mm.
long, 2-cleft, lower lip 4 mm. long, entire; banner nearly orbicular
when spread out, 7 mm. in diameter, shorter than wings, distance
between apices of banner and wings about 3 mm.; wings broadly
oblong, exceeding the keel by 2 or 3 mm.; keel strongly curved,
margins ciliate from middle to near the comparatively blunt, blue-
purple apex: pods densely villous, 2–3 cm. long and 8–9 mm.
wide; seeds 4 or 5.

A species with distinctive characteristics and peculiar aspect due to the
numerous long racemes of very small, "compact," light blue flowers with the
conspicuous brown areas on the banners. Collected only at Tabeguache
Basin on the Uncompahgre Divide, alt. 8000 ft., where it is rather plentiful
in open aspen groves or on brushy hillsides. Type no. 547, July 29, 1914.

ONOBRYCHIS SATIVA L.—Escaped from cultivation on upper
La Sal Creek, Utah; no. 437, June 16, 1914.

ASTRAGULUS COLTONI Jones.—Common in Long Park, near
Naturata among the junipers and piñons; no. 336, May 22, 1914.

Astragalus naturitensis, n. sp.—Perennial: gray with appressed
pubescence throughout: stipules large, scarious, ovate to broadly
lanceolate, free; leaves 5–7 cm. long; petioles 2–3 cm. long;
leaflets 9–11, narrowly elliptical, more densely pubescent on lower
side, 6–7 mm. long: inflorescence racemose-capitate in flower,
elongating somewhat in fruit; peduncles equaling or exceeding the
leaves, spreading: flowers few (6–10); calyx tubular, about 7 mm.
long, appressed pubescent and sparsely nigrescent, teeth short
(1–2 mm.), triangular-subulate; corolla 12–15 mm. long, conspicuously bicolored, standard white, 2 mm. longer than keel,

apical portion of lateral petals and blunt keel red: pods sessile, mottled with red or purplish blotches, horizontal or ascending, 2 cm. long, 6 mm. wide, acute, straight, linear-elliptic, flattened dorsally and on account of the intrusion of the dorsal suture very broadly cordate in cross-section.

This species is in general aspect so remarkably close to *A. desperatus* Jones that at first glance one would suppose them to be identical. Closer examination of the floral and fruit characters, however, show them to be quite distinct. In *A. desperatus* the standard, as well as the wings and keel, is red, while in the new species the flowers are conspicuously bicolored. JONES has described the pod of *A. desperatus* as being "long hairy"; the pods of the new species are short pubescent. No mention is made of any markings on the pod of *A. desperatus;* the pods of the new species are very conspicuously mottled.

Collected on dry, rocky mesas near Naturita; alt. 5400 ft.; May 27, 1914; no. 360.

Astragalus amplexus, n. sp.—Biennial or short-lived perennial: stems 3–4 dm. high, several from vertical root, sparsely pubescent, erect or spreading, more or less flexuous and often serpentine: leaves comparatively few, 8–12 cm. long; leaflets 17–21, oblanceolate, truncate or retuse at apex, those of lower leaves opposite, of upper often alternate, 9–12 mm. long, 4–6 mm. broad, green, glabrate above, sparsely pubescent below; stipules 3–4 mm. long, united only at base: racemes axillary to and not surpassing the leaves, 7–10 cm. long: flowers 12–15, horizontal, pale violet when fresh; calyx tube cylindric-campanulate, 6 mm. long, nigrescent; lobes linear, 2–3 mm. long; banner scarcely reflexed and inclosing the keel and wings; wings about 14 mm. long with slender claw half as long as the narrow blade; keel about 2 mm. shorter than wings, blade about 2 mm. wide, very blunt: pods sessile, sparsely pubescent, completely 2-celled, coriaceous, 2–2.5 cm. long and 6 mm. in diameter, neither suture intruded and hence pod nearly circular in cross-section, arcuate dorsally, mottled on dorsal surface with irregular purplish splotches.

This species evidently belongs to the MICRANTHI section of WATSON'S revision of the genus in KING'S *Report,* and to the genus *Astragalus* as restricted by RYDBERG in his segregation of the genus in Bull. Torr. Bot. Club **32**:657. 1905. Collected on dry hillsides in Long Park near Naturita, May 22, 1914; alt. about 6000 ft. (apparently infrequent); no. 335.

Psoralea aromatica n. sp.—Perennial herb, 1.5–2 dm. high: stems decumbent and sparingly branched, 2–3 dm. long, sparsely pubescent with short, rigid, appressed hairs: leaves green, rather numerous, on petioles 3–5 cm. long; leaflets usually five, petiolulate, cuneate-obovate, obtuse, mucronate, 2 cm. long, 1 cm. broad, lepidote, sparingly appressed pubescent; stipules lanceolate and scarious, persistent, 6–8 mm. long: flowers borne in axillary racemes 1.5–2 cm. long, light blue, in part at least; bracts linear-lanceolate, scarious, much shorter than the calyx; calyx somewhat accrescent in fruit, bilabiate, sparingly pubescent except on the margins where it is densely pubescent with short rigid hairs, lower lip lanceolate, nearly equaling the corolla, upper lip 2-cleft, almost one-half as long as the lower division, lateral lobes 2 mm. long; corolla 1 cm. long, blade of standard broadly elliptical; lateral petals narrow, scarcely equaling the standard; keel much shorter (7 mm. long): pods 1.5 cm. long, bearing a broad, slightly curved acute beak 8 mm. long; seed smooth, compressed, elliptical, 4–5 mm. long.

This plant is perhaps most closely related to *P. californica* Wats., from which it is easily distinguished by the glabrate calyx with its very unequal lobes. The name is given because of the delicate but pleasant odor of the dried plants. Collected on a dry, clay hillside near Bedrock, Paradox Valley, Colorado; alt. about 5000 ft.; no. 451, June 17, 1914.

EUPHORBIA FENDLERI **dissimilis**, n. var.—A well marked variety, differing from typical *E. Fendleri* T. & G. principally in the larger ovate to lanceolate leaves (7–10 mm. long) which are often acute, and in the usually entire appendages which are 1–3 times as broad as the gland.

Common on dry rocky hillsides, throughout the western half of Montrose County; no. 119, June 27, 1913, and no. 493, July 13, 1914, at Naturita, Colorado.

RHUS UTAHENSIS L. N. Gooding.—Collected on a rocky hillside near Bedrock, Paradox Valley; alt. about 5200 ft.; June 17, 1914; no. 457.

OROGENIA LINEARIFOLIA **lata**, n. var.—Leaves elliptic and quite entire or irregularly lobed and cleft at apex and then truncate, 4–6 cm. long, 1–3 cm. broad.

Growing with the species on dry flats in Tabeguache Basin; alt. 8000 ft. Intermediate forms were not seen. No. 369, June 1, 1914.

AULOSPERMUM BETHELI Osterhout.—Collected at Naturita on dry hillsides, April 27, 1914, where it is rather common. Previously known only from type locality near De Beque, Mesa County, Colorado. No. 225.

NAVARRETTIA BREWERI (Gray) Greene.—Typical plants of this species were collected in Tabeguache Basin, June 23, 1914; no. 459.

OREOCARYA TENUIS Eastwood.—Collected in flower in Long Park, near Naturita, on May 22, 1914; alt. about 6500 ft.; no. 337.

Oreocarya gypsophila, n. sp.—Densely cespitose perennial: caudex much branched, woody, and clothed with the petioles of former leaves: radical leaves crowded, narrowly spatulate, usually folded, 2–3 cm. long, obtuse, canescent with short, stiff, appressed hairs which on the dorsal surface are often pustulate at base; blade 3–4 mm. broad; petioles ciliate with coarse white bristles; stem leaves reduced upward, uppermost broadly linear, passing into inconspicuous, lanceolate bracts: stems slender, 6–10 cm. high, moderately hirsute: inflorescence short-hirsute, not becoming fulvous, thyrsoid, the relatively few flowers somewhat capitate; pedicels short (1–2 mm. long): calyx consisting of five narrowly lanceolate, bristly divisions, 6 mm. long; corolla white, salverform, tube twice as long as the calyx lobes, limb 12 mm. in diameter, consisting of five subelliptical lobes, the divisions of which do not reach to the throat; crests present but not conspicuous; anthers less than 2 mm. long, filaments very short, attached about midway on the corolla tube: nutlets broadly ovoid, white, sharply rugose transversely, all of the nutlets developing.

This species is probably most closely related to *O. cristata* Eastwood. Collected on a dry gypsum hill in Paradox Valley, Colorado; alt. slightly over 5000 ft.; no. 458, June 18, 1914.

Pentstemon cyanocaulis, n. sp.—Glabrous and more or less glaucous perennial: stems usually one, 3–6 dm. high, erect and branched, often becoming blue or purple in the inflorescence: basal leaves spatulate, 4–7 cm. long; cauline leaves oblong-spatulate, sessile, passing gradually into the much reduced lanceolate bracts: flowers in a somewhat secund, interrupted thrysus, 1–2 dm. long; sepals lanceolate, acute or acuminate, 4–5 cm. long, with a rather broad membranous margin; corolla usually horizontal, blue, scarcely ventricose-gibbous, obscurely bilabiate, 17–20 mm. long, lobes subequal; sterile filament sparsely hirsute for over half its length, slightly dilated; anthers bearing short, stiff hairs, cells not confluent, dehiscent from base to apex.

The closest relative of this species is apparently *P. strictus* Benth., from which it is easily distinguished by the shorter, broader leaves and the unusually small flowers for this group. Collected on high, dry mesas near Naturita, May 25, 1914, where it is rather infrequent; alt. about 5800 ft.; no. 348.

PEDICULARIS CENTRANTHERA A. Gray.—Rather common among junipers and piñons in western Montrose County at altitudes from 5800 to 7000 ft. Collected near Naturita, May 25, 1914; no. 345.

CHRYSOTHAMNUS FORMOSUS Greene.—This almost unknown species has apparently been collected but once before, by GREENE near Grand Junction, Colorado, in August 1899, and has been omitted from recent works on the flora of this region. When GREENE described it he did not know the floral characters and therefore I append a brief description of the species.

Low, branched from a woody base, forming dense mats 3–4 dm. high, branches permanently white tomentose: leaves narrowly filiform, somewhat revolute, tomentose but becoming greenish with age, 3–4·5 cm. long: inflorescence cymose, heads rather large and showy; involucres nearly or quite glabrous, their bracts imbricated in distinct vertical rows, 4–5 bracts in each row, ovate to oblong-linear, subacute: corollas usually five in a head, 1 cm. long, the pubescent tube gradually enlarging to form the somewhat inflated throat; tube and throat subequal, lobes short, less than 1 mm. long, scarcely acute; anthers equaling the corolla; style branches exserted, appendages subulate-filiform, three times as long as the stigmatic portion: pappus somewhat deciduous, dull white; achenes pubescent.

Collected near Naturita, alt. 5400 ft., on red clay hillsides that were more or less alkaline in character; September 8, 1914; no. 605.

Helianthella scabra, n. sp.—Stems several, 6–8 dm. high, from a woody, perennial root, somewhat cymosely branched above and bearing many rather small heads: leaves dark green, coriaceous, prominently nerved, linear-lanceolate at base to nearly linear in the inflorescence, hispidulous-scabrous (occasionally glabrate above), acute or obtuse; radical leaves 15–25 cm. long, attenuate into a petiole two-fifths the length of the blade; cauline 5–10 cm. long, short-petioled to sessile, first pair opposite: peduncles whitened with minute pubescence; involucral bracts more or less pubescent, in about three series, mostly lanceolate and shortly acute; ray flowers 8–10, neutral ray flowers bearing orange-yellow ligules over 1 cm. long, oblong, entire or cleft at apex, somewhat recurved; disk flowers purple with pubescent teeth; chaffy bracts of involucre nearly equaling disk, truncate and ciliate pubescent at apex: pappus of several short. lacerate squamellae and two awnlike elongated ones; achenes flattened but scarcely winged, from densely long pubescent in the marginal flowers to sparsely pubescent in central ones.

This species seems to be most closely related to the little known *H. microcephala* Gray; from it, however, it is apparently easily distinguished by the very much larger size, the relatively longer ligules, and the different character of the pappus of the new species. Collected at Naturita, August 11, 1914, where it grows in gulches and ravines; no. 591.

CHAENACTIS SCAPOSA Eastwood (*Chamaechaenactis scaposa* [Eastw.] Rydb.).—Nos. 267 and 305 of my collection may best be referred to this species, at least until more is known of the variability of *C. scaposa*. My plants differ principally from the type as described by Miss EASTWOOD in having leaves that are pubescent but not at all hirsute on the upper surface; pappus that is *quite as long* as the achene; and an achene that is appressed long-pubescent rather than villous. The original description does not mention the existence of rows of many small, superficial dark dots on the achene such as the Naturita specimens possess. Collected on gypsiferous hillsides, April 30 and May 15, 1914. It is interesting to note that the flowers of this plant have a peculiar, "heavy" fragrance that reminds one somewhat of *Abronia fragrans*.

UNIVERSITY OF WYOMING
LARAMIE, WYOMING

MORPHOLOGY OF THE LEMNA FROND

Frederick H. Blodgett

(with plate xiv and one figure)

In the case of structures simplified by reduction, it is sometimes necessary to trace the development of the parts through their immature stages in order to understand the morphology of the mature individuals. This has especial interest when the normal method of reproduction depends upon the development in rapid succession of vegetative propagules, as in the Lemnaceae. The character of such greatly reduced bodies has been variously interpreted, being regarded as stem, leaf, or both fused together, by different writers; while some use a term like "disklike bodies." Since GRAY[1] used *Lemna* as the type of "frondose stem or frond," this name will be used for convenience through the paper. In but few of the articles mentioning the character of the frond has any attempt been made to indicate the basis for the choice of terms, even HEGELMAIER[2] depending largely upon academic argument rather than upon structural details in his discussion of the frond. CALDWELL[3] devoted most of his attention to the development of the flower parts, and did not attempt more than a review of the general character of the frond as presented by others. In the following notes an effort is made to show what structural units contribute to the formation of the frond and the part taken by each in the development of the successive vegetative individuals.

Inception of frond rudiment

The *Lemna* frond is characterized by a sheath or pouch on each side of the base. Within each of these pouches a frond of the next vegetative generation arises, but the two do not develop at the same time. When flowers are produced each one grows in the

---

[1] GRAY, ASA, Structural botany. New York. 1879 (p. 67).

[2] HEGELMAIER, F., Die Lemnaceen. Leipzig. 1868.

[3] CALDWELL, OTIS W., Life history of *Lemna minor*. BOT. GAZ. 27:37–66. 1899.

position in which the frond ordinarily grows, although a young frond may later develop within the same pouch. Behind the growing frond or flower there is a bud normally present. The rudiments of this bud appear very early in the development of the frond, in the axil of which it is located. Such a bud is shown in fig. 1. The frond subtending this bud was slightly smaller than that shown on the same scale in fig. 10. The development of the bud involves only elongation at first (figs. 3 and 4), but at an early stage there is a thickening of the basal region. The limitations of space soon force the elongating bud to turn from the nearly erect position to one more or less oblique and finally to the horizontal. This is less rapidly accomplished when the bud lies behind a flower, since in this case the developing seed by its bulk spreads the forward part of the pouch, and leaves immediately behind itself a space in which an erect development of the bud may go on for a considerable time; but even in this case the frond becomes horizontal while still minute.

Differentiation of parts

When the bud is about twice as long as the diameter of its base, the cells of the root initial may be distinguished. They are usually the fourth cell layer from the ventral surface, and develop a cleavage along the contact region of the third and fourth series of cells (figs. 5 and 8). The cells of the fourth layer and deeper develop the root cap and the root structures within the cap; those of the second and third layers are crushed by the elongation of the root tissue; but the outer layer elongates with the forward growth of the root and forms the "root sheath" which commonly persists about the base of the mature root.

At nearly the same time when the root develops there is formed upon the dorsal surface of the bud a slight elevation or mound of cells. This is nearly opposite the group of root initials and marks the origin of the bud for the next frond in the series (fig. 3). Very soon a second mound is visible, close to the insertion of the bud upon the parent tissue, just behind the bud of the daughter frond (figs. 7 and 7a). This is the inception of the bud axillary to the frond the development of which is being followed. This bud will

develop but slowly, as compared with the bud anterior to it,
which represents the daughter frond; thus when the latter is as
shown in fig. 10, the axillary bud is only as shown in fig. 2, or in
figs. 6 and 10.

The next stage marks the beginning of the upper wall of the
inclosing pouch by which the bud will be protected until large
enough to project beyond the pouch margin. This first appears
as a sharp projection just anterior to the bud of the daughter frond,
as shown in fig. 6. At first the bud and the protective ridge are
both in the median region of the young frond, and nearly opposite
the root rudiment; but with the continued growth of all parts,
the space available for vertical enlargement becomes insufficient
and there is a lateral shifting of the dorsal structures. The bud
itself becomes divided into two groups of cells, each a potential
growing point. The development of the pouch wall is shifted from
the position of a covering which is directed posteriorly over the
bud, to that of an outgrowth which is directed obliquely backward
from either side of the original position.

A further change follows this shifting of the bud. The over-
growing wall is displaced along with the inclosed bud, and thus
there are formed two pouches at the base of the frond. The pouch
walls, which are first recognized as a slight ridge anterior to the
bud rudiment, extend around this as a flattened ring having its
insertion along the side of the elongating base or stem of the frond
and laterally on the widening frond. At the ends there is but
little projecting growth of this ridge, but from the portion just
above and below the inclosed bud a considerable development
takes place. The pouch walls merge into the body of the frond
completely at their insertion.

A vascular connection remains between the two buds after
they are separated by the upward growth of the supporting tissue,
as is shown in fig. 9. No distinctly vascular elements are developed
at this stage, but the elongated cells here shown are their rudiments.
This figure also indicates the unequal development of the opposite
buds. In *Lemna trisulca* there is less difference in development
than in most species, but the general sequence of stages is
similar.

Later stages of development

The young frond now has present all the structural units which will be visible in the mature free frond, the later developments being in the manner of enlargement of parts present and the completion of structures already initiated. The two buds in their respective pouches will follow the same stages as the frond developing from an axillary bud, such as is being discussed; while the root shows little of special interest. Two changes which take place, however, should be mentioned; the rapid elongation of the region anterior to the bud-root region, and the thrusting forward of the whole structure by the development of the basal region into a stalk or stipe. In *Lemna* this basal region is attached marginally to the main portion of the frond, but in *Spirodela* it is inserted upon the ventral surface some distance from the edge, and in *Wolffia* it is almost completely lacking.

The axillary bud which was at first upon the dorsal surface of the stipe is forced into a lateral position, as was the case with the frond bud which was divided. In this case the lateral shifting seems to be due to the more rapid growth of the stipe to one side of the median line, resulting in a tipping of the bud and raising its base as the stipe thickens. In the frond this stage is accompanied by the development of air cavities in the anterior region which is elongating rapidly. The individual cells become highly vacuolated, and the stomata are differentiated. The vascular strands appear at this time in the anterior region, which in all essentials assumes the structural characteristics of a foliage leaf.

The vascular cells in the stipe become more clearly defined as a single strand centrally located in the tissues of the connecting structure. This strand ends in a confused group of cells at the point of origin of the vascular tissues of the root, the lateral buds, and the anterior region. This region corresponds to the mat of vascular elements in the base of a bulb as discussed by CHRYSLER[4], and may be considered as locating closely the stem apex of the plant. In bulbs the apical region usually gives rise to a single main bud, with some lateral buds axillary to bud scales. In *Lemna* the

[4] CHRYSLER, M. A., Development of the central cylinder in Araceae and Liliaceae. BOT. GAZ. 38:161–184. 1904.

terminal bud is divided by vertical pressure and constraint, and appears as a pair of lateral buds except in the initial stages; but the axillary bud is developed in its normal position, behind the foliar outgrowth.

Homologies of the Lemna frond

The *Lemna* frond is a structure of reduced character, especially adapted to vegetative propagation under particular conditions. The two forces which act upon the plant as possible determining factors in the resulting structure are the tension of the water film and the buoyant effect of the liquid. The film tension tends to attenuate in all directions any elastic structure floating upon it, while the upward thrust of buoyancy tends to lift any deep-lying portion into the same plane with the thinner parts. The thin frond responds to the conditions imposed, and may be considered as a result of the forces involved. The roots act as drags or brakes against too free movement in the water, and to return the dorsal surface to the air after the frond has been immersed by any sudden shock, the drag of the projecting roots tending to orient the frond as it rises to the surface by its buoyancy. Inverted plants of *Spirodela* will thus orient themselves in a 3-inch dish of water, but complete inversion would but rarely occur under natural conditions.

The runner of *Erythronium* is a structure also of reduced nature and also developed to meet particular conditions. In this case the forces involved are the outward thrust of growing tissues and the resistance of the surrounding soil. The root type of structure is manifestly best fitted to this condition, having the region of elongation close to the apex of the structure. An investigation of the development of these runners[5] showed that the zone of elongation is so located, and that the component parts of the runner are adapted to the conditions and factors present, namely, the offsetting of propagules, and their movement through soil. A longitudinal section of a runner is shown in text fig. 1 (I). The bud scale forming the anterior wall of the hollow tip elongates, as do leaves, by basal growth; the axial portion of the runner which thrusts the terminal bud forward elongates by growth near the tip. The two

5 BLODGETT, F. H., Origin and development of bulbs in *Erythronium*. BOT. GAZ. 50:340–372. 1910.

growing regions harmonize completely, and the inclosed bud is
thrust forward through the soil, with no serious injury from antagonistic growth forces.   The pointed cylinder meets the conditions
of progress through a solid medium as does the flat frond for floating
upon a fluid surface.

The *Lemna* frond may be closely compared to the *Erythronium*
runner in structural details.   If the outer scale of the terminal

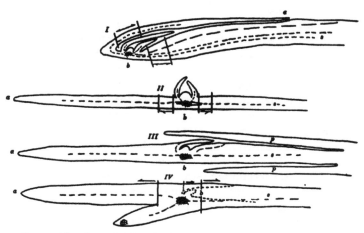

FIG. 1.—Homology of *Lemna* frond and *Erythronium* runner: *a*, tip of first leaf;
b, stem apex of bud; *s*, shoot axis; *pp*, walls of pouch.

bud of the runner could be turned forward on its base as a hinge,
a structure would result of the character shown at II in the text
figure.   This has a terminal foliar region, a central one from which
a dorsal bud and ventral root originate, and a basal stem or cauline
region,[6] but the bud is erect.   The growing regions are still at the
base of the foliar and near the tip of the cauline portions of the
whole.   If this were to be formed within a vertically restricted
space, the dorsal bud would be flattened, and probably somewhat

---

[6] These correspond respectively to the "upper internode," the "node," and the
"basal internode" of CALDWELL (*op. cit.*, p. 43), and supports his use of the word
"shoot"; but the shoot consists of a terminal leaf, a node with bud tissues, and a
stem, rather than of an "undifferentiated upper internode" and specially modified
"basal and nodal regions."

fused to adjacent parts (as in III of the text figure), as is the case on the posterior surface of the bud scale in the runner bud shown at I. The vertical thickening of the tissues at the insertion of the bud would tend to split the meristem group of cells so that a pair of lateral buds would finally result, as in the *Lemna* structure.

The protective inner scale of the bud, which is represented in diagram III as bent backward under the overlying pouch wall, in the *Lemna* frond shown at IV, is fused with the posterior margin of the foliar tissues of the frond, and laterally to the side of the stem. This pouch wall arises as a ring about the base of the bud which it will finally inclose, and is therefore homologous with the development of the scale leaf about the stem tip of a bulb, or a bud of the *Erythronium* type. Under the adjustment to space and pressure, the ring is flattened to a narrow eclipse, and is divided by the same conditions as those which divide the bud within.

In the region immediately below the stem apex in bulbs, and the corresponding point in *Lemna*, the root initials are developed; this region is indicated in the diagrams by cross-hatching, adjacent to *b* in I, II, III. In *Lemna* only one root is regularly developed, but in *Spirodela* several are formed. The activity of the adjacent apical meristem evidently continues for a considerable period, allowing for the successive formation and development of new fronds within the same parent pouch. The old fronds are released by the separation of the cells close to the insertion of the stem upon the parent tissues, apparently by the breaking down of the middle lamella of a band transverse to the stem. The stump left in such a case is seen at *s* in fig. 2. By the enlargement of the axillary bud this stump will be obliterated before the new frond is far advanced.

Conclusions

The *Lemna* frond is a propagative structure consisting of a terminal leaf; a bud inclosed by a flattened bud scale, the base of which is fused to the base of the leaf and laterally to the stem; and an apical region from which new fronds are developed. Two buds are formed through the splitting of a single bud rudiment by vertical pressure during early stages of growth. The frond meets the conditions of a floating habitat in which the tension of the surface film apparently is an active factor. Through the lack of space for vertical succession, the several outgrowths from the apical region are liberated as a horizontal series, the overlapping of successive individuals forming an element of confusion in an examination of their structure.

TEXAS AGRICULTURAL EXPERIMENT STATION
COLLEGE STATION, TEXAS

EXPLANATION OF PLATE XIV

FIGS. 1, 2.—Early stages in formation of axillary buds: magnification, indicated by scale of 0.05 mm., uniform for all figures except fig. 9, which has its special scale indicated.

FIG. 3.—Median section of older bud: *b*, daughter bud rudiment; *a*, axillary bud.

FIG. 4.—Similar section at next stage: buds displaced by thickening of base; 4*a*, details of root from adjacent section.

FIG. 5.—Adjacent sections of bud in axil of nearly grown seed (to left), showing daughter bud and root regions; axillary bud not yet defined; intermediate in age between figs. 2 and 3.

FIGS. 6, 7.—Daughter buds from opposite sides of median section shown in fig. 4; 7*a*, axillary bud the edge of which shows at *a* in fig. 7.

FIG. 8.—Sagittal section of frond, showing origin of inclosing scale leaf or pouch (*p,p*) about daughter bud (*b*); below the root is pushing out, and to the right the leaf is elongating.

FIG. 9.—Transverse section through apical region of frond, showing displaced halves of daughter bud, now unequally developed, and connecting cells; compare figs. 2 and 3; low power.

FIG. 10.—Young frond in sagittal section: leaf forms right half; basal buds partly inclosed by bud scale or pouch (at 10*a* and 10*c*); axillary bud at 10*b*, with edge of buds (*a* and *c*) and of pouch wall (*p*).

PLATE XI

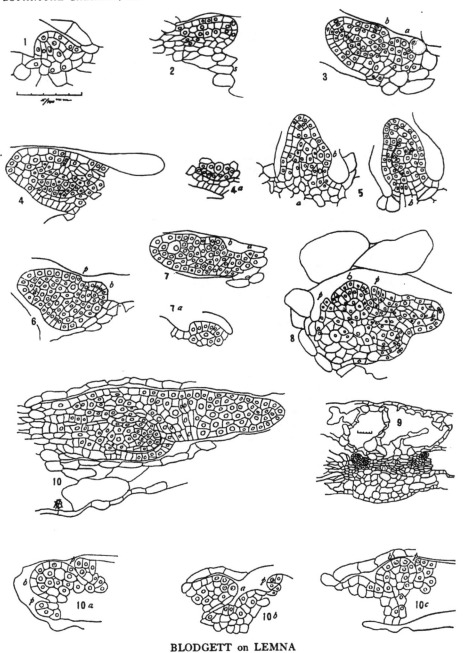

BLODGETT on LEMNA

NOTES ON NORTH AMERICAN WILLOWS. III

CARLETON R. BALL

The first[1] of a projected series of papers under this title was published in this journal in November 1905. Since that date, the writer's contributions on Salices have been confined to treatment of the genus for two floras covering widely different sections of the country, namely, the Rocky Mountains[2] and the Central Atlantic states.[3] In the ten-year period, however, thousands of specimens have been determined for collectors and herbaria, extensive field studies and collections have been made, and a considerable fund of data accumulated, some of which are presented in the third paper[4] of this series and in the following paragraphs.

1. The willows of the Black Hills

The first list of the plants of this district, of which the writer has knowledge, was published by RYDBERG[5] in 1896. He collected at some 20 points in the Black Hills, during the period from May 27 to August 18, 1892, visiting some places twice. Four species of *Salix* are recorded (p. 523): *S. fluviatilis* Nutt., *S. cordata* Muhl., *S. discolor* Muhl., and *S. Bebbiana* Sarg. Of these, *S. fluviatilis* is listed only from Rochford; *S. cordata* only from Custer and Lead; *S. discolor* only from Custer; while *S. Bebbiana* was collected at 5 points. No data on frequency of occurrence are given for the willows, though such notes are given commonly throughout the paper. It is probable, however, that the three species first named were met with at points other than where collections were made.

[1] BALL, CARLETON R., Notes on North American willows. I. BOT. GAZ. 40:376–380. *pls. 12, 13.* 1905.

[2] ———, *Salix* in COULTER and NELSON, New Manual Rocky Mt. Bot. pp. 128–139. 1909.

[3] ———, *Salix* in TIDESTROM, Elysium Marianum 3:19–37. *pls. 4–9.* 1910.

[4] ———, Notes on North American willows. II. BOT. GAZ. 60:45–54. *figs. 3.* 1915.

[5] RYDBERG, P. A., Flora of the Black Hills of South Dakota. Contrib. U.S. Nat. Herb. 3:463–536. 1896.

While the Black Hills have always been recognized as belonging to the Rocky Mountain system, and as having a flora allied
to that of the Rocky Mountain region, it will be noted that the
four willows reported by RYDBERG are all eastern species. They
range from the North Atlantic and the New England states westward across the great interior valley. *S. Bebbiana* extends commonly and *S. fluviatilis* less commonly into the Rocky Mountain
region itself, but primarily they are eastern species. The question
of the identification of the specimens collected is not raised at this
point.

Another list which includes the Black Hills district was published
by WILLIAMS,[6] presumably in 1895. The willows of the state,
10 in number, include the 4 of RYDBERG's Black Hills list and also
3 others credited to the Black Hills. It does not appear from this
publication that herbarium specimens of all the species were obtained. WILLIAMS refers frequently to RYDBERG's collections from
the Black Hills and was doubtless familiar with a manuscript copy
of his list. The 3 additional species credited to the Black Hills
are as follows:

"18. YELLOW WILLOW (*Salix flavescens* Nutt.).—A shrub found in the
Black Hills. It may be recognized by the nearly entire leaves which are
downy, or smooth and dull green above, and pale with a rufous pubescence
beneath" (p. 106). No reference is made to any collections.

· "19. HAIRY WILLOW (*Salix glauca villosa* Anders.).—. . . . Found at
various places in the Black Hills and on the Yellow Bank in Grant County"
(p. 106).

"20· PRAIRIE WILLOW (*Salix humilis* Marsh.).—A small shrub reported
from the Black Hills by Mr. RYDBERG" (p. 107). In no case is herbarium
material cited. Again the question of identification is waived for the time.

A third list covering the territory of the Black Hills in South
Dakota was published by SAUNDERS.[7] The striking thing about
this list is that it does not include two of the three additional willows
credited to the Black Hills by WILLIAMS, namely, *S. flavescens*
and *S. glauca villosa*. It does list *S. humilis*, but as occurring only

[6] WILLIAMS, T. A., Native trees and shrubs. S.D. Exper. Sta. Bull. 43. 1895
(pp. 105–107).

[7] SAUNDERS, D. A., Ferns and flowering plants of South Dakota. S.D. Exper.
Sta. Bull. 64. 1899 (pp. 134, 135).

in the eastern part of the state. The reasons for omitting these three willows can be guessed only. Either specimens were not found in the herbaria examined or they proved to belong to other species than those listed. It is a fair guess that *S. humilis* was included by WILLIAMS on the strength of RYDBERG'S[8] identification of his no. 1019 from Rochford, of which he says:

> A few specimens with thick leaves as in *S. humilis* and *S. tristis* were collected at Rochford, altitude 1700 m., July 12. Even these have been referred to *S. rostrata* by Mr. BEBB (no. 1019).

At any rate, the list of Black Hills willows was again reduced to four species. So far as publications are concerned, the number seems still to rest there, though in 1912 PERISHO and VISHER[9] reported *Salix lutea*, a western species, as being "fairly common along brooks" in Washington County, which adjoins the Black Hills on the southeast.

On September 19, 1908, the writer studied and collected the willows occurring along Whitewood Creek from the railroad station upstream for a mile or more. Five species were found. Two of them, *S. lutea* Nutt. and *S. prinoides* Pursh, were very common in the floodplain of the stream. Of each of the remaining three, *S. Fendleriana* Anderss., *S. Scouleriana* Barr., and *S. Bebbiana* Sarg., only a single plant was found. All three were obtained in the edge of the city. An examination of RYDBERG'S specimens showed that *S. lutea* was the species reported as *S. cordata* by him, while *S. prinoides* was the basis of his record for *S. discolor*. The finding of *S. Scouleriana* bears out the statement quoted from WILLIAMS under *S. flavescens*, which is a synonym. This is the first recorded collection of *S. Fendleriana* in the Black Hills. It occurs commonly in the canyons of the Rocky Mountains and westward at elevations of 5000 feet and over. The number of species was now increased to six.

Collections of *S. amygdaloides*, *S. fluviatilis*, and *S. cordata* were made also at Bellefourche on the same date, but as Belle-

[8] *Op. cit.*, p. 523.

[9] PERISHO, E. C., and VISHER, S. S., The geography, geology, and biology of south central South Dakota. State Geol. and Biol. Survey Bull. 5. p. 90. 1912.

fourche properly is not in the Black Hills, *S. amygdaloides* is not added.

On September 14, 1910, the writer again visited Deadwood for a short time and studied the willows found on the northern flank of the White Rocks, principally around a spring lying just below the wagon road. Here *S. Bebbiana* and *S. prinoides* were found abundantly and *S. lutea* less commonly on the seepy slope below the spring. Two or three stunted plants of *S. Scouleriana* occurred in dry ground just across the road above the spring.

On August 23, 1913, the writer collected *S. lutea* at the railroad station at Mystic, and *S. Bebbiana* and *S. Scouleriana* again at the spring north of the White Rocks at Deadwood. All these collections are cited under the respective species at the end of this article.[10]

In the spring and summer of 1910, Mr. JOHN MURDOCH, JR., of the United States Forest Service, collected an extensive series of specimens in various parts of the Black Hills. A portion of the series was sent to the writer for identification through the United States Forest Service, on April 1, 1913. Seventeen numbers were flowering specimens of *S. prinoides* and *S. Bebbiana*, one was *S. fluviatilis*, and one proved to be a young pistillate specimen of *S. Nelsonii* Ball, a rather rare member of § PHYLICIFOLIAE, known previously from the mountains of Colorado and Wyoming. A more complete set from the private herbarium of the collector was submitted to the writer by Dr. GEO. B. SUDWORTH, dendrologist of the Forest Service, on January 14, 1915. It was found to contain, besides numerous specimens of *S. prinoides* and *S. Bebbiana*, the specimen of *S. Nelsonii* (no. 4039) mentioned above, collected at Redfern Plantation no. 2, and also a specimen bearing mature foliage (no. 4233), collected at the same place in July.

---

[10] Since this paper was prepared, the writer has seen a series of short papers by VISHER, entitled Additions to the flora of the Black Hills of South Dakota (Torreya 9:186–188. 1909; Muhlenbergia 8:135–137. 1913; *ibid.* 9:33–39. 1913). In the first paper he records *S. Scouleriana* as "Frequent, forming trees, in deep woods, well up on Custer's Peak." The collection was made in August 1908, just a month before it was collected by the writer at Deadwood

In one of another series of papers (Additions to the flora of South Dakota. II. Muhlenbergia 9:69–77. 1913) VISHER reports *S. Fendleriana* as "Common in boggy soil in the forest reserves and along the Little Missouri River" of Harding County, north of the Black Hills, in the northwest corner of the state.

There was also one specimen each of *S. amygdaloides* and *S. lutea*. Very interesting also were a specimen of *S. Fendleriana* (no. 4375), with mature foliage, from Rochford; three foliage specimens of *S. Scouleriana*, also from Redfern Plantation no. 2; a foliage specimen of *S. Geyeriana* (no. 4367) from Bear Butte Creek; and finally a single collection of what is probably *S. monticola* Bebb (no. 4374) from South Rapid Creek, at an elevation of 6100 feet. By this collection four species, *S. amygdaloides*, *S. Geyeriana*, *S. monticola*, and *S. Nelsonii*, not previously reported, were added to the flora of the Black Hills. All except the first are strictly plants of the western mountains. The specimen identified as *S. monticola*, which consists of autumn-colored foliage collected September 16, is certainly not referable to either *S. cordata* or *S. lutea*, and with equal certainty is a species not previously recorded for the Black Hills. Two of the sheets of *S. Scouleriana* show only small but mature leaves, suggesting a dwarfed habit, and the data indicate that they come from dry situations.

In the early spring of 1913, some fragmentary specimens collected in 1912 by Messrs. T. C. SETZER and N. E. PETERSEN, of the Merritt Ranger Station in the Black Hills Forest Reserve, were sent the writer for identification. On examination these were found to include such Rocky Mountain species as *S. glaucops* Anderss., *S. chlorophylla* Anderss., *S. Nelsonii* Ball, and *S. Geyeriana* Anderss. Of these, the first two had not been collected previously in the Black Hills. At the request of the writer, Mr. W. H. LAMB, of the Forest Service, kindly made immediate arrangements to have better material collected in 1913. As a result, there were received for identification, on January 21, 1914, another series of these same species. Careful examination confirmed the previous determinations and hence added two more species to the list of those known to be native to the Black Hills. This list now contains 12 species, all native. Three out of the 12, namely, *S. amygdaloides*, *S. fluviatilis*, and *S. Bebbiana*, are found nearly across the continent. One, *S. prinoides*, is an eastern species finding here nearly the western limit of its range. The 8 others are habitants of the Rocky Mountains and westward, which find the eastern limits of their ranges in the Black Hills.

Capsule glabrous
  Scales yellow
    Leaves large, lanceolate, serrate; stamens 3–5
      Leaves glaucous beneath........................1. *S. amygdaloides.*
      Leaves green beneath..........................2. *S. Fendleriana.*
    Leaves linear; stamens 2
      Leaves green beneath, remotely denticulate..........3. *S. fluviatilis.*
  Scales brownish to black
    Aments short-peduncled; twigs yellow; style less than 0.5 mm.
                                                      4. *S. lutea.*
    Aments sessile; twigs reddish; style 1 mm. or over......5. *S. monticola.*
Capsule hairy
  Style long; scales usually black, pilose
    Aments peduncled; leaves gray pubescent above.. ......6. *S. glaucops.*
    Aments sessile; leaves bright green above, glabrous   .
      Leaves broadly elliptical or oval. . ... 7. *S. chlorophylla.*
      Leaves narrowly elliptical........................ 8. *S. Nelsonii.*
  Style short or none
    Scales obovate, black, pilose; aments sessile
      Leaves elliptical, undulate-crenate, glabrous....... . 9. *S. prinoides.*
      Leaves obovate, entire, often pubescent beneath ..10. *S. Scouleriana.*
    Scales linear-oblong, yellow or the tip dark; aments short-pedunculate
      Leaves elliptic-obovate; ,pedicels long... ....11. *S. Bebbiana.*
      Leaves linear-oblanceolate; pedicels short....... ..12. *S. Geyeriana.*

SPECIMENS EXAMINED

The following specimens have been examined by the writer. Their present location in herbaria, so far as known, is indicated by the letters following the collection numbers. The key to these letters is: B, Herb. C. R. BALL; C, Herb. Canada Geol. Survey; F, Herb. U.S. Forest Service; G, Gray Herbarium; I, Herb. Iowa State College; M, Herb. JOHN MURDOCH, Jr.; N, U.S. National Herbarium; R, Rocky Mountain Herbarium.

1. *Salix amygdaloides* Anderss.

BELLEFOURCHE.—Floodplain of Bellefourche River, *Ball* (1341, B, C, N, R) 1908, (1697, B; 1699, B) 1910.

DEADWOOD.—Whitewood Creek below Smelter, alt. 1370 m., *Murdoch* (4322, M) 1910.

2. *S. Fendleriana* Anderss.

DEADWOOD.—Bank of Whitewood Creek, south edge of town, alt. about 1400 m., *Ball* (1356, B, N, R) 1908.
ROCHFORD.—Alt. 1585 m., *Murdoch* (4375, M) 1910.

3. *S. fluviatilis* Nutt.

BELLEFOURCHE.—Floodplain of the Bellefourche River, *Ball* (1340, B, C, N, R; 1344, B, N) 1908; (1696, B, N, R) 1910.
ROCHFORD.—Alt. 1675 m., *Rydberg* (1020, N) July 11, 1892.
SAVOY.—Creek banks, alt. 1525 m., *Murdoch* (4166, F, M) 1910.

4. *S. lutea* Nutt.

BELLEFOURCHE.—Floodplain of Bellefourche River, *Ball* (1694, B, N; 1695, B, C, N, R), foliage, 1910.
CUSTER.—Alt. 1690 m., *Rydberg* (1015, N) June 4, 1892.
DEADWOOD.—Floodplain of Whitewood Creek, alt. about 1400 m., *Ball* (1352, B, C, G, N, R; 1354, B), foliage, 1908; spring below White Rock, alt. about 1460 m., *Ball*, 1686, B, C, I, N, R), foliage, 1910.
FT. MEADE.—*Forwood*, flowers, May 1, 1887.
LEAD.—Alt. 1675 m., *Rydberg* (1017, N), foliage, July 7, 1892.
MYSTIC.—At railroad station, alt. 1460 m., *Ball* (1827, B, N), foliage, 1913.
REDFERN.—Plantation No. 2, alt. 1675 m., *Murdoch* (4038, M), pistillate flowers, May 13, 1910.
STURGIS.—Alt. 1065 m., *Carr* (344, B; 345, B), staminate and pistillate flowers, May 17, 1912.

5. *S. monticola*(?) Bebb

SOUTH RAPID CREEK.—Alt. 1860 m., *Murdoch* (4374, M), foliage, September 16, 1910.

6. *S. glaucops* Anderss.

MERRITT.—Ranger Station, headwaters of Jim Creek, Sec. 8, Tp. 2 N., Range 5 E., alt. about 1525 m., *Setzer* (F) 1912; *Setzer* (2, B), staminate, May 6, 1913.

7. *S. chlorophylla* Anderss.

MERRITT.—Ranger Station, headwaters of Jim Creek, Sec. 8, Tp. 2 N., R. 5 E., alt. 1525 m., *Setzer* (F) 1912; *Setzer* (1, B), staminate, May 6, and foliage, June 2, 1913.

8. *S. Nelsonii* Ball

MERRITT.—Ranger Station, headwaters of Jim Creek, Sec. 8, Tp. 2 N., R. 5 E., alt. 1525 m., *Setzer* (F) 1912; *Setzer* (3, B), pistillate, May 6, and foliage, June 2, 1913.

REDFERN.—Plantation no. 2, alt. 1710 m., *Murdoch* (4039, B, M), pistillate, May 13, 1910; *Murdoch* (4233, M), foliage, July 13, 1910.

9. *S. prinoides* Pursh

BEAR BUTTE CREEK and vicinity.—Mostly 6–8 miles southeast from Deadwood, and about the same distance northeast from Custer Peak; alt. 1525–1650 m., *Murdoch* (4002–4015 inclusive, F, M), staminate and pistillate flowers, April 7–10, 1910; (4366, 4369, both M), foliage, September 8, 1910.

CUSTER.—Alt. 1675 m., *Rydberg* (1014, N), staminate and pistillate flowers, May 31, 1892.

CUSTER PEAK.—About 3 miles E., alt. 1710 m., *Murdoch* (4001, F, M) staminate flowers, April 7, 1910.

DEADWOOD.—Floodplain of Whitewood Creek, alt. about 1400 m., *Ball* (1349, B, N, R; 1353, B, C, G, N, R; 1355, B, N, R), foliage, 1908; at spring north of White Rock, alt. about 1460 m., *Ball* (1681, B, N, R; 1684 and 1685, B, C, I, N, R), foliage, 1910.

DUMONT.—Alluvium, Thompson's Bridge, alt. 1830 m., *Murdoch* (4267, M), foliage, August 4, 1910.

STURGIS.—*Carr* (308, B), flowers, May 17, 1912.

10. *S. Scouleriana* Barratt

DEADWOOD.—Bank of Whitewood Creek, alt. about 1400 m., *Ball* (1350, B, C, G, N, R), foliage, 1908; at spring below White Rock, alt. about 1460 m., *Ball* (1683 and 1687, B, C, I, N, R), foliage, 1910, (1832 and 1833, B, N), foliage, 1913.

REDFERN.—Plantation no. 2, alt. 1740 m., *Murdoch* (4220, 4221, 4237, all M), foliage, July 13, 1910.

11. *S. Bebbiana* Sargent

BEAR BUTTE CREEK.—About 5–8 miles S.E. of Deadwood, alt. 1555 m. *Murdoch* (4010, in part, and 4022a, F, M), flowers, April 10–27, 1910; (4368, M), foliage, September 8, 1910.

CUSTER.—Alt. 1675 m., *Rydberg* (1012, N; 1013, N), May 31 and June 6, 1892.

CUSTER PEAK.—About 3 miles E., alt. 1710 m., *Murdoch* (4365, M), foliage, September 8, 1910.

ELK CANON.—Alt. 1525 m., *Rydberg* (1016, N), June 2, 1892.

DEADWOOD.—Floodplain of Whitewood Creek, *Ball* (1351, B, N), foliage, 1908; (1680 and 1682, both B, C, N, R), foliage, 1910; (1829, B, N; 1830 [pistillate], B, N; 1831 [pistillate], B), foliage, 1913.

HOT SPRINGS.—Alt. 1100 m., *Rydberg* (1018, N), August 9, 1892.

REDFERN.—Plantation no. 2, alt. 1710–1740 m., *Murdoch* (4036 and 4037, F, M), flowers, May 13, 1910; (4219 and 4232, M), foliage, July 13, 1910.

ROCHFORD.—Alt. 1675–1830 m., *Rydberg* (1019, N), foliage, July 12, 1892.

12. *S. Geyeriana* Anderss.

BEAR BUTTE CREEK.—Alluvium, Sec. 19, Tp. 4 N., R. 4 E., alt. 1645 m., *Murdoch* (4367, M), foliage, September 9, 1910.

MERRITT.—Ranger Station, headwaters of Jim Creek, Sec. 8, Tp. 2 N., R. 5 E., alt. 1525 m., *Setzer* (F) 1912; (4, B), staminate flowers, May 6; leaves, June 2, 1913.

U.S. DEPARTMENT OF AGRICULTURE
   BUREAU OF PLANT INDUSTRY
      WASHINGTON, D.C.

ON THE OCCURRENCE AND SIGNIFICANCE OF "BARS" OR "RIMS" OF SANIO IN THE CYCADS

H. B. SIFTON

(WITH PLATE XV)

The "bars" or "rims" of Sanio were probably first figured by GÖPPERT (2) in 1849, as has been pointed out by Miss GERRY (1). SANIO (7), however, did not, as she states, claim to be the discoverer of them, although he was the first to describe them at all adequately, and hence they have borne his name. In studying the development of the wood cells in *Pinus silvestris*, he found thickened rings, complete or incomplete, which grew up around thin spots in the cellulose primary walls. The thin areas he designated as the "Primordialtüpfeln," and on these bordered pits were formed. The thickened rims, or "Umrissen," of these areas are the structures that have until recently been known as the "bars of Sanio." In 1913 GROOM and RUSHTON (3) applied the name "rims" to them. This term is descriptive of SANIO's theory of their method of formation and serves to distinguish them from trabeculae.

Until the last few years they did not come into much prominence, but in several recent papers they have been considered of great importance. Miss GERRY, in the paper mentioned above, gave a comprehensive account of their distribution in the conifers. She found them in all the members of this group except the araucarians, whose wood structure could in this way be distinguished from the other forms. She also suggests the extension of this distinction to the fossil forms, as the following sentence indicates: "The distribution of the bars of Sanio . . . . establishes a constant and useful diagnostic character in the determination of fossil woods" (p. 122). She would associate all the fossil conifers which have the "rims" with the Abietineae, and those which do not exhibit them with the Araucarineae, regardless of other characteristics of their tracheary structure.

Miss GERRY's suggestion has been followed in a number of recent articles on fossil plants. Miss HOLDEN (4, 5) has used it in several papers. Her statement of the importance of the "rims" is even stronger than that of Miss GERRY, making their occurrence "by far the most reliable criterion for diagnosing coniferous woods" (5, p. 252). Again, in another paper she speaks still more strongly: "Comparative examination of living and fossil forms leads to the rejection of all criteria except cellulose bars of Sanio as an infallible test for tribal affinities" (4, p. 544).

JEFFREY (6) has also attached much weight to it, especially since his discovery in 1912 of what he considers a vestigial bar in the araucarian cone axis. His conclusion from its presence here is stated as follows: "*Agathis* and *Araucaria* have obviously come from ancestors which, in accordance with the accepted principles of comparative anatomy, had . . . . bars of Sanio in their tracheids" (p. 548). The phylogenetic significance which he attached to its occurrence in this position figures prominently in his theory of the abietinean ancestry of the Araucarineae. Recognizing that his conclusion from its occurrence in the araucarian cone axis would be invalidated if the same condition were found to occur in similar regions of forms such as the cycads, JEFFREY examined the primitive regions of *Cycas*, *Zamia*, and *Ginkgo*, but found no evidence of its presence. He was therefore doubly assured of its meaning in the araucarians.

In connection with some investigations of the wood structure in the cycads, I have found a "bar"[1] of Sanio entirely similar to that in the araucarians, and since this discovery has such an important bearing on the theories above stated, it has been thought advisable to publish it separately.

The material in which the "bars" were found is from the petiole of *Cycas revoluta*. The sections of this, as well as all the others figured in the plate, were stained in the usual way with a double stain of Haidenhain's iron-hematoxylin and safranin. The bars thus appear in the sections as dark bands on a red background and stand out as distinctly as in the photographs.

----

[1] The name "bar" has been retained in this paper for the structure as found in *Araucaria* and *Cycas*, for considerations to be stated later.

Fig. 1 is from a section of the vascular bundle of the *Cycas* petiole cut so as to show the tangential walls of the tracheids of the primary wood. Spiral and scalariform elements of the xylem are shown at the left center of the figure, while the two transitional tracheids to the right of these exhibit the border-pitted condition typical of the region. Crossed pores can be seen in several of the pits, showing that the walls are not adjacent to parenchyma, but separate the lumina of two tracheids. The dark lines observable between the pits are thickenings in the primary wall and constitute the "bars of Sanio."

Fig. 3 is a portion of fig. 1 at a higher magnification and shows the part of the two border-pitted tracheids represented between (*a*) and (*b*) on that figure. The knife has passed diagonally through the wall of the left tracheid, leaving only a portion of one of the secondary layers, with parts of the primary attached to it in the upper region. In this area the bars of Sanio come out clearly, especially in connection with the fourth to the seventh pits from the top of the series.

The forking which has been mentioned by JEFFREY as occurring in the bars which he found in the araucarian cone axis is indicated here. An extreme instance of this can be seen between the fifth and sixth pits from the top. These pits are farther apart than is usual, and the bar instead of having merely its ends forked is cleft throughout its length, half clinging to each pit, as a definite "rim" of Sanio. This seems to indicate a double origin for the bars, suggesting that each may have its origin in two "rims," formed in the manner indicated by SANIO, one in connection with each pit. In the tracheid at the right, the bars are very plainly visible, though the forking is not so evident.

The "rims" which form the bars in *Cycas* are characterized especially by two features. They invariably cling closely to the borders of the pits, and they are short, never reaching across the entire width of a tracheid, but only across that of a single pit. These are the two characteristics mentioned by THOMSON (8) as distinguishing the type of bar found in the Araucarineae from that in the members of the Abietineae.

Fig. 4 is from the cone axis of *Araucaria Bidwilli* and shows the structures referred to by JEFFREY as occurring there. Here again, in the upper part of the figure, the wall has been partly cut away. The photograph is at the same magnification as fig. 3, and on comparison it will be seen that the characters of the bars in the two are identical. Here also they invariably cling closely to the pit borders and extend only across the width of a single pit. In almost every case the forking is plainly seen. A very favorable place for observing this is in the left hand tracheid at about the center of the figure. Near the top of this tracheid and in the majority of cases in the other two, the pits are so closely approximated horizontally, owing to the multiseriate character of the pitting, that the ends of the forks of adjacent bars meet, so that only more or less trapeziform lighter areas are seen between them. At one place, indeed, below the center of the middle tracheid, even this is not shown. It is quite evident, however, that this is not a case of a single bar passing across two pits, but of the application of two bars to each other, end to end. This is more evident from the section than from the photograph, for in it a light area similar to those shown in the series above, though not so pronounced, is visible.

There are, therefore, bars in the primitive region of *Cycas* exactly comparable to those in the araucarian cone axis. This, as JEFFREY recognized, makes their presence in the araucarian cone axis valueless as an argument in favor of abietinean ancestry of the Araucarineae. Indeed, the evidence of a reverse relationship is apparent from a study of the characters of the "bar" in the different regions of the Abietineae themselves.

The typical condition in old stem wood of the Abietineae is illustrated in fig. 2, a radial section of *Abies amabilis*. Where the pits are closely approximated in vertical series, as for example in the left tracheid of the figure or in the lower part of the tracheid to the right, the "rims" unite to form bars. Where the pits are scattered, as in the central part of the middle tracheid, each "rim" stands by itself. In neither case, however, are the "rims" closely appressed to the pit borders, as is always true in the araucarian and cycad form. They are not, moreover, limited in length to the width of a single pit. This is especially striking in the tracheid to the left, where opposition and even multiseriate pitting is in evidence.

In fig. 5 is shown a section of the wood of the cone axis of *Pinus resinosa* near the primary xylem. It was difficult to get a good photograph on account of the irregularities in the tracheid walls. The figure illustrates the type of "rim" found in this region. The structures usually adhere closely to the pit borders, appearing as dark arcs, and not extending beyond the pits. One long bar, however, is shown just below the four pits at the top of the tracheid, passing across nearly the width of two pits, and thus presenting to some extent the appearance of the bar found in specialized abietinean wood. Such structures are exceptional.

Fig. 6, from a radial section of the root of *Pinus Strobus*, shows a condition quite often found in the secondary wood close to the primary. Although the pits are large, the "rims" of Sanio, which are plainly in evidence, present the same characters as those of fig. 5. As one passes out from this region into the outer root wood, transitions are seen between this rim and the type shown in fig. 2. The rims become more and more separated from the pits. They also elongate and exhibit a tendency to fusion. The mature bar is thus often a complex structure and very specialized in character. Similar transitions are seen in the cone axis in passing outward from the center.

It is generally admitted that root and cone axis retain ancestral characters longer than the stem or branches, and the earliest formed parts of these are also accounted more primitive than the parts formed when the plant is older. If in this case the accepted reasoning holds true, the Abietineae must have been derived from ancestors which had bars of the araucarian or cycadean type.

The discovery of "bars" or "rims" of Sanio in the primitive region of the cycads must either nullify their value as evidence of the derivation of the araucarians from the Abietineae, or indicate that the Abietineae are also ancestral to the cycads, a position which can scarcely be assumed. Moreover, the study of the rims in different regions of the Abietineae indicates that the ancestral type of bar, as found in the cycads and araucarians, has become

SIFTON on BARS OF SANIO

specialized in the higher conifers, the original type persisting only in the primitive regions, the cone axis and the root.   Thus the discovery of the bar in the cycads has, besides confirming the primitive character of this structure in the araucarians, made apparent its significance in the different regions of the Abietineae.

This work has been carried on at the suggestion of Professor R. B. THOMSON, to whom I wish to express my indebtedness for his constant assistance.

UNIVERSITY OF TORONTO

LITERATURE CITED

1. GERRY, MISS E., Bars of Sanio in the Coniferales. Ann. Botany 24:119–123. 1910.
2. GÖPPERT, H. R., Monographie der fossilen Coniferen. Leiden. 1850.
3. GROOM, PERCY and RUSHTON, W., Jour. Linn. Soc. Bot. 41: 457–490. 1913.
4. HOLDEN, MISS R., Jurassic coniferous woods from Yorkshire. Ann. Botany 27:243–254. 1913.
5. ———, Fossil plants from eastern Canada. Ann Botany 27:533–545. 1913.
6. JEFFREY, E. C., The history, comparative anatomy, and evolution of the Araucarioxylon type. Proc. Amer. Acad. 48:531–561. 1912.
7. SANIO, K., Anatomie der gemeinen Kiefer. Jahrb. Wiss. Bot. 9:50–126. 1873.
8. THOMSON, R. B., On the comparative anatomy and affinities of the Araucarineae. Phil. Trans. Roy. Soc. London B 204:1–50. 1913.

EXPLANATION OF PLATE XV

FIG. 1.—*Cycas revoluta:* petiole; tangential section of primary wood; ×225.

FIG. 2.—*Abies amabilis:* radial section of secondary wood of stem, showing "rims" and "bars" of Sanio; ×225.

FIG. 3.—*Cycas revoluta:* a portion of fig. 1 more highly magnified; ×445.

FIG. 4.—*Araucaria Bidwilli:* radial section of secondary wood of seed cone axis, showing "bars" of Sanio; ×445.

FIG. 5.—*Pinus resinosa:* radial section of secondary wood of seed cone axis close to the primary wood, showing primitive "rims" of Sanio; ×445.

FIG. 6.—*Pinus Strobus:* radial section of secondary wood of root close to primary xylem, showing primitive "rims" of Sanio; ×445.

THE DISTRIBUTION OF BEACH PLANTS

EDITH A. ROBERTS

The following observations were made of the distribution of
plants on the lower and middle beaches of the coasts of the
Elizabethan Islands and Falmouth. The Elizabethan Islands and
Falmouth are situated in eastern Massachusetts, Falmouth being
a part of the mainland, and the islands extending in a chain into
the ocean, bordered by Buzzard's Bay on the west and by Vineyard
Sound on the east.

The facts here noted were observed while I was collecting
material for the investigation of the osmotic pressure of succulent
plants in the above-named localities. A more or less definite
grouping of the succulents was found, and, when present, certain
plants were always in advance, where they would the more often
be covered by the salt water. Later investigation will attempt to
show whether the different locations on the beaches present more
or less difficult situations for maintenance of growth, and, if so,
whether there is a greater or more variable osmotic pressure among
the forms to account for their being able to occupy more adverse
situations. Competition of species cannot account for the presence
or absence of forms, for GANONG (6) has shown that the ground is
not already occupied, therefore the struggle must be with the
physical environment.

Thirteen locations were selected as affording types of all shores
on the Islands and Falmouth where succulents might be found.
These may be divided into sea beach and sea cliff types. The
coasts of the sea beach type are readily divided into lower, middle,
and upper beaches according to COWLES (4). The upper beach
in all cases is a low dune, back of which there is always found a
pond. The dune at Chappaquoit Point is the highest.

The sea cliffs may be divided into two classes: those at whose
base the rise and fall of water varies so little that a zone of *Zostera
marina* is present and those at whose base this plant is absent.
The sea beach type was found at Chappaquoit Point, Falmouth

Beach, Nobska Beach, Lackey's Bay, Tarpaulin Cove, Cuttyhunk, and Quicks Hole. The sea cliff without *Zostera marina* was found at Nobska Cliff and Falmouth Cliff. The sea cliff with *Zostera marina* was found at Lackey's Bay (inside), Ram Island, Cuttyhunk Pond, and Deer Island.

The following 9 plants were the characteristic succulents found on the middle beach: *Atriplex arenaria*, *Salsola Kali*, *Chenopodium rubrum*, *Arenaria peploides*, *Cakile edentula*, *Artemisia Stelleriana*, *Solidago sempervirens*, *Euphorbia polygonifolia*, and *Lathyrus maritimus;* while *Ammophila arenaria* has been included in the chart as a tenant of the middle beach because it plays a part in the zonal distribution of some of the succulents. SNOW (7) speaks of the upper part of the middle beach as the low *Ammophila* zone, thus subdividing the middle beach into the succulent zone (lower half) and low *Ammophila* (upper half). The drift spoken of as limiting the middle beach at the front is the daily tide line. The spring tide line seems to sweep back the daily tide débris, together with its own gatherings, to a line back of the daily tide. This irregular line of débris at the front of the lower middle beach, determined by the spring tide, marks the extent forward of the succulent zone, while the summer storm tide line marks the beginning of what I term the transitional zone, corresponding, I believe, to the low *Ammophila* zone of SNOW. Succulents and *Ammophila* tenant the region together, with the *Ammophila* dominating just back of the winter storm tide line or on the upper beach, the term *Ammophila* being thus used to apply to the upper beach. The plants exclusive of *Ammophila arenaria* were found in the order given as outpost plants as shown on the chart.

At Chappaquoit Point all the 10 characteristic plants were found. *Atriplex arenaria* was the forerunner, as it always is when present. BERGEN (1) states that *Salsola Kali* is the most tolerant of the plants he observed. It ranges here from *Artemisia Stelleriana* to just back of *Atriplex arenaria*, with some plants occasionally at the top of the dune. *Chenopodium rubrum* is confined to a region just back of *Salsola Kali*. *Arenaria peploides* grows in masses and monopolizes a wide zone. *Cakile edentula* grows more in bunches and is closely associated with *Artemisia Stelleriana*,

which often finds its best location at an *Ammophila arenaria* outpost. The *Artemisia Stelleriana* presents more or less of a solid
front and only thin strands of *Ammophila* are found beyond; in
fact, it might be said to define the border line of upper and middle
beaches. *Euphorbia polygonifolia* becomes established on the
wind-swept dunes, where there is nothing else, but never is found
toward the ocean beyond the limit of *Ammophila* runners. *Lathyrus maritimus* occasionally extends into the transitional zone, but
always in company with *Ammophila*.

At Falmouth there is an illustration of what is common at
several of the beaches; at the region of the incoming daily tide
there is a mass of dry eel grass and immediately back of that a

zone of small shingle, doubtless brought there by the stronger
spring tide. At the limit of this shingle a slight depression occurs,
then an elevation and the loose sandy zone with the succulents,
and beyond this another depression, then an elevation rises slowly
to the winter storm tide line.

At Nobska Beach *Chenopodium rubrum* is abundant, with only
a little *Atriplex arenaria* and *Salsola Kali*, but wherever all three
are present the same relationship holds, *Atriplex* being always
nearer the water than the others.

Cakile edentula forms masses in front of *Ammophila* at Lackey's
Bay. At Tarpaulin Cove much *Salsola Kali* is found in mats, and
bunches of eel grass have been caught by it, forming stable mounds.
The beach formed at Cuttyhunk Harbor and Vineyard Sound is

separated from Vineyard Sound by a high shingle mound.   Among the shingle *Cakile edentula* grows most luxuriantly and forms thick, hedgelike masses.   They are the most abundant and thrifty plants seen.   Here the entire shingle strand must be washed often by the summer storms.

The relationship of the plants found at the base of Nobska Cliff and Falmouth is the same as that found on the beaches, but the tide lines are nearer together there, and the zones are condensed.   The telescoping comes in the zone between the summer storm tide line and the winter drift, so that there is a more sudden transition from the succulent zone to the *Ammophila* zone, practically eliminating the transitional zone.

The low sea cliffs, at whose base the lower beach is eliminated and where there is living *Zostera marina*, as at Lackey's Bay, furnish an additional grouping.   Here, on what corresponds to the plantless lower beach, is a grouping of *Zostera marina* and *Spartina glabra pilosa*.   Their advance is checked by a mass of dry eel grass three feet in depth, but on the top of the eel gras, as on the substratum of the lower middle beach, is a grouping of succulent plants with exceptionally long roots.   This is a strictly succulent zone, for the *Ammophila* does not advance beyond the eel grass débris. Why succulents can grow on the eel grass substratum and *Ammophila* not is an interesting question.

Along the cliffs where *Zostera marina* is present and the lower beach is covered by water, the following new forms are found: *Suaeda maritima, Salicornia europaea, Limonium carolinianum,* and *Ligusticum scoticum*.   Their ecological relationship is shown on the chart.   The last three divisions of the chart show what GANONG (5) speaks of as a *Salicornia-Suaeda* association.   CHRYSLER (3) gives a list of plants arranged in a descending scale with respect to salt-resisting capacity, and in this *Salicornia europaea* is next to *Limonium carolinianum*.   GANONG found that the root hairs of *Salicornia europaea* can endure 90 per cent of sea water without plasmolysis, and those of *Suaeda maritima* 60–70 per cent. CANNON (2) refers to salt-containing places as having marked zonal distribution; several species of *Atriplex* and *Suaeda* are nearer the center of salt spots, where the concentration of salt is highest.

Summary

1. The plants on the lower and middle beaches of some of the coasts of the Elizabethan Islands and Falmouth are found to have in general definite zonal distribution.

2. The shores may be divided into sea beach and sea cliff types; the latter may be subdivided into cliff types with *Zostera marina* and those without.

3. The following plants were selected as most representative of the beach type: *Ammophila arenaria, Atriplex arenaria, Salsola Kali, Chenopodium rubrum, Arenaria peploides, Cakile edentula, Artemisia Stelleriana, Solidago sempervirens, Euphorbia polygonifolia, Lathyrus maritimus,* and *Suaeda maritima. Salicornia europaea, Limonium carolinianum, Ligusticum scoticum, Zostera marina,* and *Spartina glabra pilosa* were taken as representatives of the sea cliff type.

4. The upper half of the middle beach is called the transitional zone.

5. When the lower beach is covered most of the time by water, *Suaeda maritima, Salicornia europaea,* and *Limonium carolinianum* are found in the order given, from ocean to shore.

6. The forms of the middle beach are found in definite zones, and when not present their places are not filled; for instance, *Atriplex arenaria, Salsola Kali, Chenopodium rubrum,* and *Arenaria peploides* are always found in the order given on the lower half of the middle beach, while *Cakile edentula, Artemisia Stelleriana, Solidago sempervirens,* and *Euphorbia polygonifolia* are related to *Ammophila* in distribution.

7. As to the cause of this distribution I hope to offer some suggestions in the near future.

MOUNT HOLYOKE COLLEGE

LITERATURE CITED

1. BERGEN, J. Y., Some littoral spermatophytes of the Naples region. BOT. GAZ. **41**:327–333. 1906.
2. CANNON, W. A., Some relations between salt plants and salt spots. Leland Stanford Junior Univ. Publ. Dudley Memorial vol. 123–129. 1913.

3. CHRYSLER, M. A., Ecological plant geography of Maryland.   Maryland
   Weather Service 3:148–197. 1911.
4. COWLES, H. C., The ecological relations of the vegetation on the sand dunes
   of Lake Michigan. BOT. GAZ. 27:94–117, 167–202, 281–308, 361–391.
   1899.
5. GANONG, W. F., Vegetation of Bay of Fundy marshes. BOT. GAZ. 36:
   161–186, 280–302, 349–367, 429–455. 1903.
6 ———, The nascent forest of the Miscou beach plain. BOT. GAZ. 42:81–
   106. 1906.
7. SNOW, L. M., The ecology of the Delaware coast. BOT. GAZ. 34:284–306.
   1902.

BRIEFER ARTICLES

THE INFLUENCE OF ETHER ON THE GROWTH OF
ENDOTHIA

The writers found it necessary to conduct a series of experiments on the influence of ether on the growth of *Endothia*, preliminary to more extensive studies on the influence of the tannin content of the host plant on *Endothia parasitica* and related species. Owing to a change in plans this phase of the work was abandoned. However, it has been considered advisable to publish the results of these experiments.

The medium used was as follows: water 1000 cc., glucose 20 gm., peptone 10 gm., potassium phosphate (dibasic) 0 25 gm., magnesium sulphate 0 25 gm. The organisms used were *Endothia parasitica*, *E. radicalis* (American and European strains), including one from *Quercus alba* which was sent to us by Dr. GEORGE P. CLINTON. This last organism was a much slower grower than the others. In order to prevent the evaporation of the ether, the cotton plugs were pushed a short distance into the tubes and close-fitting corks inserted.

In the first experiments, approximately 10 cc. of the medium was placed in each tube and two or three drops of ether added by means of a graduated pipette (that is, not more than 0 1 cc.). These tubes were then inoculated with the organisms. In all cases the checks grew more slowly than those to which the ether had been added. These results made it necessary to conduct a more careful series of experiments, in which definite but varying amounts of the ether should be used.

In this second experiment, 0 1 cc. ether was added to certain tubes of the medium and 0 4 cc. to others. These were inoculated and kept as above. The results confirmed the first experiments, but made it necessary to conduct a still more extensive experiment.

This experiment was as follows. To 10 cc. of culture solution, ether was added in quantities of 0 05, 0 1, 0 2, 0 3, 0 4, 0 5, 0.6, 0 7, 0 8, 0 9, and 1 per cent. In this experiment only one species was used, *E. parasitica*. On the third day the check showed a good normal growth; that having 0 05 per cent ether was somewhat accelerated and that having 0 1 per cent was noticeably so. The next two members of the series showed slight germination. The next day but little change was noticeable except that germination was apparent up to and including the tubes containing o 6 per cent ether. The second day the culture containing o 3, o 4, and o 5 cc. ether showed accelerated growth, while the one with o 6 per cent showed good growth. On the tenth day after inoculation, the cultures with the higher percentages of ether showed signs of more vigorous growth, except the two highest members of the series, which never germinated. From this time on the results do not agree in all details with the earlier stages of the experiment, owing to the unavoidable escape of varying amounts of ether from the different tubes and the consequent change of percentages. Therefore, it was not practicable to keep these cultures containing ether under observation for more than ten days or two weeks, a time too short for pycnidial formation in liquid media.

These results are represented in the accompanying table, in which the check is rated 5, and the acceleration of retardation computed on this basis.

TABLE I

GROWTH OF *Endothia parasitica* IN LIQUID MEDIA TO WHICH THE INDICATED AMOUNTS OF ETHER WERE ADDED (AMOUNT IN CC.)

Days after inoculation	o o	o o5	o 1	o 2	o 3	0.4	o 5	o 6	o 7	o 8	o 9	1 o
3.......	5	7	9	2	1	0	0	0	0	0	0	0
4.......	5	7	9	4	3	2	1	2	0	0	0	0
5.......	5	5	6	5	4	4	3	3	0	0	0	0
6.......	5	5	6	5	5	5	4	3	0	0	0	0
7.......	5	6	7	6	5	5	4	3	1	0	0	0
8.......	5	6	6	6	5	5	4	3	2	1	0	0
10.	5	7	6	6	5	5	4	3	2	3	0	0

It appears that small quantities of ether have a stimulating effect on the fungus, quantities of from o 2 per cent up retard germination, and quantities from o 4 per cent up have injurious effect on the growth of the fungus. While it is possible to grow the mycelium under the influence of ether, the volatile nature of this chemical makes it impossible to keep such cultures intact long enough for pycnospores to appear. After two to two and one-half weeks the mycelium showed signs of dying and pulling away from the glass in the cultures containing the higher percentages of ether.—MELVILLE T. COOK and GUY W. WILSON, *Rutgers College and New Jersey Agricultural Experiment Station, New Brunswick, N.J.*

WIND BURN IN AMORPHOPHALLUS

Wind burn, or, as it is more commonly called, "sunburn," does not seem to be conspicuous in plants. An interesting case was observed during the past typhoon season at Los Baños, Philippine Islands. During July and August a strong southwest wind is continuous and may often become violent. This wind is frequently warm and dry.

In the latter part of August 1914, the southwest wind was particularly and continuously heavy. By the third day of the heavy wind, large red blotches were noticed on the petioles of plants of *Amorphophallus campanulatus* (Roxb.) Blume (Araceae). Investigation of many plants revealed the fact that only in those plants which were exposed to the wind was the red color present. Light green areas on the petioles were more affected than dark areas. When wind and sun acted in combination, the red color was more apparent, occurring on both blades and petioles. Plants growing in ravines and in places otherwise sheltered from the wind, even though fully exposed to the sun, did not show red coloring. Opening up the brush to allow the wind access resulted in the development of the red coloring within two to three days.

The red color from this cause, however, was not a permanent effect, for in the rainy period which succeeded the heavy wind the red color entirely disappeared within ten days.—FRANK C. GATES, *Los Baños, P.I.*

CURRENT LITERATURE

NOTES FOR STUDENTS

General biology of rusts.—KLEBAHN[1] has published a series of observations on rust fungi made during 1912 and 1913. The report covers three phases of work: a study of the factors which bring about the termination of the rest period in teleutospores, cultural work, and some observations on the mallow rust in relation to ERIKSSON's mycoplasma theory.

In order to study the factors influencing the termination of the rest period in teleutospores, the author subjected teleutosporic material of *Puccinia graminis*, *P. Phragmites*, and *P. Magnusiana* to various treatments such as exposure to low temperatures, alternate wetting and drying, and continuous immersion in water. Apparently the changes in the spore which make germination possible depend not so much upon exposure to low temperatures as upon the action of moisture. Spores suspended in a dry shed and subject only to atmospheric moisture failed to germinate. Alternate wetting and drying is most potent in shortening the rest period; spores subjected to such treatment, even at room temperature, germinated in January two to three months before the usual time. Even continuous immersions with frequent changes of water somewhat shortens the rest period, although occasional drying seems to be essential to produce any marked effect.

The second part of the paper has already been noted.[2] In the third section some experiments and observations on the mallow rust furnish a basis for discussion of ERIKSSON's views on the means of distribution of this fungus. ERICKSSON's main contention[3] is that the fungus is distributed in the mycoplasma state with the seed, and manifests itself by a general outbreak over the entire surface of the infected plant when it is about three months old. These contentions are not borne out by the experiments of KLEBAHN, who finds that different groups of plants grown from the same lot of seed become infected at different times, and the outbreak does not involve the whole plant at once. The sori are at first scattered and few, as if originating from local infections. To the conidia-like segments sometimes produced by the promycelium of this rust ERIKSSON attributes the special function of giving rise to the mycoplasma

[1] KLEBAHN, H., Kulturversuche mit Rostpilzen. Zeitschr. Pflanzenkrank 24:1–32. *fig. 1*. 1914; also Beobachtungen über Pleophagie und über Teleutosporenkeimung bei Rostpilzen. Jahresb. Verein Angew. Botanik 11:55–59. 1913 (a short statement of the main facts in the foregoing paper).

[2] Rev. BOT. GAZ. 60:245. 1915.   [3] *Ibid.* 54:431. 1912.

in the plants infected by these spores. DIETEL[4] has already shown that the production of "conidia" by the promycelium does not indicate a functional differentiation of the teleutospores, but is dependent on external conditions. In this matter KLEBAHN agrees with DIETEL, but the two investigators differ somewhat as to the factors which determine the segmentation of the germ tubes. KLEBAHN finds that spores germinating within a liquid medium produce, in place of normal mycelia, germ tubes which may become segmented or not. The number of segments produced is always four, and under some conditions the "conidia" produce sporidia. Teleutospores germinating in contact with air always produce promycelia and sporidia. DIETEL attributes the production of segmented germ tubes to a scanty moisture supply in the substratum. KLEBAHN infected a number of leaves of hollyhock with cultures of teleutospores which were producing "conidia." The groups of sori originating from these infections were cut out of the leaves which were kept under observation during the rest of the season. According to ERIKSSON, these leaves should have contained mycoplasma which should later have given rise to a general outbreak of rust on the leaves. KLEBAHN found, however, that the leaves from which the sori had been removed remained free from infection.

The mycoplasma theory is attacked from another standpoint by HAASE-BESSEL,[5] who finds in and about mature rust pustules of secondary origin on leaves of *Althaea rosea* attacked by mallow rust mycelial structures resembling the promycelium described by ERIKSSON. These structures on account of their delicate walls have the appearance of naked masses of protoplasm which conform to the shape of the intercellular spaces which they occupy. A study of their origin shows that they are the outgrowths of older hyphae at the base of the rust pustules. The author attributes to them the function of distributing the mycelium through the leaf. Since these structures, conforming in every way to ERIKSSON's promycelia, are observed originating from the mycelium of mature pustules of undoubted secondary origin, the author argues that the similar structures of ERIKSSON may be explained without the aid of the mycoplasma theory. To this conclusion ERIKSSON would object, as to the other similar arguments, that the mycoplasma does not occur in connection with rust infections originating from spores. Thus the elusive mycoplasma again escapes annihilation.

A cytological study supplemented by field observations has enabled OLIVE[4] to explain some apparently anomalous phenomena in the sequence of spore forms and the formation of aecidia in certain rusts with perennial mycelia.

[4] Rev. BOT. GAZ. 56:163. 1913

[5] HAASE-BESSEL, G , Zur Erikssonchen Mycoplasmatheorie. Ber. Deutsch. Bot. Gesells 32:393-403. *pl. 9.* 1914.

[6] OLIVE, E. W., Intermingling of perennial sporophytic and gametophytic generations in *Puccinia Podophylli, P. obtegens,* and *Uromyces Glycyrrhizae.* Ann. Mycol. 11:298-311. *pl. 1.* 1913.

The forms investigated are *Puccinia Podophylli*, *P. suaveolens*, and *Uromyces Glycyrrhizae*. Early in spring *Puccinia Podophylli* usually produces teleutospores on the bud sheaths at the base of the stems of the host plant (*Podophyllum peltatum*). These teleutospores are produced before any other spore forms and also before the plants are fully developed. Sometimes, according to OLIVE, traces of aecidia also, but no spermagonia, are found on the bud sheaths. Later, spermagonia and aecidia develop abundantly on the leaves. These aecidia have been shown by SHARP[7] to arise from a binucleate mycelium. Teleutospores are produced on the leaves in late summer. OLIVE finds that the teleutospores on the bud sheaths and the aecidia, both here and on the leaves, arise from a perennial sporophytic (binucleate) mycelium which extends throughout the plant. Intermingled with the sporophytic mycelium is a perennial gametophytic (uninucleate) mycelium which gives rise to spermagonia and to incepts of aecidia which, however, are soon invaded by the sporophytic mycelium whose hyphae give rise to basal cells and rows of binucleate aecidiospores. The mycelia of these two generations are generally commingled throughout the plant, but in the young parts the gametophytic mycelium is somewhat in advance of the sporophytic. Besides these perennial forms, local binucleate mycelia, originating from aecidiospores and giving rise to teleutospores, occur on the leaves. In *Puccinia suaveolens* on the Canada thistle and in *Uromyces Glycyrrhizae* on *Glycyrrhiza lepidota* three conditions of mycelial distribution have been observed by the author: first, perennial gametophytic mycelium giving rise to spermagonia and perennial sporophytic mycelium giving rise to secondary uredospores and to teleutospores are commingled throughout the same plant; second, perennial sporophytic mycelium giving rise to secondary uredospores and to teleutospores occurs alone; third, annual local mycelium occurs which also produces uredospores and teleutospores, but the sori in which these are produced ar rarely confluent like those of the perennial mycelium. Perennial gametophytic mycelium alone or local colonies of gametophytic mycelium have not been observed.

In all these cases no spore forms resulting from gametophytic cell fusions have been found. The aecidiospores of *Puccinia Podophylli* and the uredospores of *P. suaveolens* and *Uromyces Glycyrrhizae* are all regarded as secondary in nature. They arise apogamously, solely from sporophytic mycelia The author has not observed aecidiospores nor primary uredospores arising as the result of fusions between gametes in these forms.

DAWSON[8] has studied the distribution of the mycelium of *Puccinia fusca* and *Aecidium leucospermum* in the rhizomes, buds, and leaves of *Anemone nemorosa*. The mycelium of these fungi is found in the rhizomes and buds of the

[7] SHARP, L. W., Nuclear phenomena in *Puccinia Podophylli* BOT. GAZ. 51:463–464. 1911.

[8] DAWSON, W. J., Über das Mycel des *Aecidium leucospermum* und der *Puccinia fusca*. Zeitschr. Pflanzenkrank. 23:129–137. 1913.

infected plants, but its distribution may be more or less local. Frequently one
section of the rhizome with its buds is infected, while the adjacent sections remain free from the fungus. All the leaves arising from the infected sections
become infected. Experiments like those reported by TISCHLER⁹ for *Uromyces
Pisi*, to determine if the plants could outgrow the fungus, were not conducted.
The mycelium is found in all the tissues except the phloem and xylem; in the
buds even the cells of the dermatogen are invaded. In this respect the distribution of the mycelium in the growing points agrees with that of *Uromyces
Pisi* in *Euphorbia Cyparissias* as observed by DE BARY and TISCHLER. There
is apparently a marked difference, however, in the behavior of the mycelium
of these two fungi in the embryonic tissue. TISCHLER observed the interesting
fact that no haustoria were formed by the hyphae of *U. Pisi* in the strictly
meristematic cells, while DAWSON states that haustoria are formed in all parts
of the plant containing mycelium, even "close" to the growing point. Unfortunately neither the description nor the imperfect sketches make this physiologically important point entirely clear.

TISCHLER⁹ has published a further contribution to his interesting studies
on the relation between *Uromyces Pisi* and its aecidial host *Euphorbia Cyparissias*. In his former account the fact was brought out that under certain conditions the upper portions of infected shoots of this plant might produce normal
leaves to all appearances free from the fungus. In such cases it was found that
the growing point had became emancipated from the fungus, which was thereafter unable again to invade the strictly embryonic tissues. The behavior
of these shoots is now contrasted with that of shoots from dormant buds
of infected rhizomes. Such shoots produce only typically deformed leaves
characteristic of plants infected by the fungus. If, however, the infected
rhizomes are kept permanently in a vegetative state by pruning or other means,
the new shoots produced show no external signs of the fungus and bear only
normal leaves. If infected plants, which as a result of such treatment have
produced only normal leaves for a season, are allowed to remain dormant for
a time, only infected shoots grow out from the dormant buds. It appears,
therefore, that buds that develop on the rhizome while the plant is kept continually in a vegetative state, and which start into growth without having
undergone a distinct period of rest, are not subject to the formative influence
of the fungus; but buds which have undergone a period of dormancy become
subject to such influence and produce only deformed leaves. Nevertheless,
the apparently sound shoots are not free from the fungus, for the mycelium
is found in all parts except the meristematic tissue of the growing point; this
it is entirely unable to invade in shoots which have once become emancipated.

---

⁹ Rev. BOT. GAZ. **56**:161. 1913.

¹⁰ TISCHLER, G., Über latente Krankheitsphasen nach *Uromyces*-Infektion bei
Euphorbia Cyparissias. Bot. Jahrb. **50**:95–110. *figs. 6.* 1914

Wherein lie the causes of this peculiar difference of behavior of the two kinds of buds the author was unable thus far to determine.

HAAK[11] has published a somewhat lengthy account of a series of general observations and experiments on the biology of *Peridermium Pini*, and particularly on the question of the propagation of this rust by aecidiospores. From the observation that no probable telial host is coextensive in its distribution with *Pinus sylvestris* and its blister rust (*Peridermium Pini*), and also because of the failure up to this time to find the telial phase of this rust, HAAK infers the likelihood that this fungus is propagated on the pine by aecidiospores. This assumption contains in itself nothing anomalous, since several rusts with repeated aecidial generations are known. The experiments by which it is endeavored to substantiate this assumption furnish only probable evidence. Cultures on sound trees having failed to give satisfactory results (out of 72 trees with about 200 sowings only one infection resulted), subsequent cultures were made only on trees which showed the presence of *Peridermium*. The diseased branches were removed and numerous small twigs were infected in wounds. Although the number of *Peridermium* cankers resulting from these infections was small compared with the number of infected twigs, the proportion of diseased twigs to sound twigs was much greater on the infected branches than on the uninfected ones. The circumstantial evidence, therefore, seems to indicate that this fungus can be propagated on the pine by means of aecidiospores. A histological examination to determine whether the mycelium under the aecidia was sporophytic or gametophytic was not made. Two further points of interest in the biology of this fungus are brought out by the author. First, natural infections take place only in green twigs still bearing needles; no infections occur in older twigs. Even old cankers on trunks and large branches, where the origin has not been completely obliterated, can be traced to infections of small twigs from which the disease has gradually spread. Secondly, when a tree is once infected, the disease spreads by new infections to other branches during succeeding years. This fact is in accord with the theory that the aecidiospores are capable of reinfecting the pine.

BAUDYS[12] adds a contribution to the question of the maintainance of rusts by means of wintering uredospores or mycelium. The observations were made during the winters of 1910–1911 and 1911–1912 in Bohemia, where the temperature ranged as low as −16° C and −21° C. respectively during the two seasons. From 60 to 100 per cent of the uredospores in the winter sori of *Puccinia glumarum* and *P. dispersa* on rye, and *P. bromina* on *Bromus sterilis* were found to retain their capacity for germination throughout the winter. These rusts

---

[11] HAAK, ———., Der Keinzopf *Peridermium Pini* (Willd.) Kleb. Zeitschr. Forst u. Jagdwesen 46:3–46. *pls. 2*. 1914.

[12] BAUDYS, E., Ein Beitrag zur Überwinterung der Rostpilze durch *Uredo*. Ann. Mycol. 11:30–43. 1913.

were kept under observation from early autumn until the general outbreak of rust the following spring. It was found also that young undeveloped sori in the leaves continued to develop when temperature conditions were favorable. The mycelium in the neighborhood of the sori was found to be alive. Uredospores of *Puccinia dispersa* retained their germinative power as long as 100 days, although the percentage of viable spores decreased during that time. The uredospores of a number of other rusts were found to be capable of germinating during the winter, but observations on these forms were not carried through the entire season. Uredospores of *Puccinia simplex* on *Hordeum murinum* germinated on December 29; those of *Uromyces Anthyllidis* on *Anthyllis vulneraria* on December 11; those of *U. Ervi* on *Vicia hirsuta* on January 29; and those of *Puccinia Lolii* on *Lolium perenne* on March 5.

STAKMAN[13] has investigated for the conditions of Minnesota some of the problems relating to the specialization of physiological races of the cereal rusts and the resistance of different varieties of wheat to rusts. The results are in general in accord with the observations of other investigators, and serve to emphasize the fact that particular strains of *Puccinia graminis* inhabiting the common cereals are more or less specifically adapted to particular species, but that this adaptation is not entirely rigid. Of all the forms, that on the oat is most strictly specialized. The forms on wheat, barley, and rye migrate more or less readily from one cereal to another. Treatment of the plants with anaesthetics makes them more susceptible to infection. Resistance to rust seems to depend more on the nature of the variety than upon any external factors to which the plant is subjected. Certain phenomena, such as length of the incubation period of the rust within the plant, size of pustules and of spores, seem to be correlated with the degree of resistance of the host. The observations of MARSHALL WARD that in resistant plants the rust mycelium usually killed small infected areas of the leaves and thereby brought about its own destruction was confirmed.

Miss SAHLI[14] has continued the investigations begun by FISCHER[15] on the susceptibility of hybrids and chimaeras of mixed immune and susceptible parentage to the attacks of fungi to which one parent is susceptible. A number of cultures with teleutospores of *Gymnosporangium Sabinae*, *G. clavariaeforme*, *G. confusum*, and *G. tremelloides*, all inhabiting species of *Crataegus*, *Sorbus*, and *Mespilus*, were made on hybrids and chimaeras of these plants. In general, there appears to be no constant relation between the state of immunity of the parents and that of the hybrids or chimaeras. For the details concerning each case, the paper itself should be consulted.—H. HASSELBRING.

[13] STAKMAN, E. C , A study in cereal rusts. Physiological races. Tech. Bull. 138. Agric Exp Sta Univ. Minn. pp. 56. *pls. 9* 1914.

[14] SAHLI, GERTRUDE, Die Empfünglichkeit von Pomaceen-Bastarden und Chimaren für Gymnosporangien. Mycol. Centralbl. 3:10-11. 1913.

[15] Rev. BOT. GAZ. 56:163. 1913

Germination.—MÜLLER[16] has done a very important piece of work on the breaking of seed and fruit coats in germination. The work deals with four general topics: (1) source of energy for breaking the coats; (2) structural features aiding in freeing the embryo; (3) measurement of pressure exerted against the coats by the growing contents; and (4) pressure required for bursting the coats in germination.

1. The author finds four different types of seeds and fruits as to the source of the pressure for bursting the coats: (a) swelling of the seed or fruit contents resulting from water absorption (swelling endosperm of *Ipomaea* and others, swelling embryo of *Chenopodina maritima*); (b) growth of endosperm (*Pinus Pinea*, *Ricinus communis*, and several others); (c) growth of cotyledons (*Prunus persica* and several other species of *Prunus*, *Juglans regia*, *J. nigra*, *J. cinerea*, and *Corylus avellana*); (d) growth of primary root or hypocotyl. In type (d) the author recognizes five different groups: (1) radicle borne between the subtending ends of the cotyledons and exerting its pressure upon the coat through these; (2) embryo located centrally in the endosperm and exerting its pressure upon the coat through this tissue (*Rumex* species, *Ruta graveolens*, *Nigella damascena*, *Viola tricolor*, etc.); (3) no nutrient tissue, embryo directly against the coat, hypocotyl in wedge-shaped cavity (*Helianthus annuus*, *Cucurbita Pepo*, etc.); (4) hypocotyl peripherally placed and in a wedge-shaped channel formed by integuments or by integuments and endosperm or cotyledons (*Ervum lens*, *Phaseolus multiflorus*, *Saponaria vaccaria*, *Potomogeton* species, etc.); (5) root or root-bearing portion of cotyledon generally cylindrical, pushing out through hole stopped by plug, lid, or valves (mostly monocots, as Marantaceae, Musaceae, Cannaceae, Phoenicaceae, etc.).

2. Under structural features of fruit or seed coats related to freeing of the embryo, the author discusses five classes: (a) free openings through which embryo grows without resistance (*Coix lacrymae*); (b) a special hole for exit of the embryo, the hole being stopped by a plug, cap, or valves (*Canna*, *Tradescantia*, *Commelina*, *Sparganium*, etc.); (c) a preformed dehiscent line (*Prunus*, *Pinus*, *Juglans*, *Amygdalus*, etc.); (d) zones of slight resistance at which coats break (Polygonaceae, some Lepidocaryeae); (e) no special mechanism (*Trigonella foenum graecum*, *Ligustrum vulgare*, *Capsicum annuum*, *Allium* species).

3. The pressure exerted against the coats by growing contents were measured for three forms. The growing cotyledon of *Corylus avellana* develops an average maximal outer pressure on the coat of 4759.42 gm., 34 14 gm. per mm.^{2}, or 3.305 atmospheres. The growing endosperms of *Ricinus communis* and *Pinus Pinea* develop a maximal average pressure of 4539 07 gm.

[16] MÜLLER, GOTTFRIED, Beiträge zur Keimungsphysiologieuntersuchungen über die Sprengung der Samen- und Fruchthüllen bei der Keimung. Jahrb. Wiss. Bot. 54:529–644. 1915.

and 4106.96 gm., 32 06 and 38 gm. per mm.², or 3,104 and 3 678 atmospheres respectively.

4. The hard seed and fruit coats of *Corylus* and *Pinus* are greatly weakened at the dehiscent line merely by water absorption. The breaking resistance offered by the soaked coats in these forms is generally one-half to one-sixth that offered by the dry coat. This weakening is not due to the dissolution of materials from the dehiscent line, for the old breaking strength is recovered upon drying. Soil agencies (bacterial, fungal, and acid effects) and enzymes from the living contents play no part in the dehiscence except where cellulose is present, as in *Juglans*. In *Pinus*, if rings of the coat were buried in soil for a period considerably exceeding germination time, the dehiscent zone was considerably weakened by soil agents. In these experiments, however, the agents can attack the dehiscent zone from every side. Water absorption also greatly lowers the breaking strength of brittle coats like *Ricinus*, as well as leathery or skinlike coats, as in *Vicia Faba*. It seems that the breaking strength of a great number of organic substances (*Laminaria* thallus, filter and parchment paper, etc.) falls greatly with water absorption and rises with water loss.

One must doubt the correctness of a fundamental assumption of Müller in determining whether rupture of the coat is due merely to swelling or mainly to growth of the contents. If the bursting occurs at 0° C., Müller pronounces it due to swelling; if only at a considerably higher temperature, to growth. The coats to be broken are gels of various consistency, which vary in viscosity and other physical characters, including no doubt breaking strength, with temperature, H+, OH−, and salt content. Again, the actual force with which the colloidal contents absorb water varies with conditions. We have found[17] cases where the coats are broken by swelling of the embryo and endosperm, but it occurs only at 15° C. or above, unless the coat is weakened by various treatments, when it occurs at 0° C. It is likely that there are cases where the seed or fruit contents exert much greater pressure upon the coats than those reported by Müller (3–4 atmospheres). Indeed, Crocker and Davis[18] showed by a very different method that the swelling embryo of *Alisma Plantago* exerts a pressure of about 100 atmospheres against the coat. Soil and other natural and applied agents play a much greater part in lowering the breaking strength of seed or fruit coats (*Alisma, Sagittaria, Amaranthus*, etc.) than Müller found for the seeds upon which he worked. It seems odd that Müller uses the micellar theory in his explanation rather than applying the modern colloid conception. There is no doubt that this leads him to certain incorrect inferences, such as cited above. It is very doubtful whether any generalizations can be drawn from Müller's measurements, since they cover so few forms and in the main the same types of seeds and fruits. For this

[17] Unpublished work by Davis and Crocker.

[18] Bot. Gaz. **58**:285–321. 1914.

reason his pioneer work should be followed by a similar study of a great variety of seeds and fruits.—WILLIAM CROCKER.

Light and growth.—VOGT[19] has shown that the effect of light upon the growth rate of the coleoptile of *Avena sativa* is very complex, in contrast to the older statement that light always inhibits growth. Temporary illumination of sufficient intensity is shortly followed by a considerable decrease, which in turn is immediately followed by a greater and longer enduring increase in growth rate. This increase is due to the action of light alone, and not to the combined action of light and darkness. The initial inhibitory effect of light on growth is not due to increased transpiration, as BLAAUW[20] suggests. In sufficient illumination the rate and total amount of growth of this organ is reduced. The greater amount of illumination causes a second inhibitory effect, which greatly exceeds the former acceleration, so that the total effect is a marked reduction in rate and amount of growth. This total effect of light was all that was definitely studied before JACOBI'S[20] work in 1911. In this total effect, like amount of light gives like effect, even within wide variation of the intensity and duration factors. For a given reduction in growth, *light intensity×duration of illumination* is a constant. This is the quantity of stimulus law which has been shown to have rather general application in growth and movement stimuli in plants. It seems then that there are three effects of light: inhibition, followed by a greater acceleration upon temporary illumination, and a second inhibition largely determining the total effect in long or enduring illumination of sufficient strength. These results tally in general with those of BLAAUW on sporangiophores of *Phycomyces*, except that BLAAUW has not found important the initial inhibitory effect of light.

While all the work of JACOBI, BLAAUW, and VOGT will greatly modify the statements current in our texts, there is one set of experiments by VOGT which is especially interesting in this respect. When the organ was alternately illuminated (100 M.K. or less) and darkened in 15, 30, or 60-minute periods, the illuminated periods gave less growth only in the 30 and 60-minute periods, and not in the 15-minute periods. In the longer periods, VOGT has shown that the greater growth in darkness is due largely to the stimulation effect of the previous illumination, and the slower growth in light is not due in the main to inhibitory effect of light. This quite reverses the former interpretation given to such results.

Red light acts as very weak diffuse light. This tallies with the known existence in the red of photo-stimulation power. A sudden change of 10–12° C. in the temperature has effects very similar to temporary illumination. This leads VOGT to inquire whether the effect of temporary illumination is not due

---

[19] VOGT, E., Über den Einfluss des Lichts auf das Wachstum der Koleoptile von *Avena sativa*. Zeitschr. Bot. 7:193–270. 1915.

[20] BOT. GAZ. 59:67–68. 1915.

to a purely physical change, perhaps a temperature change. Some will object to the distinction between physical and physiological changes, for is not the latter, as well as its physical aspects, largely unknown? In this connection it might be mentioned that we know little about the effect of light upon "hydration" of colloids in general, and of cell colloids especially. Recently we have been coming to see that this is quite as important in growth as is turgor pressure.—WILLIAM CROCKER.

Embryo sac of Myosurus.—In 1913 NAWASCHIN and FINN[21] concluded that chalazogams are primitive and derived from gymnosperms, the conclusion being based largely upon the reduction series shown by sperms. They found that the generative cell of *Juglans* reaches the embryo sac as a binucleate cell, which means that nuclear division had occurred, but that distinct male cells had not been organized. This condition, carried over from gymnosperms, was regarded as an intermediate one between a well developed sperm and a naked sperm nucleus. TCHERNOYAROW[22] has now found the same condition in *Myosurus minimus*, and concludes that this is probably a feature of the "Polycarpicae," and that this phylum and the chalazogams are two independent primitive branches from the gymnosperm stock. The paper also contains a detailed account of the events from the development of the pollen tube to the act of fertilization. In general the events are of the usual kind, but attention may be called to the fact that the author lays special stress upon the idea that there is some coordination which assures the discharge of the pollen tube at the moment of maturity of the embryo sac.—J. M. C.

Northern plains forests of Canada.—A recent report by CONNELL[23] deals with the forest region lying north of the prairies in Manitoba, Saskatchewan, and Alberta. It extends from the contact line with the Laurentian pine plain near Lake Winnipeg in the east, to the Rocky Mountains in the west. The northern limits are not determined. If the portion of northern Alberta most carefully studied be taken as typical of the entire region, it is made up as follows: (1) boulder clay slopes, which comprise about 50 per cent of the area, more than half of these slopes being covered with a poplar association, a little less than one-fourth with a mixed spruce-poplar forest, and less than 2 per cent with a pure spruce association; (2) sand ridges, which amount to 18 per cent of the area and are covered with nearly pure *Pinus Banksiana*, *P. Murrayana* appearing in mixture with it in the western portion of the region; (3) swamps, which occupy the remainder, showing how poorly developed the drainage is. In the swamps there is a stunted growth of *Picea mariana* and *Larix laricina*, the former predominating.—GEO. D. FULLER.

[21] Rev. BOT. GAZ. 57:162. 1914.

[22] TCHERNOYAROW, M., Les nouvelles données dans l'embryologie du *Myosurus minimus* L. Mém. Soc. Nat. Kiew 24:95-170. 1915.

[23] CONNELL, A. B., Some aspects of the nothern plains forest region of Canada. Forestry Quarterly 13:31-34. 1915.

NEW

NICAL GA

Editor: JOHN M. COULTER·

DECEMBER 1915

VOLUME LX

NUMBER 6

THE

BOTANICAL GAZETTE

DECEMBER 1915

MASS MUTATION IN OENOTHERA PRATINCOLA[1]

HARLEY HARRIS BARTLETT

(WITH FIFTEEN FIGURES)

Introduction

Of the several small-flowered wild evening primroses thus far examined by the writer for mutability, no other has yielded as valuable data as *Oenothera pratincola*. Certain mutations of this species have been treated in a former article,[2] of which this one is in effect a continuation. To recapitulate very briefly, it may be recalled that *O. pratincola*, a species found wild at Lexington, Kentucky, gives rise in successive generations to a small proportion of mutations, belonging to several distinct types. Of these the most conspicuous in the young condition is mut. *nummularia*, which originates in every generation from seven of the eight independent strains which have been studied. The eighth strain, designated in the former article as Lexington E, shows the phenomenon which the writer has elsewhere designated as mutation *en masse*.[3] Mutant species in *Oenothera*, as typified by *O. Lamarckiana*, give rise to few mutations. The frequency of mutations in

[1] From the Bureau of Plant Industry, U.S. Department of Agriculture, Office of Plant Physiological and Fermentation Investigations. Published by permission of the Secretary of Agriculture.

[2] BARTLETT, H. H., Additional evidence of mutation in *Oenothera*. BOT. GAZ. 59:81–123. 1915.

[3] ———, Mutation *en masse*. Amer. Nat. 49:129–139. 1915.

O. Lamarckiana is about 2 per cent. Some of its mutations are themselves mutable and give 4 per cent or even 6 per cent of secondary mutations.[4] The ordinary strains of *O. pratincola* produce about 2 per cent of mutations,[5] and the true *O. biennis*, of Holland, only about 0.45 per cent.[6] In strains which show mass mutability, however, the number of mutations rises to 50 per cent or even 100 per cent. This is the case with Lexington E.

Lexington E differs from all the other strains of *O. pratincola* (1) in that it gives rise to a characteristic group of four mutations, (2) in that these characteristic mutations occur in such large numbers as to justify the use of the designation mass mutation for the phenomenon, and (3) in that it does not give rise to mut. *nummularia* and certain other mutations which are characteristically produced by the strains which do not show mass mutation.

Turning now to resemblances, we find absolutely no morphological characteristics to indicate even the slightest difference between Lexington E and the other strains of *O. pratincola*. Moreover, Lexington E shows also the ordinary type of mutability, in that it gives rise to small numbers of some of the same mutations which the other strains produce. These mutations are in no wise distinguishable morphologically from the same forms occurring in the other strains, but all the evidence at hand shows that when derived from the strain showing mass mutation the non-characteristic mutations themselves show mass mutability. The mutations characteristic of mass mutability are themselves mutable and throw as secondary mutations other members of the characteristic group.

In view of the extraordinary interest of the phenomenon of mass mutation, it is hoped that the reader will pardon the presentation of this confessedly preliminary report. Many of the genetic relationships between the mutations and the parent species remain to be worked out, and, as will be very obvious, a cytological study of the whole subject is imperative.

[4] DeVries, H., Gruppenweise Artbildung. pp. 312–315.

[5] An estimate based upon the results tabulated in Bot. Gaz. 59:105–109. 1915.

[6] DeVries, H., The coefficient of mutability in *Oenothera biennis* L. Bot. Gaz. 59:169–196. 1915.

The characteristic mutations of the mass mutating strain

The four mutations, which by occurring in such large numbers characterize the hereditary behavior of Lexington E, form a group with a common structural peculiarity which sets them apart from typical *O. pratincola* and from all the other mutations. This peculiarity lies in the narrow, strongly revolute, veiny leaves, of which the midrib is frequently but not always prolonged from a point below the apex as a setiform appendage. This appendage strongly resembles the infra-terminal calyx tips of certain species of *Oenothera*, and suggests the translocation of a character from one organ to another which does not typically display it. The setiform appendage is exceptionally a centimeter long but is entirely absent on some leaves of each individual mutation. It may be said that the four mutations are characterized by the possibility of manifesting the appendage under favorable conditions rather than by its invariable presence. The revoluteness, narrowness, and venoseness of the leaves, however, are characters which are always distinctive.

The four characteristic mutations may be contrasted as follows:

Plants as tall as *O. pratincola* f. *typica*, with a much-branched terminal inflorescence; fruiting freely by self-pollination and producing a normal number of viable seeds....................................mut. *formosa*

Plants semi-dwarf; leaves whitish, broader and thicker than in the last; inflorescence often simple and bearing a few thick-tissued, usually cleistogamous flowers; producing by self-pollination large, apparently normal capsules which contain very few seeds.................. ....mut. *albicans*

Plants semi-dwarf; leaves green, narrower than in mut. *formosa;* inflorescence densely branched and many-flowered; ovaries almost sterile, producing no capsules by normal self-pollination and only shriveled capsules with few seeds by artificial pollination...............................mut. *revoluta*

More extreme dwarfs, with narrowly linear leaves; inflorescence-bearing branches with broader leaves than the rest of the plant, simple, with thick-tissued, usually cleistogamous flowers which produce large normal fruits but very few viable seeds by natural self-pollination ....mut. *setacea*

The four mutations do not form a linear series showing successive degrees of reduction. Mut. *formosa* and mut. *revoluta* are very similar and might be interpreted as successive reduction stages. They differ in size and fertility, but have many morphological characters in common. Before flowering they can be distinguished with certainty only when the environmental conditions are the same for both. Thus, mut. *revoluta* grown in a rich, moist soil is as large as mut. *formosa* grown in a dry, sandy soil. Under such conditions they might be indistinguishable until they flowered, when the latter would produce large capsules, filled with good seeds, and the former would produce few seeds or none in the shriveled ovaries. Grown under identical conditions, however, the two mutations differ at every stage of development. In some features mut. *setacea* also appears to be a reduction derivative in the same series with mut. *formosa* and mut. *revoluta*. In comparison with the latter, however, it shows a partial resumption of fertility. Its large, strong capsules are well filled, although the seeds are for the most part empty. In this characteristic, as also in its simple inflorescences and thick-tissued buds and flowers, it resembles mut. *albicans*. Mut. *setacea* is different from the other three mutations in its strong tendency to produce dimorphic foliage. The rosette leaves and young cauline plants have narrowly linear, grasslike leaves, which are succeeded above and on the inflorescence-bearing lateral branches by leaves much like those of mut. *revoluta*, which nearly always show the setiform terminal appendage. Like the latter mutation, mut. *setacea* responds greatly to environmental changes. In dry sandy soil it flowers and fruits when only 10 cm. high, but in moist loam it becomes 50 cm. high and has quite a different aspect. The comparatively broad-leaved mut. *albicans* is totally unlike the other mutations at every stage of development.

THE F_1, F_2, AND F_3 PROGENIES OF FORMA *typica*

The original wild mother plant designated as Lexington E did not give a progeny in any way peculiar when it was first grown in 1913. A casual inspection of the F_1 seedlings disclosed no mutations. The majority of the plants of this first culture were discarded as very young seedlings and only 30 were brought to maturity. These 30 plants were entirely typical. In 1914 the F_2 of the strain was found to show mass mutation. The remaining seeds of the original collection were therefore sown, in order to detect any mutability which, on account of the use of insufficiently rigorous experimental methods, might have been overlooked the year before.   The results from the new F_1 cultures of 1914 are set forth in table I.

TABLE I

ANALYSIS OF THE F_1 SEEDLING CULTURES OF LEXINGTON E

Culture	Seeds planted	Total plants	Forma *typica*	Mut. *albicans*	Mut. *setacea*	Mut. *latifolia*	Mut. *graminea*
2........	200*	162	160†	o	1 (no. 35)	1 (no. 36‡)	o
3........	199*	175	172	1 (no. 34‡)	o	o	2 (nos. 32‡ and 33‡)
4........	203*	62	62	o	o	o	o
5........	185*	123	123	o	o	o	o
Total	787	522	517†	1	1	1	2

* Indicates seeds from one capsule.
† 25 plants of f. *typica* from culture 2, grown to maturity, were uniform.   The remaining plants of f. *typica* were discarded in the rosette stage.
‡ Indicates that the mutation was grown to maturity.

It is clear from table I that the F_1 did not point to Lexington E as a specially mutant strain.   There were only 5 mutations in a progeny of 522 plants.   Moreover, 2 of the 4 types obtained, mut. *latifolia* and mut. *graminea*, were common to the other strains of *O. pratincola*.

In 1914, F_2 progenies were grown from 3 plants of *O. pratincola* f. *typica* belonging to strain E, and the progeny of a fourth was grown in 1915.   The results are summarized in table II.

The F_2 shows a decidedly greater degree of variability than the F_1.   One progeny only, that from Lexington E-5, shows mutations in excess of the number of typical plants; the other three progenies indicate a degree of mutability more comparable with that of certain derivatives of *O. Lamarckiana*, such as *O. scintillans*.   The F_1, however, was, if anything, less mutable than *O. Lamarckiana* itself.   Successive generations seem to show an increasing degree of mutability.   Only one F_3 progeny from f. *typica* has been studied. The parent belonged to the progeny of Lexington E-5, that is, it was selected from the most mutable line.   The analysis of the F_3 culture is shown in table III.

The salient fact shown by the data for the F_1, F_2, and F_3 progenies is that the number of mutations varies inversely with the number of seeds per capsule.   The F_1 progeny, with few mutations, came from capsules with about 200 seeds.   (Perhaps the capsules

TABLE II

ANALYSIS OF THE F_2 CULTURES OF LEXINGTON E, FROM 4 F_1 F. *typica* PARENTS

Parent	Culture	Seeds planted	Total plants	Forma typica†	Mut. formosa	Mut. albicans	Mut. revoluta	Mut. setacea	Other mutations	Total mutations	Percentage of mutations
Lex. E-5.	1	156*	131	109	0	0	3	17	2 (no. 208, *angustifolia*)	22	
"	2	96*	72	26	0	6	3	36	1 (no. 210, *chimaera*)	46	
"	3	106*	75	33	2	6	1	33	0	41	
"	4	75*	57	15	1	3	5	33	0	42	
"	5	104*	69	10	2	1	4	51	1 (no. 238, *gigas*)	59	
Total..	1-5	537	404	194	4	15	15	172	4	210	52
Lex. E-19..	1	340*	270	261	0	0	4	5	0	9	
"	2	350*	239	224	0	0	4	11	0	15	
"	3	219*	150	142	0	0	4	4	0	8	
"	4	299*	230	219	0	0	4	7	0	11	
"	5	378*	143	108	0	1(no. 67)	7	27	0	35	
Total..	1-5	1586	1032	954	0	0	24	54	0	78	7 5
Lex. E-25.	1	187*	136	130	0	0	3	3	0	6	
"	2	225*	147	142	0	0	1	4	0	5	
"	3	186*	145	138	0	0	4	3	0	7	
"	4	180*	147	134	0	1	3	9	0	13	
"	5	144	72	66	0	0	1	4	1 (no. 34)	6	
Total..	1-5	922	647	610	0	1	12	23	1	37	5 7
Lex. E-43..	1	410‡	275	238	0	0	1	36	0	37	
"	2	424‡	362	328	0	2	4	27	1 (no. 70§)	34	
Total...	1 and 2	834	637	566	0	2	5	63	1	71	11

* Seeds from one capsule.

† Plants of f. *typica* were grown to maturity as follows: Lex. E-5, 49 plants from cultures 2, 3, 4, and 5, including all of the *typica* plants shown in figs 1, 2, 3, 4, and 5; Lex. E-19, 18 plants from cultures 1 and 5; Lex. E-25, 23 plants from culture 5; Lex. E-43, 29 plants from culture 2. The remainder were classified in the seedling stage and discarded.   The mutations were all retained and classified at maturity, except that some of the weaker specimens of mut. *setacea* died at various stages of development

‡ Seeds of two capsules.

§ Lex E-43-70 was a new mutation combining characters of mut. *nitida* and mut. *angustifolia*, two frequent derivatives of *O. pratincola*.

had dehisced at the apex and lost part of their seeds, as frequently happens.   A normal capsule of *O. pratincola* contains 300 seeds, more or less.) The F₂ progenies of 3 mother plants whose capsules contained an average of about 250 seeds gave about 8 per cent of mutations, the upper limit of ordinary mutability, as far as experience goes. Another F₂ progeny from a mother plant with about 110 seeds to the capsule gave over 50 per cent of mutations. Turning to the very striking F₃ progeny, we find that a mother plant

TABLE III

ANALYSIS OF F₃ CULTURES OF LEXINGTON E, FROM LEX. E.-5-229, F. *typica*

Culture	Seeds planted	Total plants	Forma typica	Mut. formosa	Mut. albicans	Mut. revoluta	Mut. setacea	Other mutations	Total mutations	Percentage of mutations
1........	39	30	9	0	1	1	19	0	21	70
2........	86*	56	6	2	4	4	39	1 (*angustifolia*)	50	89.3
3........	91*	71	17	0	3	8	43	0	54	76.1
4........	100*	74	22	0	3	5	42	2 (*gigas?*)	52	70.3
5........	80*	48	20	3	3	4	18	0	28	58.3
6......	97*	69	24	0	4	2	39	0	45	65.2
7 ...	95*	51	19	0	3	1	28	0	32	62.7
8......	104*	73	21	0	4	3	44	1(?)	52	71.2
9.....	105*	68	30	0	3	0	35	·0	38	55.9
10	142*	102	8	0	0	5	89	0	94	92.2
11.	59*	39	4	0	5	2	28	0	35	89.8
12 .	79*	65	13	0	5	3	43	1 (*gigas?*)	52	80.0
13. ...	83*	65	8	1	4	6	46	0	57	87 7
14 .	85*	42	18	1	0	1	22	0	24	57.3
15.....	72*	59	19	2	6	2	30	0	40	67.8
16......	73*	51	15	1	7	2	26	0	36	70.6
17	87*	73	13	0	2	5	53	0	60	82 2
Total..	1477	1036	266	10	57	54	644	5	770	74 3

* Seeds from one capsule; the entire progeny was classified from the young seedlings; 30 plants of f. *typica* and all the mutations except the weaker individuals of mut. *setacea* were retained.

with only 90 seeds to the capsule gave almost 75 per cent of mutations. It would be necessary to have much more complete data to establish any exact relationship between progressive sterility and mutability. Nevertheless, it is beyond question that the decrease in the number of seeds has gone hand in hand with the increase in mutability.

In *Oenothera pratincola*, therefore, the phenomenon which I have termed mutation *en masse* is associated with the failure of a large number of zygotes to develop, for the number of ovules in all capsules appears to be about the same. Probably the zygotes which fail to develop into embryos represent the weaker individuals of mut. *setacea*, or perhaps some still more reduced mutation which is incapable of development. The problem, however, must be attacked by cytological methods. It is interesting to observe that the F_3, with an average seed germination of 70 per cent, contained 75 per cent of mutations. If all the seeds had germinated, and the additional plants had all been f. *typica*, there would still have been more than 50 per cent of mutations. It seems far more likely, however, that the seeds which did not germinate were either empty or else that they were the weak mut. *setacea*.

A comparable degree of mutability to that of Lexington E is known only in the case of *O. Reynoldsii*, in which mass mutation was first described. In *O. Reynoldsii*, also, the great increase in mutability is associated with an enormously increased degree of sterility. The data in regard to the latter species will soon be published elsewhere.

A number of photographs were made to record the appearance of the mutations at various stages of growth. Figs. 1–5 show a portion of the F_2 progeny from F_1 parent Lexington E-5, recorded in table II. All of the characteristic mutations are shown, as well as several plants which became the parents of subsequent cultures. Fig. 6 shows 6 rosettes of mut. *setacea*, one of which will be found in fig. 3. Fig. 7 shows two mature plants of the same mutation, of which one is shown in fig. 5. Fig. 8 shows 4 rosettes of mut. *revoluta*, 3 of which are likewise shown in figs. 4 and 5. A mature plant is shown in fig. 9. Fig. 10 shows 6 rosettes of mut. *albicans*, 3 of which will be found in figs. 2, 3, and 4. Fig. 11 shows two cauline plants of the same form. The one on the left is just beginning to flower; the one on the right lingered in the rosette condition and would therefore have matured as a stronger plant than its sister. Rosettes of mut. *formosa* are shown in fig. 12. The main stem of the mature plant is shown in fig. 13, and the identical cross mut. *formosa*×f. *typica* in fig. 14. Fig. 15 shows young rosettes of f. *typica* and mut. *gigas*. The *gigas* plant was the particular individual in which E. G. ARZBERGER determined the chromosome number to be 28. For a figure showing the mature stem of f. *typica* the former article on *O. pratincola* in this journal[2] should be consulted.

Lexington E-5 Pan 5
← 10cm →

FIG. 1.—Progeny of Lexington E-5, pan 5 (part of culture 2; see table II); the pan contained 23 rosettes which were classified, most of them at maturity, as follows:

Row	1	2	3	4	5
1 .	typica	typica	setacea	setacea	setacea
2	albicans	setacea	typica	setacea	setacea
3......	setacea	albicans (no 187, see fig 11)	typica	setacea	setacea
4 . .	typica	setacea	setacea	typica	setacea
5 ..	setacea	setacea	typica		

Mass mutability of the non-characteristic mutations

It has already been stated that Lexington E gives rise to certain mutations which are common to the other strains of *O. pratincola* from Lexington, and that these non-characteristic mutations, as

FIG. 2.—Progeny of Lexington E-5, pan 6 (part of culture 3; the remainder is shown in fig. 3, see table II), the pan contains 49 plants, classified as follows:

Row	1	2	3	4	5	6	7
1 .	typica	typica	typica	setacea	typica	setacea	typica
2 .	typica	setacea	setacea	setacea	setacea	typica	setacea
3 .	typica	setacea	setacea	typica	setacea	setacea	albicans (no 190, see fig 10)
4........	typica	setacea	setacea	typica	setacea	albicans	setacea
5........	typica	setacea	setacea	typica	typica	setacea	typica
6 .	typica	typica	typica	typica	typica	typica	typica
7........	setacea	typica	albicans	setacea	setacea	typica	setacea

they may be called for convenience, exhibit mass mutability superposed upon their ordinary behavior in heredity.   In order to present
the data in regard to this point it will be necessary to anticipate
somewhat the publication of the next article of this series.

Among the mutations of *O. pratincola* which have been obtained
both from Lexington E and from the strains showing only ordinary

FIG. 3.—Progeny of Lexington E₋₅, pan 7 (part of culture 3; the remainder is
shown in fig. 2; see table II); the pan contains 26 plants, which were classified at
maturity as follows:

Row	1	2	3	4	5	6
1	formosa	setacea	albicans (no 191, see fig 10)	setacea	setacea	
2	typica	albicans	setacea	setacea	typica	
3	typica	setacea	setacea	albicans	typica	
4	typica	setacea	typica	typica	setacea	
5	setacea	setacea	setacea	setacea	setacea	setacea (no 88; see fig 6)

mutability, two, mut. *angustifolia* and mut. *latifolia*, have been carried into an F_1 generation from parent plants derived from both sources. A third non-characteristic mutation, mut. *gigas*, has appeared in Lexington E and also, apparently, in the other strains,

FIG. 4.—Progeny of Lexington E-5, pan 8 (part of culture 4; see table II); the pan contains 49 plants, which were classified (most of them at maturity) as follows:

Row	1	2	3	4	5	6	7
1	typica	setacea	setacea	typica	typica	setacea	setacea
2	setacea	albicans	typica	setacea	formosa	setacea	setacea
3	typica	setacea	typica	revoluta (no. 203, see fig 8)	setacea	typica	revoluta (no. 202; see fig. 8)
4	setacea	typica	setacea	typica	typica	setacea	setacea
5	setacea	typica	setacea	typica	setacea	albicans (no. 195, fig. 10)	setacea
6	setacea	setacea	setacea	setacea	setacea	setacea	setacea
7	setacea	revoluta	setacea	setacea	setacea	setacea	setacea

but no progenies have yet been grown or chromosome counts made, except in the case of one plant, belonging to the mass mutant strain.

Lexington E-5 Pan 11
10 cm

FIG. 5.—Progeny of Lexington E-5, pan 11 (part of culture 5; see table II); the pan contains 49 plants, which were classified (most of them at maturity) as follows:

Row	1	2	3	4	5	6	7
1...........	setacea	gigas (no. 238; see fig. 15)	typica (no. 239; see fig. 15)	setacea	setacea	setacea	setacea
2...........	setacea	setacea	setacea	setacea	setacea	setacea	setacea
3...........	setacea	typica	typica	setacea	setacea	typica	setacea
4...........	setacea	setacea	setacea	formosa (no. 206; fig. 12)	setacea	revoluta	setacea
5...........	setacea	setacea	setacea	setacea (no. 164; see fig. 7)	setacea	formosa (no. 207; see fig. 12)	revoluta (no. 172; see fig. 8)
6...........	setacea	setacea	albicans	setacea	setacea	revoluta	setacea
7...........	typica	setacea	setacea	setacea	setacea	setacea	setacea

(This plant is shown in figs. 5 and 14.) All three mutations are discussed here, but full data and illustrations are reserved for a paper entitled "Certain mutations and hybrids of *Oenothera pratincola*," to appear later in this journal.

Mut. *angustifolia.*—It has been found that mut. *angustifolia* from ordinary strains gives no descendants resembling itself; aside from the usual small proportion of other mutational types, the progeny consisting of f. *typica* only. The variation, although striking and entirely discontinuous, appears to be somatic. In this connection it is interesting to note that a perfect branch of f. *typica* has been observed as a bud sport on mut. *angustifolia*. As would be expected, mut. *angustifolia* crossed reciprocally with f. *typica* gives f. *typica* together with the usual few mutations. The hereditary behavior may be stated:

mut. *angustifolia*×mut. *angustifolia* → f. *typica*
mut. *angustifolia*×f. *typica* → f. *typica*
f. *typica*×mut. *angustifolia* → f. *typica*

The behavior of mut. *angustifolia* from Lexington E is most remarkable. The parent plant was a sister of the *typica* plant whose progeny is analyzed in table III, but the degree of mutability proved to be much more extreme than in the case of the *typica* sister. Only two plants in the progeny, out of a total of 505, were f. *typica;* the other 503 plants, 99.6 per cent of the progeny, were mutations belonging to the group characteristic of Lexington E. The results are summarized in table IV.

The cross mut. *angustifolia*×f. *typica* and the reciprocal gave respectively 100 per cent and 91.5 per cent of mutations. By comparison with table III it will be seen that each of the reciprocal crosses tends to show the same degree of mutability as the female parent. To be sure, there would seem to be a considerable discrepancy between 74.3 per cent, representing the mutability of f. *typica*, and 91.5 per cent, representing the mutability of f. *typica* ×mut. *angustifolia*. No significance can be urged for this discrepancy, however, when we consider that one of the cultures from a single capsule of f. *typica* contained 92.2 per cent of mutations among 102 plants, as compared with 91.5 per cent of mutations

FIG. 6.—Mut. *setacea:* 6 rosettes of the F_2 progeny of f. *typica*, from F_1 parent Lexington E-5; the plant in the upper left-hand corner, Lexington E-5-88, is shown also in fig. 3.

among 118 plants of the cross. As a matter of fact, the significance of ratios in *Oenothera* can be maintained only with great caution, in view of the enormous elimination of gametes during maturation and the subsequent failure of large classes of zygotes to develop. Nevertheless, the absence or almost complete absence of a strong zygote such as f. *typica* in the progenies of mut. *angustifolia* and mut. *angustifolia*×f. *typica* is strong evidence for the view that the composition of the progeny among the mass mutating strains is conditioned by the female gametes. The failure of a

TABLE IV

ANALYSIS OF F₁ CULTURES OF MUT. *angustifolia*, LEXINGTON E-5-208, SELF-POLLINATED AND RECIPROCALLY CROSSED WITH F. *typica*, LEXINGTON E-5-229 (THE PLANT WHOSE PROGENY IS ANALYZED IN TABLE III)

The mutation was a sister plant of the *typica* plant with which it was crossed; for position in pedigree see table II, culture 1 from Lexington E-5

Parent	Seeds planted	Total plants	Forma typica	Mut. formosa	Mut. albicans	Mut. renovia	Mut. selaces	Other mutations	Total mutations	Percentage of mutations
Mut. *angustifolia*..	651*	505	2	4	2	21	475	1 (no. 1)	503	99.6
Mut. *angustifolia* ×f. *typica*......	199†	173	0	0	0	4	168	1 (no. 1)	173	100
F. *typica*×mut. *angustifolia*.....	182‡	118	10	0	1	4	99	4	108	91.5

* The 651 seeds were from 11 capsules, containing respectively 51, 75, 53, 62, 44, 59, 82, 101, 46, 33, and 45 seeds.

† The 199 seeds were from 3 capsules, containing respectively 73, 47, and 79 seeds.

‡ The 182 seeds were from 2 capsules, containing respectively 79 and 103 seeds.

class of strong zygotes to appear has much greater evidential value than any fluctuation in the proportion of weak zygotes. From other sources the evidence is unusually strong that the female and male gametes of *O. pratincola* are not equivalent, and that many characters are not carried by the male gametes.

In conclusion: mut. *angustifolia* ordinarily gives a progeny containing nearly 100 per cent of f. *typica;* in a strain exhibiting mass mutation many of the *typica* plants are replaced by mutations of the characteristic group. Presumably other individuals of mut. *angustifolia* could be found which would be less mutable than the one tested, just as different individuals of f. *typica* show widely varying degrees of mutability.   The progenies of crosses indicate that mass mutability is conditioned by the female gametes.

Mut. *latifolia*.—In contrast with mut. *angustifolia*, mut. *latifolia* reproduces itself in part of its progeny.   Its descendants include roughly 50 per cent f. *typica* and 50 per cent mut. *latifolia*, the proportion varying within rather wide limits.   Moreover, mut. *latifolia* gives progenies of the same type whether self-pollinated or

Fig. 7.—Mut. *setacea:* 2 mature plants from the F₂ progeny of f. *typica*, from F₁ parent Lexington E-5; the right-hand plant, Lexington E-5-164, is shown also in fig. 5; note particularly the dimorphic foliage.

cross-pollinated with f. *typica*.   The reciprocal cross, with f. *typica* as the pistillate parent, consists only of f. *typica*, aside from the usual low proportion of mutations, among which mut. *latifolia* may or may not happen to occur.   These relations are as follows:

mut. *latifolia*×mut. *latifolia* → f. *typica*+mut. *latifolia*
mut. *latifolia*×f. *typica* → f. *typica*+mut. *latifolia*
f. *typica*×mut. *latifolia* → f. *typica*

The type of heredity here exemplified is shown by several mutations from *O. Lamarckiana*. *O. lata* DeVries provides the classic case. The heredity of *O. scintillans* DeVries and *O. oblonga*

FIG. 8.—Mut. *revoluta:* 4 rosettes of the F_2 progeny of f. *typica*, from F_1 parent Lexington E.-5, the upper right-hand plant, Lexington E.-5-172, will be found in fig. 5; the two lower plants, nos. E.-5-202 and 203, will be found in fig. 4.

DeVries is in essentials the same.[7] Another case is provided by
O. stenomeres mut. *lasiopetala*.[8] HERIBERT-NILSSON[9] has recently
described several new mutations from *O. Lamarckiana* ("hetero-
game Kombinanten" *dependens, undulata, stricta*, etc.) which
probably show the typical *lata* type of inheritance, although he
erroneously concludes that the repeated segregation of *O. Lamarckiana* from these mutations in each generation is due to the existence of two types of functioning gametes on the male rather than
on the female side.

It must not be inferred from the similarity of the names that
mut. *latifolia* is a parallel variation to *O. lata* DeVries. Such is
not the case. Its characters are quite different. Both mut.
latifolia and mut. *angustifolia* will be described and illustrated in a
future article.

As in the case of mut. *angustifolia*, the progeny of mut. *latifolia*
from the mass mutating strain contained the expected types, f.
typica and mut. *latifolia*, together with the characteristic mutations.
The latter did not show differences among themselves which would
enable one to classify them as modified *typica* or modified *latifolia*,
as the case might be.   The mother plant belonged to the F_1 generation from Lexington E, and showed about the same degree of mutability as the *typica* sister plant, Lexington E-5 (see table II).   The
data for mut. *latifolia* are summarized in table V.

Mut. *gigas*.—E. G. ARZBERGER's discovery that this mutation
has 28 chromosomes has already been announced.[10] The count
has been made only in one plant, Lexington E-5-238 (figs. 5 and 14),
belonging to the mass mutant strain. An apparently identical
mutation in one of the other strains has appeared this summer
(1915), but its heredity is unknown. Mut. *gigas* is treated, therefore, as a non-characteristic mutation. Only 196 seeds were

[7] For the latest treatment of these mutations see DEVRIES, Gruppenweise
Artbildung. pp. 244–267.

[8] BARTLETT, H. H., The mutations of *Oenothera stenomeres*. Amer. Jour. Bot. 2:
100–109. 1915.

[9] HERIBERT-NILSSON, N., Die Spaltungserscheinungen der *Oenothera Lamarckiana*.
Lunds Universitets Arsskrift. N.F. Avd. 2. 12: no. 1. pp. 132. 1915.

[10] BARTLETT, H. H., The experimental study of genetic relationships. Amer.
Jour. Bot. 2:132–155. 1915 (see p. 143).

obtained from 15 capsules of the primary mutation, Lexington E-5-238. The progeny consisted of 25 plants, only 16 of which survived transplanting from the seed pan. None of the progeny resembled the parent. All were extreme dwarfs which resembled, but were not identical with, mut. *revoluta* and mut. *setacea*. They differed mainly in the thicker leaves, which in 4 plants were narrow but not markedly revolute. Although a very nondescript lot, differing much among themselves, 5 most resembled mut. *revoluta*, and 7 mut. *setacea*. The result of this culture might almost have been predicted. The mass mutability was inherited by mut. *gigas* from f. *typica*. In view of the dependence of the *gigas* characters upon

TABLE V

ANALYSIS OF AN F₁ CULTURE FROM SELF-POLLINATED MUT. *latifolia*, LEXINGTON E-36

The parent plant belonged to the F₁ from Lexington E, f. *typica;* see table I, culture 2, for position in pedigree

Parent	Seeds planted	Total plants	Forma *typica*	Mut. *latifolia*	Mut. *albicans*	Mut. *revoluta*	Mut. *setacea*	Other mutations	Total mutations‡	Percentage of mutations‡
Lex. E-36, mut. *latifolia*........	375*	182	95	48	5	2	29	3†	87	47 8

* Seeds from 7 capsules, containing respectively 54, 72, 49, 30, 38, 70, and 62 seeds.
† Nos. 32 and 43, a new mutation; no. 51 mut. *gigas* (?), morphologically identical with Lexington E-5-238 which had 28 chromosomes, see table II, culture 5 from Lexington E-5, for position of latter plant in pedigree.
‡ Excluding, of course, mut. *latifolia*.

the double complement of chromosomes, which would in general be handed on to any secondary mutations, it follows that the mutations occurring *en masse* would not be identical with those from f. *typica*. Furthermore, chance irregularities in chromosome distribution might increase the polymorphism of the progeny. In such a highly modified germ plasm irregularities would be expected.

The cross mut. *gigas*×f. *typica* yielded 160 seeds in a single capsule, of which 10 germinated. The plants were all extreme dwarfs, of the most nondescript nature, hardly any two alike. All had thick leaves, some plane, others revolute. No mutation of

f. *typica* could be identified among them. The reciprocal cross yielded no seeds.

Inheritance and mutability of the characteristic mutations

Of the group of characteristic mutations, including mut. *formosa*, mut. *albicans*, mut. *revoluta*, and mut. *setacea*, only the first is both normally fertile and vigorous. The second is vigorous, but produces few good seeds. The third is almost sterile, and the fourth is not only difficult to cultivate, but like mut. *albicans* gives very few good seeds. All of the forms were self-pollinated and reciprocally crossed with f. *typica* in 1914, but, except in the case of mut. *formosa*, the resulting F_1 cultures were very fragmentary or entire failures. The other three forms bloomed in September, when only a few weak, belated flowers of f. *typica* were available for the crosses. The results of the cultures are summarized in table VI.

Mut. *formosa*.—The entirely satisfactory cultures of this form show that it is constant in the sense that it gives no reversions to f. *typica* in its progeny. Moreover, there is no

FIG. 9.—Mut. *revoluta:* a mature plant, Lexington E.-19-21 (for position in pedigree see table II).

TABLE VI

ANALYSIS OF F_1 CULTURES IN THE GROUP OF MUTATIONS CHARACTERISTIC OF MASS MUTABILITY

Parentage culture	Seeds planted	Total plants	Forms *typica*	Mut. *formosa*	Mut. *albicans*	Mut. *revoluta*	Mut. *setacea*	Other mutations	Total mutations
Mut. *formosa* Lex. E-5-199 — 1	244*	196	0	171	0	1	24	0	25
2	269*	163	0	137	0	0	26	0	26
3	201*	177	0	149	0	2	26	0	28
4	216*	174	0	130	1	0	43	0	44
5	217*	193	0	150	0	1	42	0	43
6	224*	179	0	146	0	0	33	0	33
Total 1–6	1371	1082	0	883	1	4	194	0	199
Mut. *formosa* Lex. E-5-199 × f. *typica* Lex. E-5-217	365†	309	0	218	0	0	91	0	91
Mut. *formosa* Lex. E-5-206	177‡	146	0	130	2	0	14	0	16
Mut. *formosa* Lex. E-5-206 × f. *typica* Lex. E-5-229	233§	177	0	151	0	3	23	0	26
F. *typica* Lex. E-5-229 × mut. *formosa* Lex. E-5-206	246‖	133	7	0	1	0	121	4	126
Mut. *albicans* Lex. E-19-67 (seeds of 4 capsules)	386	265	0	0	36	3	226	0	229
Mut. *albicans* Lex. E-5-182 (seeds of 4 capsules)	173	69	0	0	7	3	59	0	62
F. *typica* Lex. E-5-229 × mut. *albicans* Lex. E-5-182	22*	14	1	0	0	1	12	0	13
Mut. *revoluta* Lex. E-5-190 (seeds of 14 capsules)	85	23	0	0	1	17	5	0	6
Mut. *setacea* Lex. E-5-17 (seeds of 4 capsules)	625	140	0	0	0	0	140	0	0
Mut. *setacea* Lex. E-5-20 (seeds of 16 capsules)	1997	8	0	0	0	0	8	0	0
Mut. *setacea* Lex. E-5-66 (seeds of 2 capsules)	114*	31	0	0	0	0	31	0	0
Mut. *setacea* Lex. E-5-135 (seeds of 3 capsules)	463	14	0	0	0	0	14	0	0

* Seeds from one capsule.
† Seeds from 2 capsules; 215+150.
‡ Seeds from 2 capsules; 48+129.
§ Seeds from 2 capsules; 64+169.
‖ Seeds from 2 capsules, 115+131.

Fig. 10.—Mut. *albicans:* 6 rosettes of the F₂ progeny of f. *typica*, from F₁ parent plant Lexington E-5; the right-hand plant in the middle row, Lexington E-5-190, will be found in fig. 2; the two plants in the lower row, E-5-191 (left) and 195 (right), will be found respectively in figs. 3 and 4.

difference between the progenies resulting from self-pollination and those resulting from pollination with f. *typica*. In other words, mut. *formosa* is dominant over f. *typica* if it enters the cross as a female gamete, but is not even dominant over the weak mut. *setacea* when the *formosa* gamete is male. The scheme of heredity is:

mut. *formosa* × mut. *formosa* → mut. *formosa*
mut. *formosa* × f. *typica* → mut. *formosa*
f. *typica* × mut. *formosa* → f. *typica*

It would have been instructive to cross mut. *formosa* reciprocally with f. *typica* from a non-mass mutant strain. (Such crosses have been made this year and will be grown next year.) From the data at hand, concerning only crosses within the mass mutant strain, it appears clear that the external features of all the characteristic mutations are determined by the female gametes. The female and male gametes are not equivalent. Thus, the progeny obtained from f. *typica*, Lex. E-5-229, by pollination with mut. *formosa* is not significantly different from the progeny obtained by self-pollination (cf. tables III and VI). The characteristic mutations occur with their usual frequency regardless of which pollen is used. We know that this particular individual of f. *typica* gave about 1 per cent mut. *formosa* when grown in large cultures from self-pollinated seeds. That pollination with pollen of mut. *formosa* does not increase the proportion of this mutation in the progeny is strikingly shown by the absence of even a single individual among the 133 plants of the cross. In a culture of this size from self-pollinated seed the chances are about even that an individual of mut. *formosa*, with a frequency of 1 per cent, would or would not turn up. If the use of *formosa* pollen had appreciably increased the frequency of this form in the progeny, a culture of 133 plants might have been expected to show it. The results can be interpreted in only one way, that is, the female gamete carries all the factors which determine the visible characters of the several forms, not only of the 4 mutations, but of f. *typica* as well.

Both parent plants of mut. *formosa* showed a high degree of mutability themselves, and gave rise to the other 3 characteristic mutations.   As in the case of progenies from f. *typica*, mut. *angusti-folia*, and mut. *latifolia*, the predominating form among the mutations was mut. *setacea*.

FIG. 11.—Mut. *albicans:* 2 cauline plants, Lexington E.-5-196 (left) and 187 (right), from the F₂ progeny of f. *typica;* no. 187 is shown also in fig. 1.

Mut. *albicans*.—This mutation reproduces itself in only a small proportion of its progeny, but can be said to come true in the sense that it gives no reversions to f. *typica*. All of the aberrant plants in the cultures, both from self-pollination and from pollination with f. *typica*, are mutations belonging to the characteristic group. As in the case of mut. *formosa*, most of the secondary mutations were mut. *setacea*.

The small culture of f. *typica*✕mut. *albicans* emphasizes the fact that the composition of the culture is conditioned by the female gamete. As in the case of the analogous cross f. *typica*✕mut. *formosa*, the progeny is just what we should expect from self-pollination of the *typica* parent.

FIG. 12.—Mut. *formosa:* 2 rosettes, Lexington E-5-206 and 207, from the F₂ progeny of f. *typica;* both are shown in fig. 5.

Mut. *revoluta.*—Only one small progeny was obtained from this nearly sterile mutation. It showed that the form reproduces itself except for throwing other mutations of the characteristic group. None of the crosses made with mut. *revoluta* were successful, but there can be little doubt, from collateral evidence, that mut. *revoluta*, as well as mut. *albicans*, follows the same type of inheritance as mut. *formosa*.

Mut. *setacea.*—So far as can be determined, this form comes entirely true from seed, and represents the most extreme modification which can take place in the direction followed by the group of characteristic mutations. Although the crosses with f. *typica* have so far not been successful, it is probable that this extreme reduction phase would also be dominant when introduced into the cross as the female gamete.

The numerical data for mut. *setacea* in all cultures have had to be based largely on the determination of very young plants, for many weak plants do not succeed in forming new roots after being transplanted. There is no difficulty in growing to maturity practically every individual of mut. *formosa* that germinates, and most of those of mut. *albicans* and mut. *revoluta*. It is the rule rather than the exception, however, to lose three-fourths or more of the *setacea* plants. They show some variation among themselves which may possibly indicate that mut. *setacea* is itself mutable and that more than one type is covered by this name. If so, only one type survives in the part of the cultures which reaches maturity.

The phenomenon of mass mutation

From the results of the crosses between f. *typica* and muts. *formosa* and *albicans*, as well as from the insignificant

FIG. 13.—Mut. *formosa* (Lexington E-5-206-51): the setiform leaf appendages show very clearly; the position of the plant in the pedigree may be determined from table VI.

variation in the composition of cultures showing mass mutation regardless of the source of the pollen, it appears clear that the factors responsible for the mutational characters are carried in the female gametes. So far, there is no evidence that the pollen of any of the characteristic mutations differs from that of f. *typica*.

FIG. 14.—Mut. *formosa* × f. *typica* (Lexington E-5-206 × E-5-229, one of the F₁ progeny): this cross is identical with mut. *formosa* itself (cf. fig. 13); the constitution of the F₁ progeny is given in table VI, the progenies resulting from self-pollination of the parent plants are recorded in tables III and VI.

It follows that mass mutation in *O. pratincola* must be due to the wholesale modification of female gametes. The relations have not been worked out in the case of *O. Reynoldsii*, which also shows mass mutability.

There can be no doubt that mass mutation is not Mendelian segregation, although the two phenomena have points of resemblance. HERIBERT-NILSSON's hypothesis to account for the mutability of *O. Lamarckiana* depends upon the segregation of plural factors for the same character, and involves such complications as the elimination of all zygotes which are homozygous with regard to the presence of any of the numerous plural factors. Needless to say, he has also relied upon the doctrine of the equivalence of male and female gametes. His last paper bears evidence that his faith in the equivalence of gametes is beginning to waver, although he has formerly trusted so implicitly that he has made crosses only one way. If he had studied the reciprocals of his crosses it is safe to assume that he

would never have advanced his Mendelian explanation of mutability. As far as his results extend, his derivatives of *O. Lamarckiana* fall, for the most part, into two classes, which conform in hereditary behavior to the two main classes of mutations which have been obtained from *O. pratincola*.

Class I.—The mutation breeds true, in the sense that it gives no reversions to the parent form. The reciprocal crosses with the parent species are matroclinic. The progeny conforms to the type which supplies the female gamete.

Class II.—The mutation gives a progeny consisting of the parental and mutational types in greatly varying proportions. The progenies from reciprocal crosses are mixed if the mutation supplies the

FIG. 15.—Mut. *gigas* (above) and f. *typica* (below): rosettes from the F_2 progeny of f. *typica;* the rosette of mut. *gigas*, Lexington E.-5-238 had a darker color and more conspicuous pubescence than the sister plant of f. *typica*, but the difference does not appear in the photograph; both plants are shown in fig. 5.

female gametes, but consist of the parental type only if the mutation supplies the male gamete.

Several mutations of each class have been studied by the writer in more or less detail, and the results will soon be published. As already announced,[11] the interesting mut. *nummularia* belongs to class I, as do also all of the mutations characteristic of Lexington E. Mut. *latifolia* is a typical member of class II. There are mutations, of course, which show neither type of behavior, but they need not be involved in the present discussion.

HERIBERT-NILSSON's hypothesis demands the recessiveness of mutations of class I, regardless of which way they are crossed with the parent. This condition is not fulfilled. It demands that the female gametes of the mutations of class II should be of one kind, and the pollen of two kinds. Neither is this condition fulfilled. His hypothesis makes no provision for the appearance of mutations in excess of one-third of the progeny. In this respect it is quite inadequate. On Mendelian grounds it is as difficult to account for too many mutations as for too few. His assumption is that after a homozygous and recessive condition has been attained in *O. Lamarckiana*, except for one of the plural factors which produce the *Lamarckiana* phaenotype, monohybrid splitting will take place. The one-fourth of dominant homozygotes will be eliminated, and therefore the progeny will consist of heterozygotes and recessives (mutations) in a 2:1 ratio. He has not attempted to explain how more than one-third of a progeny can consist of mutations, although he states in a vague and general way that the discovery of highly mutable strains is an argument in favor of his thesis. Nothing, he says, has made the mutation phenomena appear so exceptional as the low frequency of mutations. In his opinion, the high mutability of *O. Reynoldsii* has rendered the mutation fiction an absurdity.

Further comment on this opinion is rendered unnecessary by the serious discrepancies between HERIBERT-NILSSON's hypothesis and the facts. It can do no harm to point out, however, that even if mutations appeared through the operation of Mendelian segregation, as no one denies may sometimes be the case, it is still necessary to account for the origin of heterozygosis in the parent strain. The writer believes that mutations may often appear as a result of segregation, but that the antecedent heterozygosis has its origin in a mutative change. To attempt to account for the hetero-

[11] Amer. Jour. Bot. 2:146. 1915.

zygosis by hybridization leads to such absurdities as the denial that new forms have ever originated except by hybridization and recombination.

It is perhaps unwise to hazard even a guess at the nature of the modification of the female gametes which results in mass mutation. At one time the writer was inclined to believe that the modification had involved the cytoplasm rather than the nucleus, and that cytoplasmic inheritance might account for the matroclinic crosses. However, there are now adequate data at hand to show that similar matroclinic crosses in other cases cannot be explained by cytoplasmic inheritance. The reason for discarding this hypothesis will be explained in a future paper, since it involves data which cannot be touched upon here.

Mendelian expectations require that the largest class in a progeny showing mutation shall consist of the parent phaenotype. No explanation of the high mutability of mass mutating strains can be accepted which requires the elimination of zygotes of this phaenotype, which according to all other experience are strong and viable. If a deficiency in any class of zygotes were to be expected in a mass mutant strain, it would be the class of weakest mutations; in the case of *O. pratincola*, for example, it would be mut. *setacea*. Yet this mutation is the very one which occurs in the largest numbers.

Mass mutation is neither more nor less easily explained than ordinary mutation. It seems to be due to sudden mutative transformations of certain female gametes, and to be apparent in the zygotes without the necessity of subsequent segregation because of the fact that the factors involved have no counterparts in the male gametes. There is no real distinction between mass mutation and ordinary mutation except that in the former type large numbers of gametes may be simultaneously affected, whereas in the latter only a few are affected.

Summary and conclusions

1. Mass mutation consists in the production of unexpectedly large numbers of mutations, in some cases amounting to 100 per cent of the progeny.

2. The phenomenon is known in two species of *Oenothera: O. Reynoldsii*, in which it was first described, and *O. pratincola*, the subject of this paper.

3. It cannot be explained by HERIBERT-NILSSON's Mendelian hypothesis.

4. The mutations of the mass mutant strain of *O. pratincola* are: (A) common to other strains of the species; the non-characteristic mutations are not produced in unexpected numbers and show mass mutability superposed upon their ordinary behavior in heredity; (B) characteristic of the mass mutant strain.

5. The characteristic mutations are constant in that they do not throw the type form of the species, but, except in the case of the most reduced member of the group, are themselves highly mutable.

6. As far as tested, the characteristic mutations adhere to the following scheme of inheritance:

mutation × mutation → mutation
mutation × parent → mutation
parent × mutation → parent

7. They belong to a group with certain structural characters in common, but do not seem to form a linear reduction series.

8. They seem to result from the mutative modification in the female gametes of factors which have no counterparts in the male gametes.

9. Mass mutation is associated with a high degree of sterility, which manifests itself in the production of a greatly reduced number of seeds or in the production of many empty seeds.

UNIVERSITY OF MICHIGAN

FERTILIZATION IN ABIES BALSAMEA

CONTRIBUTIONS FROM THE HULL BOTANICAL LABORATORY 210

A. H. HUTCHINSON

(WITH PLATES XVI–XX AND ONE FIGURE)

A general account of fertilization and the related phenomena in this species has been given by MIYAKI (19). A number of supernumerary nuclei were noted in the micropylar end of the egg at the time of fertilization. These were generally regarded as derived from the male gametophyte, while in some cases another nucleus, that of the ventral canal cell, was added to the number. Repetition has been avoided, inasmuch as the present account is restricted to special problems, while the general account has received little attention.

The problems suggested by the regular occurrence of four nuclei in the egg cytoplasm, near the micropylar end, at the time of the 4-nucleate proembryo, and by the unusual grouping of chromosomes during the division which follows the conjugation of the egg and sperm nuclei, have led to this investigation. Special attention, therefore, will be given to the fertilization of the nucleus of the ventral canal cell and to the cytological features connected with the fertilization of the egg.

The material was collected in Ontario, Canada. I am greatly indebted to Professor W. R. SMITH for collections made at Lake Joseph. On June 25 and 26 I obtained ovules showing fertilization at N. lat. 44°, W. long. 79°12′, and on July 2 and 3 ovules showing similar stages at N. lat. 45°30′, W. long. 78°32′. The altitude in the first case was 900 feet; in the second 1800 feet.

In a former paper on the male gametophyte of *Abies* (10), the excessive number of prothallial cells has been recorded. One of these polar cells, during mitosis, is shown in fig. 1. Fig. 5, which was drawn from a pollen grain lodged in the micropyle, shows the division of the body cell nucleus, to form two male nuclei, taking place before the tube breaks through the spore coat. At this time the cytoplasm surrounding the tube nucleus is extremely vacuolate, that of the body cell very dense. The tube nucleus begins to disintegrate before the pollen tube emerges from the exine; usually during the pollen tube stage it appears only as an irregular aggregation; seldom, if ever, does it enter the egg.

A number of gametophytes have been found similar to that shown in fig. 2. The first division has cut off a polar ("prothallial") cell which later has disintegrated; the second division has given rise to two equivalent cells; which of the two might have been regarded as prothallial, under other circumstances, is impossible to determine. Evidently there are two antheridia. The conclusions based on a study of *Picea* (9), namely, that under favorable conditions any of the cells resulting from the three primary divisions of the male gametophyte may be antheridial, are supported by the facts as found in *Abies*.

The period of time between pollination and fertilization is from four to five weeks. During the greater part of this time the pollen grains lie dormant on the nucellus, or lodged in the micropyle. The course of the pollen tube is direct, and the motion rapid. Although the rate is difficult to determine, it is believed that the passage time of the pollen tube does not exceed two days, and probably may be measured in hours.

Fertilization of the ventral canal cell

The division of the central cell to form the egg and the ventral canal cell is similar to that in *Pinus* (figs. 7, 8, 9; compare, 3, 5, 6, 19); the nuclei formed are similar in size and form. The egg nucleus at once begins to move toward the center of the egg cytoplasm (fig. 7). What the attractive force may be is unknown, but the nucleus of the ventral canal cell responds to the same force and moves in the same direction. This nucleus breaks through the cell wall and enters the cytoplasm of the egg (figs. 11, 12); here it increases in size until it reaches a length, in some cases of 80 μ, which is approximately one-half of the greatest length attained by the egg nucleus. The structure of the ventral nucleus is very similar to that of the egg, which will be described later. The latter is surrounded by a dense granular layer which is not present in the case of the former.

The fertilization of the nucleus of the ventral canal cell has been seen in several instances. One of the male nuclei fuses with the ventral nucleus (figs. 15, 17, 21); the stalk nucleus also may be in close proximity (figs. 17, 18). Sometimes two tubes enter the archegonium, in which case male nuclei from different gametophytes may fuse with the ventral nucleus and egg nucleus respectively (fig. 21). The chromatin of the fusion nucleus condenses near its center (fig. 18), and the first division takes place. Two successive divisions (figs. 16, 23, 24, 26) result in four nuclei, which as might be expected, are generally arranged in pairs (figs. 22, 25). The nuclei of this ventral proembryo range in diameter from 40 to 50 μ; those of the proembryo proper from 60 to 65 μ; otherwise the similarity is very marked (figs. 32, 53, 54).

Nuclear changes

The changes in size of the nuclei located in the egg cytoplasm are the most readily measured of all the modifications; moreover, the increase or decrease in volume will serve to indicate the extent of the qualitative changes which occur. Immediately after the division of the central cell the egg nucleus measures about 30 μ in diameter (fig. 7). At the time of fusion its length approximates 100 μ (fig. 19); during the approximation of the chromatin groups 160 μ is the maximum measurement (fig. 27), and during anaphase of the first division the diameter is again reduced to 50 or 60 μ (fig. 46). The changes in size of the daughter nuclei are less marked; the diameter varies from 20 μ at telophase (fig. 52) to 65 μ in the resting stage   It is not surprising that the egg nucleus should vary greatly in structure while increasing to 60 times its original volume, and again decreasing to one-tenth its attained volume.

During the early stages of the first division, and even before the chromatin groups have united, four differentiations of the "nuclear" material are evident. The chromatin group or groups occupy less than one-tenth of the space within the nuclear membrane (figs. 27, 28, 42). The spindle fibers are intranuclear in origin (fig. 28). Large, vacuolate, irregular, deeply staining masses are distributed throughout the whole area. The greater part of the nuclear cavity is pervaded by slender filaments, which include small granules. An attempt has been made to trace these structures from their origin to their fate, in order that something regarding their nature and function might be determined. This complexity of structure is in contrast with the prophase of an ordinary mitosis, where only chromosomes, at most chromosomes and the nucleolus, are inclosed by the nuclear membrane.

The dark vacuolate masses are most conspicuous just after fusion (fig. 27). In the early stages small refractive globules are scattered throughout the granular egg nucleus; later, these become inclosed in a gelatinous network (fig. 19), and next appear as previously described. During the first mitosis they accumulate into several globular, vacuolate bodies (fig. 34). The latter decrease in volume or become distributed throughout the nucleus during metaphase (fig. 46). At anaphase the irregular masses once more become conspicuous (fig. 51). When the daughter nuclei are formed, these bodies are not included, and soon disappear. The fact that they are extruded would seem to indicate that they are not fundamental nuclear material. The fact that they are stored up during the growth of the nucleus and decrease in amount during mitosis would suggest that they are simply stored food bodies.

Two successive groups of intranuclear fibers become differentiated during the processes of fusion, and the first division after conjugation. The first is concerned with the approximation of chromosomes to be described later; the second with the first mitosis. While the two chromatin groups, from the egg and from the male nucleus, are still distinct, fibers which penetrate and surround these groups are organized (fig. 28). A union of the two groups of fibers results in the formation of a single spindle made up of large complex strands (figs. 29–33). The spindle drawn was 60 μ in length. After the approximation of the chromosomes into pairs the spindle broadens; the fibers become less conspicuous, and finally disappear, leaving a group of irregularly arranged chromosomes within the now much reduced nuclear space (figs. 43–45). Meanwhile the chromosomes migrate to the center of the nucleus, and soon the second set of fibers is formed (fig. 46). At first these are restricted to the region in which the chromosomes are situated; as the chromosomes move to the poles, the whole nucleus becomes pervaded by fibers radiating from irregular centers (fig. 51). In the egg of *Abies* there is an excessive development of intranuclear fibers. It seems reasonable to suppose that they originate from cytoplasmic material which has entered the nucleus during its growth period, and which may be differentiated into the fibrous form under physiological conditions not yet determined.

The slender filaments pervaded by small granules are scattered throughout the greater part of the nucleus. They are most definitely organized at the time when the spindle fibers are most conspicuous (figs 27, 33, 42). When the fibers disappear, the filaments become disorganized, resulting in a granular mass (figs. 43, 46); when the spindle fibers appear the second time, the whole nucleus becomes more or less fibrous (fig. 51); when the daughter nuclei are formed, these bodies are not included, but form a matrix for the nuclei (figs. 49, 50). We may conclude that these filaments also are cytoplasmic, resembling the spindle fibers in nature, and becoming differentiated under similar conditions.

Pairing of chromosomes in fertilization

The succession of events which occur in connection with fertilization and the first division of the zygote has been traced not only by a consideration of the stages in the approximation and redistribution of chromatin bodies, but also by tracing parallel series of changes in the size of the nucleus, in the modifications of the deeply staining food bodies, and in the formation and dissolution of spindle fibers. A study of the chromatin, involving as it does the union of the male and female elements and distribution in the daughter nuclei, is of primary importance. Emphasis has been given first to a study of the related phenomena, already described, thereby eliminating, in so far as is possible, the possibility of a misinterpretation of the order of events.

The approximation of male and female nuclei has been described for a number of conifers. In general the process as found in *Abies* agrees with that of these descriptions. A few features may be noted. No cytoplasm could be detected adhering to the male nucleus as it approached the egg nucleus. There is a great disparity in the sizes of the pairing nuclei; that of the egg, as shown in fig. 19, is 120 μ by 80 μ, while the male nucleus is 45 μ by 15 μ. The membranes of the nuclei are resorbed at the place of contact, and the contents of the male nucleus pass into the interior of the female nucleus, thereby leaving a protuberance containing very little cytoplasm.

In *Abies* the chromatin could not be detected definitely until the formation of two groups in the micropylar end of the egg nucleus takes place. While the groups are still distinct, the individual chromosomes become separate (fig. 28), each group containing the haploid number of chromosomes. As the two spindles unite, the chromosomes become paired (figs. 21–33); at first the individuals of a pair approximate side by side (figs. 31, 32); soon they twist about one another and jointly loop into the form of a C (figs. 30, 32, 33). The chromosomes are very large, in some cases exceeding 20 μ in length. There is abundant evidence that this is a pairing, not a longitudinal splitting of chromatin elements. First, the number of pairs is haploid. This is the number which would necessarily result from a pairing of the double number of chromosomes already present; a splitting would, of course, result in $2x$ pairs. Also, the twisting of the chromosomes about one another is identical with their behavior in what is generally regarded as a pairing during the prophase of the first reduction division. Moreover, there follows a transverse segmentation. If the diploid number of chromosomes should undergo two divisions, one longitudinal and one transverse, an $8x$ number of chromosomes would necessarily result. The facts cannot be explained by the supposition that there is a longitudinal split; they are readily explained by regarding this paired appearance as a true pairing.

The segmentation.—The bending into C or V-shaped forms is followed by a segmentation of each component of the pairs; in other words, a transverse fission at the angle of the bent chromosomes (figs. 34, 40). The resulting $4x$ daughter segments are approximately 10 μ in length, or one-half the length of the pairs before and during the looping process. At first the segments remain more or less twisted about one another (figs. 34, 37, 41, 42), and for some time retain a paired relation (figs. 43, 45). They may be in the form of X's, or V's, or parallel rods. At the time when the second set of spindle fibers begins to be differentiated, the $4x$ number of chromosomes are indiscriminately intermingled (figs. 46–48). Half the number pass to each pole to form the daughter nuclei.

The chromosome count.—Repeated chromosome counts in the sporophyte and gametophyte confirm one another in fixing the x and $2x$ numbers for *Abies balsamea* as 16 and 32. The individual chromosomes which appear during the prophase of the division in the central cell are shown in figs. 8 and 9. The count is 16. The division of the body cell in anaphase gives 39 segments. A reconstruction of parts separated by the microtome knife results in a count of 32, or 16 passing to each pole. At the time of approximation of the chromosomes in the egg, there are 16 pairs (figs. 29–33). When segmentation takes place, 32 pairs of segments are present (figs. 34–40). In the nucleus represented in figs. 43–45, there are 72 chromosome pieces; figs. 46–48 show 63 almost complete chromosomes, besides a number of ends. Undoubtedly we have in each case the $4x$ number, or 64 chromosomes.

The daughter nuclei.—During telophase the chromatin strands remain remarkably distinct (figs. 49–52). They elongate greatly, and become irregularly looped (figs. 50, 52). It would seem that each is in contact with the periphery somewhere throughout its length. Contraction is followed by an increase in the size of the nuclear space; the latter is accompanied by a vacuolization of the chromatin (fig. 55). The nuclear outline is still lobed, the lobes corresponding to the loops of chromatin. The nuclear membrane forms late. As the nucleus continues to enlarge, the chromatin becomes still more discontinuous, but the outline of the strands may still be readily traced.

It is to be noted that of all the material which was inclosed by the membrane of the egg nucleus, only the chromatin is included in the newly organized daughter nuclei. The large vacuolate darkly staining bodies, the filaments pervaded by granules, and the fibers are all excluded when the membranes inclose the daughter nuclei. We may conclude that these materials, although they may be found within the nuclear membrane, are not essentially nuclear and are not directly concerned in mitosis. They are, at most, cytoplasmic inclusions within the nucleus. The chromatin is the fundamental nuclear substance.

Discussion

The ventral canal cell.—The general tendency among the Coniferales is toward the reduction of the ventral canal cell. In the Abietineae a cell is cut off by a cell wall; in the Taxineae and Cupressineae, as groups, a ventral nucleus is formed, but no cell is organized; while in *Torreya* (15) there is no ventral cell, the nucleus of the central cell becoming the egg nucleus. In *Abies* the nucleus of the ventral canal cell functions as an egg. In *Pinus Laricio* (3, p. 278), "while the ventral canal cell nucleus usually disappears soon after it is formed, in some cases it persists, and its nucleus becomes as large as that of the oosphere, passing through a similar developmental history. New support is thus given to the theory that the ventral canal cell is the homologue of the egg."

NICHOLS (21) describes (fig. 90) "two nuclei resulting from the division of the ventral canal cell nucleus" in *Juniperus*. The ventral canal cell "is fairly persistent in *Tsuga* (20). When division is complete, its nucleus is equal in size and similar in structure to the nucleus of the egg, and for some time shows the same stages of development."

The most extreme development recorded is in the case of *Thuja*, described by LAND (14). "A number of the writer's preparations of *Thuja* lead him to believe that both the ventral nucleus and the egg, in the same archegonium, may be fertilized. In fig. 17 the proembryo is well advanced, while the ventral nucleus has formed a group of four cells. Another preparation shows eight cells with indications that walls are soon to appear. The probability of such a fertilization is strengthened by finding occasionally in the same ovule embryos growing upward into the nucellus, as well as downward into the endosperm" (p. 224). These facts and those already described for *Abies* remove any doubt that the ventral canal cell is potentially an egg.

The cytoplasmic "mantle" about the egg nucleus is present in most species of Coniferales. NICHOLS (21) describes it in *Juniperus* as follows: "The mass of cytoplasm and starch derived from the male cell gradually surrounds the conjugating nuclei so that there is never any possibility of mistaking the fusion nucleus for an unfertilized egg nucleus." And NORÉN (22) states: "Nachdem die Kopulation der Kerne erfilgt ist legt sich das Plasma der Sperma-zelle und deren Stärke wie ein Mantel um die beiden Kerne herum." The description is similar for *Taxodium* (4), *Torreya* (15), *Sequoia* (16), and *Thuja* (14). In *Ephedra* (13), a dense cytoplasmic mass develops about the egg nucleus and extends downward in the cyto-plasm of the egg. In *Abies* a finely granular cytoplasmic layer develops about the egg nucleus during its movement toward the center of the egg (figs. 7, 14). This area extends along the path of the nucleus in the form of a short streamer. The male nucleus penetrates the mantle, but there is no appreciable addition to its mass by cytoplasm accompanying the same (fig. 19). At the telo-phase of the first division, the excess material from the egg cyto-plasm is added to the "mantle" (figs. 49, 50, 55).

The darkly staining bodies described in *Abies* as globular or irregularly shaped and vacuolate, according to conditions, have been variously interpreted. NICHOLS (21) writes: "In the egg nucleus frequently the entire chromatin content of the nucleus seems to have resolved itself into nucleoli and pseudo-nucleoli. Yet even after a study of a large number of preparations one is unable to formulate any satisfactory conclusions as to the nature of these structures." Describing the egg nucleus of *Taxodium*, COKER (4) says: "In addition to the reticulum and plastin nucleoli there are also present numbers of chromatin nucleoli." And LAWSON for *Libocedrus* (17a) states that "It is impossible to dis-tinguish the true chromatin from the nucleoli and other irregularly shaped bodies which stained like chromatin, and seemed to be closely associated with the latter." These bodies seem to resemble very much the karyosomes which SHARP (24) has described in the resting nuclei of *Vicia Faba*. "They appear in connection with the chromatic network and resemble the latter in staining quality. They seem to represent an elaboration product of a process actively going on during rest." A study of *Abies* has led to the conclusion that the darkly staining bodies are storage materials derived from the cytoplasm.

A cytological study of fertilization in conifers has been made for a number of species: CHAMBERLAIN, *Pinus Laricio* (3); BLACKMAN, *Pinus silvestris* (2); Miss FERGUSON, *Pinus* (6); MURRILL, *Tsuga canadensis* (20); NORÉN, *Juniperus* (22); and NICHOLS, *Juniperus* (21). "After the male pronucleus is within the oosphere nucleus the chromatin of the two pronuclei appears as two distinct masses in the spirem stage" (3). "Es scheint als würde jede der beiden Chromatingruppen zuerst ihre eigen Kernspandel ausbilden, die sich dann zu einzigen vereinigen" (22). On the fibers "the long bent and twisted chromosomes appear" in *Tsuga* (20). Miss FERGUSON (6) states: "When the chromosomes are being oriented at the nuclear plate, the maternal and paternal elements can no longer be distinguished." The number of chromosomes at this stage was found to be $2x$. No count is recorded during the phase of pairing (or splitting, as it has generally been regarded). Miss FERGUSON's figures confirm what has been described for *Abies*. Her figs. 236 and 237 may be compared with figs. 29–33; her fig. 238 with figs. 34–39; and her fig. 241 with fig. 51. As illustrated by diagram *B*, a number of stages in the process of fertilization have not been described heretofore. It is not surprising that, without these stages and not having the chromosome count throughout, the pairing should be interpreted as a longitudinal split, and that evidence of segmentation should not be found. Because of the complete series found in *Abies* and the extreme size of the chromosomes, it has been possible to discover the facts which center about the pairing, followed by the transverse segmentation.

Stages in fertilization.—A study of the union of egg and sperm in plants and animals makes it evident that there are several phases in the process of fertilization. The primary phase is illustrated by Uredineae. Here the fusion is evidently incomplete during the binucleate stages. "In the young aecidium the nuclei become paired and divide together in very close association. The teleutospores in the young state are binucleate, but when mature become uninucleate by the fusing of the two paired nuclei" (1). It is only after many separate but simultaneous divisions of the pairing nuclei that this second phase of the process is accomplished. HARPER (8) believes that bivalent chromosomes are formed in

Ascomycetes. "The time and degree of the combination of the sexual chromosomes is a variable matter. If the prochromosomes can remain in one nucleus with the double chromosome number, or in two distinct nuclei through part or all of the sporophyte generation, it is also possible that they may combine in one nucleus into bivalent chromosomes, and maintain their identity in this

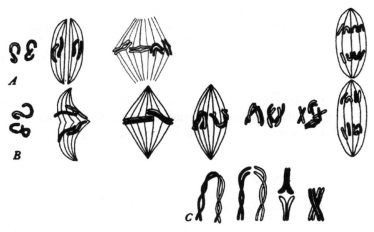

FIG. 1

Diagram *A*.—An interpretation of fertilization in some animals and in *Pinus*, according to the accounts heretofore recorded.

Diagram *B*.—An interpretation of the phenomena already described   the chromatin originating from the egg nucleus is shown in solid black, that from the male nucleus is outlined and barred; the argument upon which such an interpretation is based is included in the description given in the text; the facts of spindle formation are also indicated.

Diagram *C*.—A copy from GRÉGOIRE's (7) "schéma de l'interprétation metasyndetique des tétrades-crois," as found in certain lower animals; compare fig. *B*

condition through the sporophyte generation until a true reduction occurs in spore formation." In the well known case of *Cyclops*, studied by HAECKER, the parental chromosomes do not mingle, but persist as individuals and maintain their separation into two groups through several cell generations after "fertilization." We may consider the union of egg and sperm as consisting of three stages: (1) the two nuclei enter the same cytoplasmic mass; (2) the two groups of chromosomes enter the same nuclear membrane, but retain their former grouping; and (3) the chromosomes pair. The first is illustrated by the paired nuclei in Uredineae; the second by the chromatin groups during the first divisions of the *Cyclops* embryo; the third takes place soon after the fusion of the egg and sperm in *Abies*. It seems altogether probable that the chromosomes do not actually pair until the prophase of the reduction division in such animals as *Cyclops*. There is a striking similarity between the pairing process as described in *Abies* and that concerned with tetrad formation in some animals (compare diagram *A* with diagram *C*). To what extent we are justified in suggesting that the processes are identical and that they are both concerned with the pairing of corresponding chromosomes from the egg and from the sperm, only further research can definitely determine.

Relationships

Abies, as compared with *Pinus*, shows a number of primitive characters:

1. The male gametophyte: (*a*) an excessive polar ("prothallial") tissue; (*b*) the equality of the male nuclei; both may function.

2. The female gametophyte: (*c*) the large number of neck cells; as many as five tiers (MIYAKI 19); (*d*) the persistence of the ventral canal cell; its nucleus may function as an egg nucleus.

3. The ovulate strobilus and ovules: (*e*) the almost complete separation of scale and bract; (*f*) the development of a rudimentary pollen chamber; (*g*) the comparatively free integuments.

4. The staminate strobilus: (*h*) the staminate strobili are borne on ordinary branches in the axes of ordinary leaves.

5. The arrangement of leaves: (*i*) the spiral arrangement of leaves on ordinary branches (THOMSON 25; LLOYD 18).

6. Vascular anatomy: (*j*) the general absence of ray tracheids (THOMPSON 26, 27), which in *Pinus* are present in the mature wood, but absent in the seedling and strobilus axis.

It may be noted, however, that resin canals are not found in the woody axis of the *Abies* stem except as traumatic responses (JEFFREY 11). JEFFREY states that the presence of resin ducts is an ancient character which has persisted in *Pinus*. "On account of the reduced foliage of the abietineous conifers, this [that is, resinous secretion] was a very serious drain on the assimilatory apparatus. Gradually the more economical tendency arose of forming resin passages in the case of need only." So regarded, *Pinus* would be more primitive than *Abies*. Whether or not this argument is sufficient to overbalance the numerous ancient characters of *Abies* previously tabulated is a matter of judgment which we do not presume to decide.

Many of the foregoing characters are such that they tend to relate more closely the two ancient groups of Coniferales, the Abietineae and the Araucarineae.

Summary

1. *The male gametophyte.*—The polar ("prothallial") cells may divide mitotically. The body cell divides to form the male nuclei while within the spore coat. Under favorable conditions a "prothallial" cell may develop as an antheridial cell, a biantheridial gametophyte resulting. The male nuclei are equivalent; one fuses with the egg nucleus and frequently the other fuses with the ventral canal cell nucleus.

2. *The ventral canal cell and ventral proembryo.*—The ventral canal cell nucleus breaks through its wall into the egg cytoplasm, enlarges, and fuses with one of the male nuclei. A ventral proembryo is formed by two successive divisions.

3. *The egg nucleus.*—The egg nucleus enlarges to 60 times its original volume. At the time of fertilization (fig. 28) irregular, darkly staining, vacuolate masses, slender filaments pervaded by small granules, spindle fibers, and chromatin are differentiated within the nuclear membrane. The chromatin is the fundamental nuclear substance; the other bodies are accretions gained during the growth of the nucleus and excluded from the nuclei of the proembryo.

4. *Fertilization and the first division.*—After fusion two chromatin groups appear at the base of the egg nucleus; in each $2x$ chromosomes become separate; the two original spindles unite; the chromosomes approximate to form x pairs; they twist the one about the other and become looped; each of the components of a pair

segments medianly, that is, at the apex of the loop; $2x$ pairs of segments result; these separate to form $4x$ chromosomes; a new spindle is formed and $2x$ chromosomes pass to each pole.

5. Fertilization may be regarded as having three phases: (1) sex nuclei enter a common cytoplasm; (2) the two groups of chromosomes enter a common nuclear membrane; (3) the chromosomes approximate in pairs. The first phase may be prolonged, as in Uredineae; the second may be prolonged, as in some animals; or the three phases may follow one another in rapid succession, as in *Abies*. Attention is drawn to the similarity existing between the phenomena connected with pairing in *Abies* and tetrad formation in animals. It is suggested that they may be like processes occurring at different stages of the life history.

I wish to express my thanks for many suggestions and helpful criticisms given by Professor J. M. COULTER and Professor C. J. CHAMBERLAIN, under whose direction the investigation was pursued.

UNIVERSITY OF CHICAGO

LITERATURE CITED

1. BLACKMAN, V. H., On fertilization, alternation of generations, and general cytology of the Uredineae. Ann. Botany 18:323–326. 1904.

2. ———, On the cytological features of fertilization and related phenomena in *Pinus silvestris*. Phil. Trans. Roy. Soc. B 190:395–426. 1898.

3. CHAMBERLAIN, C. J., Oogenesis in *Pinus Laricio*. BOT. GAZ. 27:268–279. 1899.

4. COKER, W. C., On the gametophyte and embryo of *Taxodium*. BOT. GAZ. 36:1–27. *pls. 1–11*. 1903.

5. FERGUSON, MARGARET C., Contributions to the life history of *Pinus*, with special reference to sporogenesis, the development of the gametophyte, and fertilization. Proc. Wash. Acad. Sci. 6:1–202. *pls. 1–24*. 1904.

6. ———, The developement of the egg and fertilization in *Pinus Strobus*. Ann. Botany 15:435–479. 1901.

7. GRÉGOIRE, V., Les cinèses de maturation dans les deux règnes. La Cellule 26:223–422. 1910.

8. HARPER, R. A., Sexual reproduction in the organization of the nucleus in certain mildews. Pub. Carnegie Inst. Wash. no. 37. 1–92. *pls. 1–7*. 1905.

9. HUTCHINSON, A. H., The male gametophyte of *Picea canadensis*. BOT. GAZ. 59:287–300. *pls. 15–19. fig. 1*. 1915.

10. HUTCHINSON, A. H., The male gametophyte of *Abies balsamea*. BOT. GAZ. 57:148–153. *figs. 15.* 1914.

11. JEFFREY, E. C., The comparative anatomy of the Coniferales and Abietineae. Mem. Boston Soc. Nat. Hist. 1905.

12. ———, Comparative anatomy of Abietineae. Mem. Boston Soc. Nat. Hist. no. 6. 1904.

13. LAND, W. J. G., Fertilization and embryogeny in *Ephedra trifurca*. BOT. GAZ. 44:273–292. 1907.

14. ———, A morphological study of *Thuja*. BOT. GAZ. 34:249–259. *pls. 6–8.* 1902.

15. COULTER, J. M., and LAND, W. J. G., Gametophytes and embryo of *Torreya taxifolia*. BOT. GAZ. 39:161–178. 1905.

16. LAWSON, A. A., On the gametophytes, fertilization, and embryo of *Sequoia sempervirens*. Ann. Botany 18:417–444. 1904.

17. ———, The gametophytes, fertilization, and embryo of *Cephalotaxus drupacea*. Ann. Botany 21:1–23. 1907.

17a. ———, The gametophytes and embryo of the Cupressineae, with special reference to *Libocedrus decurrens*. Ann. Botany 21:281–301. 1907.

18. LLOYD, F. E., Morphological instability, especially in *Pinus radiata*. BOT. GAZ. 57:314–319. 1914.

19. MIYAKI, K., Contribution to the fertilization and embryogeny of *Abies balsamea*. Bot. Centralbl. 14:134–145. 1903.

20. MURRILL, W. A., The development of the archegonium and fertilization in the hemlock spruce, *Tsuga canadensis*. Ann. Botany 14:583–607. *pls. 31, 32.* 1900.

21. NICHOLS, G. E., A morphological study of *Juniperus communis* var. *depressa*. Bot. Centralbl. 25:201–241. 1910.

22. NORÉN, C. O., Zur Entwicklungsgeschichte des *Juniperus communis*. Uppsala Univ. Arks. pp. 64. *pls. 4.* 1907.

23. OVERTON, J. B., On the organization of the nuclei in the pollen mother cells of certain plants, with especial reference to the permanence of chromosomes. Ann. Botany 23:19–60. 1909.

24. SHARP, L. W., Somatic chromosomes in *Vicia*. La Cellule 29:297–331. *pls. 2.* 1913.

25. THOMSON, R. B., The spur shoot of pines. BOT. GAZ. 57:362–385. *pls. 20–23. figs. 2.* 1914.

26. THOMPSON, W. P., On the origin of ray tracheids in the Coniferae. BOT. GAZ. 50:101–116. 1910.

27. ———, Ray tracheids in *Abies*. BOT. GAZ. 53:331–338. 1912.

EXPLANATION OF PLATES XVI–XX

FIGS. 1–5.—Pollen grains: male gametophyte.

FIG. 1.—Division of polar cell; ×510.

FIG. 2.—Biantheridial gametophyte; ×510.

FIGS. 3–5.—Division of body cell; ×865.

FIG. 6.—Central cell and two neck cells; ×510.

FIG. 7.—Ventral canal cell and egg nucleus; ×510.

FIGS. 8, 9.—Mitosis of nucleus of central cell; ×865.

FIG. 10.—An archegonium: egg nucleus, ventral canal cell with wall broken and escaped nucleus in the cytoplasm of the egg; ×85.

FIGS. 11, 12.—Nucleus of ventral canal cell escaping.

FIG. 13.—Egg nucleus near a male nucleus; ventral canal cell nucleus, and near by the other male nucleus and stalk nucleus.

FIG. 14.—Detail of egg nucleus of fig. 10.

FIG. 15.—Two archegonia, one showing fertilization of both egg and ventral canal cell nucleus; ×85.

FIG. 16.—Divisions following the two fusions; ×85.

FIG. 17.—Ventral canal cell nucleus: a male nucleus and the stalk nucleus in contact, ×510.

FIG. 18.—Fusion nucleus resulting from union of ventral canal cell nucleus and male nucleus; ×510.

FIG. 19.—Egg nucleus and male nucleus in contact; ×510.

FIG. 20.—Egg and male nucleus in contact (see detail in fig. 19): the ventral canal cell nucleus as described for fig. 18; ×510.

FIG. 21.—An archegonium showing the egg nucleus, ventral canal cell nucleus, four male nuclei, a stalk nucleus, and two pollen tubes; ×510.

FIG. 22.—Postion and relation of ventral proembryo and proembryo proper; ×510.

FIGS. 23–26.—Detail of division following the fertilization of the ventral canal cell nucleus; ×865.

FIG. 27.—Nucleus of the fertilized egg; ×510.

FIG. 28.—Nucleus of the fertilized egg: two groups of chromosomes; ×510.

FIGS. 29–33.—Pairing of chromosomes; ×865.

FIGS. 34–40.—Segmentation of chromosomes; ×865.

FIGS. 41, 42.—Paired segments; ×865.

FIGS. 43–45.—Chromosome segments intermingling, still somewhat paired; ×865.

FIGS. 46–48.—Chromosomes starting to the poles; ×865.

FIG. 49.—Early telophase of first division; ×865.

FIGS. 50, 52, 55.—Late telophase of first division; ×865.

FIG. 51.—Anaphase of first division; ×865.

FIG. 53.—Resting stage of nucleus from ventral proembryo; ×865.

FIG. 54.—Resting stage of nucleus of the proembryo; ×865.

PLATE XVI

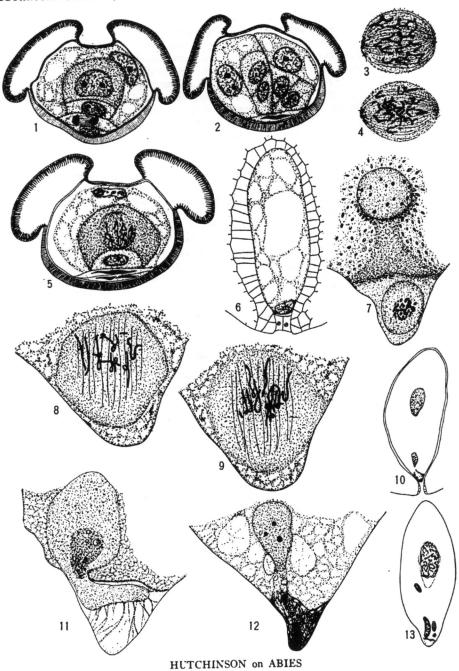

HUTCHINSON on ABIES

PLATE XVII

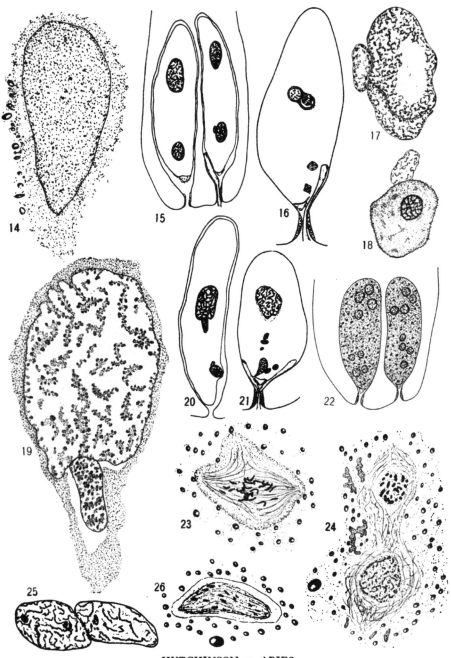

HUTCHINSON on ABIES

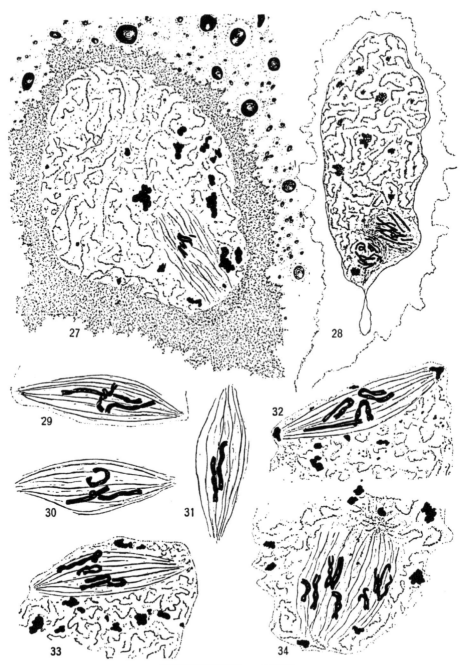

HUTCHINSON on ABIES

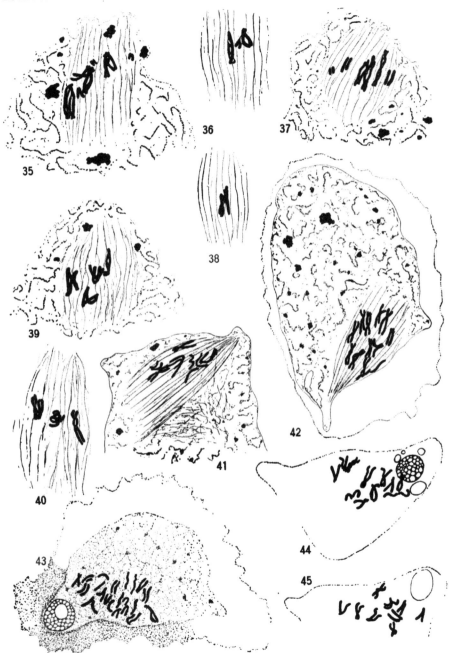

HUTCHINSON on ABIES

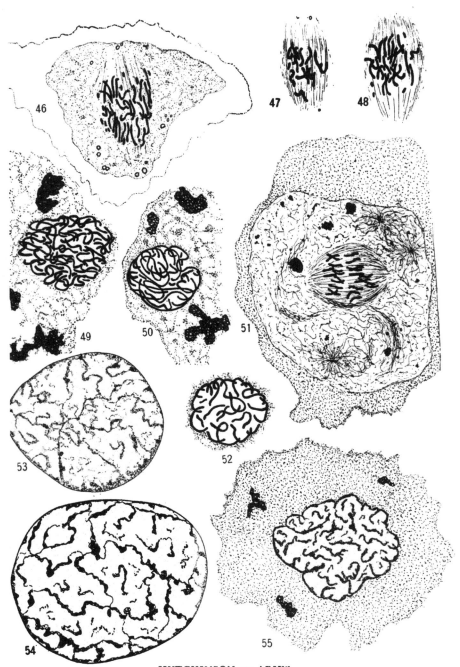

HUTCHINSON on ABIES

THE SIZE OF KELPS ON THE PACIFIC COAST OF NORTH AMERICA

T. C. FRYE, G. B. RIGG, AND W. C. CRANDALL

(WITH TWO FIGURES)

Introduction

The personal observations of the authors on the size of kelps have covered much of the west coast of North America from Cedros Island on the coast of Mexico to the Shumagin Islands on the coast of western Alaska.

The work of FRYE has been done during ten summers spent on Puget Sound, during several of which he has been director of the Puget Sound Marine Station at Friday Harbor, Washington, and one summer (1913) during which he was in charge of the expedition · sent to southern Alaska by the United States Bureau of Soils for the purpose of investigating the kelps as a source of potash fertilizer.[1] The investigations of that summer included a careful examination of the coast from Dixon Entrance to Juneau (not including the west coast of Baranof Island), together with observations on kelps at a number of points along the coast of British Columbia where opportunity offered on the way to Alaska in May.

The observations of RIGG cover six summers in the Puget Sound region, portions of three of which were spent at the Puget Sound Marine Station, and portions of two of which were spent in making a survey of the kelp beds of the Puget Sound region for the United States Bureau of Soils as a part of its investigation of those kelps as a source of potash fertilizer. RIGG's Alaskan observations were made during the summer of 1913 when he was in charge of the United States Bureau of Soils expedition to western Alaska for the investigation of the kelps of that region as a source of potash

[1] All of the kelp investigation expeditions of which the writers of this paper were in charge were a part of the general investigation of the fertilizer resources of the United States conducted by Dr. FRANK K. CAMERON of the United States Bureau of Soils.

fertilizer. A few incidental observations were made on kelps along the British Columbian coast and in southern Alaska on the way through that territory in May. Stops were made at Dixon Harbor, just north of Cape Spencer, and at Yakutat. No stop was made from Yakutat to Prince William Sound, but from this sound to the Shumagin Islands (including Cook Inlet and Kodiak Island) a reconnoissance survey was made. A careful examination of the coast was made at all points where conditions of weather and tides would permit.

The work of CRANDALL covers the coast from Cedros Island, Mexico, to Cape Flattery, Washington. It was done by him as biologist at the Scripps Institution for Biological Research and under appointments of the United States Bureau of Soils in 1911, 1912, and 1913 to investigate the kelps of the region as a source of potash fertilizer.

The maximum size of kelps on the Pacific coast of North America has been a matter of interest ever since the beginning of botanical investigation in the region. Naturally a careful watch for large kelps was kept on these expeditions and frequent soundings in kelp beds were made. No specimens of *Nereocystis* or *Macrocystis* were found that at all approached in length the figures given for these species by the earlier writers. The figures that had been reported for the length of *Alaria fistulosa* were, however, practically corroborated, and the width of that species was found to be much greater than had been reported.

Macrocystis

Macrocystis pyrifera has been reported (without locality) by KJELLMAN (7) to reach a length of 200–300 m. It has been reported by FITZROY (3) that it grew about the Falkland Islands in 30 fathoms of water, and by CAPTAIN COOK (1) that it grew at Kerguelen Island in 24 fathoms of water and that he feels warranted in saying that it reaches a length of 60 fathoms.

HARVEY quotes HOOKER as having calculated this species to reach a length of 700 feet (213 m.), and states that BORY ST. VINCENT attributes to specimens a length of 1500 feet (457 m.). HOWE (6) reports measurements of *Macrocystis pyrifera* and *M. integrifolia* from Peru.   In regard to the size of the first species he quotes (p. 64) a field note of the collector (Dr. ROBERT COKER) as follows: "The largest piece measured 10 m. and the entire single cluster weighed 30 lbs." He refers to the second species (p. 61) as "the short plant (2–3 m. long) that grows gregariously on surf-swept rocks with densely intertangled rhizomatous holdfasts. . . . ." BURD (2) found *Macrocystis* plants varying in weight from 27 to 300 pounds. The largest specimen reported from the Puget Sound region is 40 feet in length (RIGG 12).

This species was not found growing in the territory covered by the western Alaskan expedition, but beds of it were found by the southeastern expedition.   The deepest water in which it was found growing was 12.5 m.   Where the plants were growing in water of this depth, the longest portion of an attached plant floating at the surface was 13.8 m.   If this plant rose from the bottom at an angle of 45°, the under-water portion would be 17.3 m. long, and the whole would be 31.1 m.   It seems likely that it rarely if ever reaches a length of 38 m. in Alaska.

The longest *Macrocystis* plant found on the Californian coast measured 45.7 m.   Many are only 9–13 m. long.   This species with a definite holdfast has been found by the writers growing in water varying in depth from 2½ to 14 fathoms.   When it grows in shallower depths the holdfast is more like a rhizome.   A single specimen of this plant has been found on the Californian coast that weighed 136 kg.   A specimen collected at Santa Barbara, California, was 23 m. in length and weighed 22 kg.   Another one collected at the same place was 25 m. long and weighed 37 kg.   The deepest sounding made in a *Macrocystis* bed was 14 fathoms.   This was on the Californian coast.   The deepest sounding made in Alaska was 7 fathoms.

Pelagophycus

Pelagophycus porra has been found 45 m. in length on the Californian coast (fig. 1).   It has not been found in the Puget Sound region or in Alaska.   The deepest sounding made in *Pelagophycus* beds was 20 fathoms.   BURD (2) found that these plants varied in weight from 16 to 71 pounds.

Nereocystis

Statements of the length of *Nereocystis luetkeana* in various localities are shown in table I.

TABLE I

Observer	Locality	Date	Depth reported	Meters	Part measured
Mertens....	Alaska	1829	45 fathoms	82.3	Stipe only
Harvey.........	Without locality	1852	340 feet	103.0	Whole plant
Ruprecht......	Alaska	1852	250 "	76.2	" "
Kjellman ..	Without locality	1897	100 m.	100.0	" "
MacMillan. .	British Columbia	1899	80 feet	24.4	" "
MacMillan......	" "	1901	100 "	30.5	" "
Saunders........	Alaska	1901	70 "	21.3	" "
Frye ...	Puget Sound	1906	21 m.	21.0	" "
Setchell. .	California	1908	41 "	41.0	" "
McFarland. .	"	1912	100 feet	30.5	" "
Rigg...........	Puget Sound	1912	72 "	21.9	" "

SAUNDERS (15) thinks that MERTENS' statement is "probably not true," since he (SAUNDERS) "has measured many fully developed plants on the California, Oregon, and Washington coasts, as well as on the Alaska coast, and has never found one exceeding the above figures" (70 feet). FRYE (4) has indorsed SAUNDERS' view. SETCHELL (16) sees a possibility that the "45" in MERTENS' statement is a "misprint . . . . for 15, which, from all experience seems likely." He prefaces this, however, with the suggestion that MERTENS' statement "must have been made with care, as are his statements in general." He also quotes RUPRECHT, who had found specimens 25 feet long on the Californian coast, but had added "that according to trustworthy natives in the employ of the N. A. Company at Fort Ross, it reaches a length ten times greater in the vicinity of Sitka and the Aleutian Islands." SETCHELL also says that in his observations from Kodiak Island, Alaska, to California he did not find any specimens that approached the dimensions reported by MERTENS and RUPRECHT, but adds "Off the entrance to Yakutat Bay, Alaska, while at anchor during a fog I saw soundings taken close to and among *Nereocystis* plants showing that the solid portions of the stipe were 10–12 fathoms in length. Altogether the plants thus observed must have been 45–50 m. in length." SETCHELL's report of 41 m. above is based on a specimen

FIG. 1.—*Pelagophycus porra*, or elk kelp.—From Report no. 100, Bureau of Soils.

that he paced at Carmel Bay, California. MacMILLAN (9) expresses the opinion that the length stated by KJELLMAN "is not at all excessive." McFARLAND (8) says: "I have been unable to find individual plants of *Nereocystis* in the region examined by me which attained anything like the maximum dimensions given by KJELLMAN, MERTENS, SETCHELL and GARDNER, and MacMILLAN. Specimens up to 100 feet in length are met with."

It should be noted that RUPRECHT's statement was based on mere hearsay. KJELLMAN (7) does not cite any measurements in support of his statement. MERTENS' (11) statement has never been confirmed by measurements. There is practical agreement in the statements by SAUNDERS (15), FRYE (4), and RIGG (12, 13). Longer kelps than those found by the last three observers were reported by MacMILLAN (10), SETCHELL (16), and McFARLAND (8).

Not a single specimen has ever been found by the writers of this paper that reached 40 m. in length, nor has any evidence been obtained by them from soundings that plants reach this length in any of the beds examined. The largest specimen of *Nereocystis* measured on the southern Alaskan expedition was 20.7 m. This was at Point Davidson on Annette Island. The longest specimen of this species measured on the western Alaskan expedition was 18.9 m. This was at Resurrection Bay on the Kenai Peninsula. Many statements were made to the writers about extremely large kelps, but when the beds were visited, these large specimens could not be found. The longest lamina found by our expeditions is 780 cm. The greatest width of lamina found is 20 cm. The length of a pneumatocyst found at Yakutat was 462 cm. All measurements of kelps at Yakutat were from drift specimens. None were seen growing. The region was visited in May, which is too early for the new crop of kelp there. The maximum circumference of bulbs measured on our expeditions is 453 mm. This specimen was found at Danger Bay on Afognak Island. Other bulb circumferences measured were 425 mm. at Meares Passage, 367 mm. at Yakutat, 365 mm. at Sand Point, and 350 mm. at Resurrection Bay.

McFARLAND (8) has found the weight of fresh *Nereocystis* plants on the Californian coast to vary from 43 to 76 pounds. BURD found a maximum weight of 56 pounds at the season at which the sample was taken (fall).   One of us (RIGG 12, 13) has found the weight of the fronds and pneumatocyst to be 18–35 pounds on Puget Sound.   The maximum weight found by our expeditions in Alaska was 55 kg.   This included the fronds and 210 cm. of the pneumatocyst.   The fronds alone weighed 49 kg.   This specimen was collected at Resurrection Bay.   The same portions of specimens collected at other points weighed as follows: Smith Island, 24 kg.; Sand Point, 16 kg.   The slender portion of a stipe collected near Seattle, Washington, was found by the writers to weigh 87 gm. per meter.   A similar piece collected at Sand Point, Alaska, weighed 227 gm. per meter.   A holdfast at Sand Point weighed 3.4 kg.

A summary of the measurements of the largest *Nereocystis* plants found by the writers is given in table II.

TABLE II

Locality	Total length (m)	Length of fronds (m)	Diameter of bulb (mm)	Weight of plant (kg)
Pacific Grove, Cal., 1913..	34.7	1.8	93	7.0
Pacific Grove, Cal., 1913	36.5	3.6	137	13.6
Pacific Grove, Cal., 1913	38.4	1.8	100	7.5
Friday Harbor, Wash., 1911....	21.9	....	....	....
Seward, Alaska, May 1913.	19 0	7.9	116	55.4
Yakutat, Alaska, May 1913. .	.	..... .	120.5	. ...
Sand Point, Alaska, July 1913.	13.7	3.3	113.0	16.2
Pt. Davidson, Alaska, May 1913	20.6	5.4	...	....
Meares Passage, Alaska, July 7, 1913	21 2		. ... .	....

The following soundings include the deepest ones made by the writers in beds of *Nereocystis:* Resurrection Bay, Alaska, 5 fathoms; Kodiak Island, Alaska, 6 fathoms; Danger Bay, Alaska, 6.5 fathoms; Pacific Grove, California, 9 fathoms.

Alaria

The other kelp forming large beds in Alaska is *Alaria fistulosa* (fig. 2).   The maximum width mentioned by SETCHELL and GARDNER (17) is 1 m. and the maximum length is 25 m.   The number of sporophylls is given at about 200.

FIG. 2.—*Alaria fistulosa:* general appearance of a wide-leaved plant.—From Report no. 100, Bureau of Soils.

Specimens collected by GRIGGS and RIGG at Anchor Point, Cook Inlet, measured as follows: length 13 m., 21 m., 22 m., 18 m.; width 232 cm., 175 cm., 170 cm., 123 cm., respectively. The first of these specimens had 220 sporophylls, the largest of which measured 65 cm. in length and 84 mm. in width. The width of the midrib was 53 mm.

All of these plants were badly frayed at the tip, a great deal of material evidently having been worn off by the violence of the tidal currents. Regeneration from the base seems to be continuous throughout the life of the individual in this species. It is thus impossible to say whether these specimens had at any time been longer than they were when collected.

The weights found by our expeditions for portions of the fronds of *Alaria fistulosa* of the sizes given are shown in table III.

TABLE III

Locality		Length (m.)	Width (cm.)	Weight (kg.)
Cook Inlet,	Alaska............	19.0	170	8.1
" "	"	18.5	175	9.8
Red Bay,	"	19.0	135	....
Cook Inlet,	"	14.0	132	7.1
Kodiak Island,	"	15.0	25	0.9

The deepest sounding made in an *Alaria* bed was 5 fathoms (Middle Islands, Alaska).

Summary

1. The observations of the writers do not tend to confirm the earlier statements of the great length of *Nereocystis* and *Macrocystis*.

2. Their observations on *Alaria fistulosa* confirm earlier statements of its length. Its maximum width was found to be more than twice as great as reported in literature that has come to the attention of the writers.

UNIVERSITY OF WASHINGTON, SEATTLE,
AND
THE SCRIPPS INSTITUTION FOR BIOLOGICAL RESEARCH
LA JOLLA, CAL.

LITERATURE CITED

1. COOK, CAPTAIN, Report of second voyage.
2. BURD, J. S., Univ. Calif. Publ., Bull. 248, Coll. Agric. 1915.
3. FITZROY, R., Voyages of the Adventure and Beagle 2:246–247. 1836.
4. FRYE, T. C., *Nereocystis luetkeana*. BOT. GAZ. 42:142–146. 1906.
5. HARVEY, W. H., Nereis Boreali Americana. Part I. p. 82. Smithsonian Institution. 1852.
6. HOWE, MARSHALL A., The marine algae of Peru. Mem. Torr. Bot. Club 15:1914.
7. KJELLMAN, F. R., Laminariaceae. Engler und Prantl, Die Natürlichen Pflanzenfamilien 1²: 1897.
8. McFARLAND, F. M., Kelps of the central California coast. Senate Doc. 190:194–208. 1912.
9. MACMILLAN, C., Observations on *Nereocystis*. Bull. Torr. Bot. Club 26:273. 1899.
10. ———, The kelps of Juan de Fuca. Postelsia 193–220. 1901.
11. MERTENS, HENRY, Linnaea 4: 1829.
12. RIGG, G. B., Ecological and economic notes on Puget Sound kelps. Senate Doc. 190:179–193. 1912.
13. ———, Ecological and economic notes on *Nereocystis luetkeana*. Plant World 15:83–92. 1912.
14. RUPRECHT, F. J., Neue oder unbek. Pfl. a. d. Nordl. Th. des stillen Oceans. 1852.
15. SAUNDERS, DE A., Harriman Alaska Exp., The algae. Proc. Wash. Acad. Sci. 3:391–486. 1901.
16. SETCHELL, W. A., *Nereocystis* and *Pelagophycus*. BOT. GAZ. 45:125–134. 1908.
17. SETCHELL, W. A., and GARDNER, N. L., Algae of northwest America. Univ. Calif. Publ. Bot. 1:165–418. 1913.

NEW SPECIES OF ACHLYA AND OF SAPROLEGNIA[1]

A. J. PIETERS

(WITH PLATE XXI)

In the course of physiological work on *Saprolegnia* carried on during the past four years, there was occasion to make pure cultures of a large number of water molds. Altogether some 85 numbers were isolated and cultivated on flies or in artificial media or both. Among these forms were several that could not be induced to produce oogonia and which could not, therefore, be referred to any species. The author is of the opinion that in some cases such forms have completely lost the power to produce oogonia and that they should be described and named as new species, the diagnoses to rest on physiological rather than on morphological bases. However, his experience with some of these forms shows that it will be well to be cautious in drawing conclusions. In the case of a variety of *Saprolegnia monoica* to be described, the form was in cultivation for 16 months, on flies and in various media, without any sign of oogonia being produced. Later, when just the right combination of conditions was presented, oogonia were produced.

Another species, to be described as *S. Kaufmanniana*, produced oogonia sparingly on flies, many cultures not showing a single oogonium. In this form also a number of oogonia with oospores and antheridia were produced in a certain strength of haemoglobin solution, but in no other medium. These experiences show that the production of sexual organs may depend on some special combination of conditions, differing doubtless for each form. One form, which has been studied for 18 months as no. 66 and which has been tested in every way in which any of the other forms have been tested, still refuses to produce oogonia, though yielding an abundant harvest of round single gemmae. These gemmae have the shape and size of oogonia, and are commonly borne laterally on short stalks just as oogonia are in such a species as *S. monoica*. In

---

[1] Contribution from the Botanical Laboratory of the University of Michigan, no. 148.

fact, when these gemmae first form, the observer is certain that young oogonia are being produced, but in no case have oospores or antheridia been observed, though hundreds of cultures have been examined. It is possible that in this form these gemmae are nothing but arrested oogonia, and that, could we find the proper conditions, oospores and perhaps antheridia would be developed.

Certain details in regard to these and other tests will be presented in another paper; at present the writer wishes to describe a species of *Achlya*, which he has named *A. Klebsiana* in honor of Professor GEORG KLEBS. The naming of this species for Professor KLEBS seemed to the writer especially appropriate since at one time it seemed to be an exception to the rule laid down by KLEBS in 1899 that sporangia are formed only when the food supply is quickly and markedly decreased. Further study, however, showed that this apparent exception was a real and interesting proof of the correctness of KLEBS's statement. The experiments showed that while an abundance of food was present in the solution in which sporangia were formed, it was not available to the growing hyphae and consequently might as well not have been there. While working in Heidelberg the writer observed that a culture of *A. DeBaryana* Humphrey regularly produced sporangia in agar to which pea broth had been added. The sporangia were mostly borne on the large, vigorous hyphae that made a rapid growth immediately after a fresh plate of pea agar was inoculated. Later very many slender hyphae, which grew more slowly, were formed, and these did not produce sporangia. On the strong hyphae there were often 2 or 3 sets of sporangia, but always after a time the development of sporangia ceased and the outer portions of the medium became filled with many slender hyphae. One other form which never produced oogonia behaved in a similar manner. The agar here contained an abundance of food, and yet immediately after growth commenced sporangia were formed; while later, when the amount of food present might be thought to have been decreased, no sporangia were formed. At that time no definite experiments were undertaken to explain this phenomenon. Later these cultures were lost, and it was not until the fall of 1913 that another form showing this characteristic was collected. Meanwhile, *A. prolifera*, *A. race-*

mosa, and a number of species of *Saprolegnia* had been cultivated, but none of these produced sporangia in 1.5 per cent agar with pea broth.   In the fall of 1913 nos. 67, 68, and 70 were isolated, and in every case the germ tube gave rise to several vigorous hyphae which produced sporangia after attaining a length of some 5 mm.

It was also found that sporangia were sometimes developed on mycelia growing in liquid pea broth.   Since there certainly was no question here of an abundant supply of food, the thought suggested itself that KLEBS's conclusion, that sporangium formation takes place only when there is a dearth of food, would not apply to all species.   However, when mycelia were transferred to purified water, sporangia were normally produced.

The hypothesis suggested to explain the facts was that although there was plenty of food it did not reach the surface of the growing hyphae rapidly enough; these were therefore soon in an environment poor in food and then sporangia were normally and inevitably produced.

In the pea broth the proteid constituents are colloids, and these large molecules diffuse with extreme slowness, so slowly, in fact, that their movement can be considered to be practically nil.   The large, vigorous hyphae which require a relatively large amount of food would therefore after a time find themselves surrounded by a film of liquid out of which they had absorbed the food particles and to which diffusion did not carry new particles as rapidly as the hyphae used them.   This condition would result in starvation and the development of sporangia.

To test this hypothesis several series of experiments were prepared as follows.   Four Ehrlenmeyer flasks, each holding 100 cc., were half filled with pea broth, sterilized, and inoculated with *A. Klebsiana*.   One of these was so hung by a stout string that an arm clamped to a shaft and protected by a cotton pad pushed the flask aside about once a minute.   As the flask was pushed aside, it fell against the cotton pad on the end of the arm and thus the liquid was vigorously jarred.   Another flask was placed upon a shelf subject to a slight jar from a small motor, so that the surface of the liquid could be seen to tremble slightly; the other flasks were placed upon the writer's desk.

In every case where these flasks were arranged in the afternoon and examined in the morning so that those on the desk would be unaffected by the jar caused by walking, which complicated matters during the daytime, it was found that sporangia had been formed in the flasks on the desk, that the mycelium in the flask subject to frequent shaking showed no sign of sporangia, but consisted of balls of hyphae, the whole very dense and vigorous, and that on the mycelium in the flasks subject to a slight jar there were many aborted sporangia (fig. 1). It was evident that in the latter case a condition of starvation had momentarily existed, stimulating the ends of the hyphae to the production of sporangia; that before these could fully develop, the slight jar of the liquid had resulted in a fresh supply of food being brought to the hyphae, resulting in renewed vegetative growth and the abortion of the partially formed sporangia. The formation of sporangia in the pea agar is doubtless to be accounted for by the slow diffusion of food in the thick medium. The fact that only some species behave in this way doubtless indicates a more vigorous metabolism on the part of such species; the rate of metabolism exceeds the rate of diffusion and the result is starvation.

Saprolegnia Kaufmanniana has also proved interesting in that it shows great sensitiveness to the concentration of haemoglobin in solutions into which vigorous mycelia are placed. While such forms as *S. ferax*, *S. mixta*, or *S. monoica* will produce oogonia more or less freely in concentrations of haemoglobin varying from 0.075 to 0.01 per cent, *S. Kaufmanniana* persistently refused to respond to any concentration except 0.025 per cent. In this, either alone or with certain salts, oogonia containing oospores and accompanied by antheridia were regularly produced, though never in large numbers.

Achlya Klebsiana, n. sp.

This species was collected under three numbers, the plants all proving similar, and all were isolated as single spore cultures during November and December 1913. The cultures were secured from collections of algae from Bass Lake, near Ann Arbor, Michigan, and from near Coldwater, Michigan, in both cases collected by Mr. E. B. MAINS. One culture was also secured from a dish of algae in the botanical laboratories of the University of Michigan, of unknown source, but doubtless from around Ann Arbor.

Hyphae stiff, medium thick, forming a dense zone about the fly. Among these are large, coarse, branched hyphae attaining a length of 10–15 mm. or sometimes more; sporangia cylindrical, discharging and forming secondary sporangia as in *A. prolifera*, but developing in pea agar or in pea extract when this is absolutely quiet; oogonia on short lateral branches which are about as long as the diameter of the oogonia, rarely at the ends of long hyphae, but nearly always on the basal portions of strong hyphae, near the body of the fly, never intercalary; round or slightly oval in shape; oogonium wall smooth, not pitted, oospores 4–10, averaging about 25 μ in diameter, excentric; antheridia always present, of diclinous origin, partly clasping the oogonia, never clavate nor wrapped about the oogonia; gemmae produced in chains by the breaking up of the large hyphae, cylindrical, sometimes slightly branched or with one or more protuberances at one or both ends.

This species is peculiar in the fact that besides the zone of delicate hyphae which usually surrounds the fly, there were also a number of very long, thick hyphae. These commonly extended for several millimeters beyond the thick tuft of hyphae and spread out on the surface of the liquid, later becoming densely filled with protoplasm and breaking up into chains of gemmae as shown in fig. 2.

Oogonia are produced quite regularly on flies and are always clustered near the body of the fly, but so far, with one exception, I have been unable to secure oogonia in artificial media. In one test a sterilized pea on which the fungus was growing was left in an open dish of distilled water. Bacterial decay set in slowly, the water was changed from time to time, and the fungus kept on growing vigorously, eventually forming oogonia. In no solution of haemoglobin or leucin, with or without salts or sugars, have oogonia with oospores appeared, though empty oogonia have been occasionally formed in haemoglobin. In one case penetration of the oogonium was observed (fig. 3), but whether fertilization takes place is not known. This species shows affinities with *A. DeBaryana* Humphrey in the excentric oospores and smooth unpitted oogonia, but the strictly diclinous, branched antheridia and the arrangement of the oogonia distinguish it markedly from that species. In *A. DeBaryana* the oogonia are arranged in a loose raceme along the hyphae well out from the fly, while in *A. Klebsiana* the oogonia are always borne in a dense cluster near the body of the fly.

Saprolegnia Kaufmanniana, n. sp.

This species was collected from algal material in the botanical laboratory of the University of Michigan, of unknown source, but presumably from around Ann Arbor.

Vegetative growth like that of *S. ferax*, with firm stiff hyphae; sporangia freely produced and of the same size and appearance as in *S. ferax;* gemmae round, oval, or irregular in shape, mostly single, sometimes in chains and freely produced; oogonia very large, on long or short stalks, or intercalary, scattered; oval or club-shaped, very rarely almost round, the usual size being about 70–80 $\mu \times$ 100–250 μ. The smallest oogonium noted was 30$\times$70 μ; oogonium wall thin and smooth, without pits; oospores from 3 or 4 in small oogonia to very many in large ones, averaging about 20–30 oospores per oogonium; oospores average about 30 μ in diameter, contents granular without any conspicuous oil drop; antheridia nearly always present, only occasionally absent on intercalary oogonia, diclinous, of various shapes from clavate to clasping or irregular, often curving part way round the oogonium, and borne on slender antheridial branches; usually more than one on an oogonium.

This species seems to differ decidedly from all others described, especially in the large, thin-walled oogonia without pits. Rarely two oogonia were observed in series, as in fig. 5. This species may be related to *S. anisospora*, of which species little is known, though no evidence of two kinds of zoospores was found in the present species. Besides its marked morphological characters, *S. Kaufmanniana* is interesting from the fact that it is especially sensitive to the concentration of haemoglobin. Oogonia were but sparingly produced on flies, many cultures having none, and no culture having more than a few. Tests were made by transferring vigorous mycelium to haemoglobin solution, and it was found that only where the haemoglobin had a concentration of 0.025 per cent were oogonia formed. Of the drawings, fig. 7 is from fly cultures, the others from haemoglobin 0.025 per cent.

S. MONOICA var. vexans, n. var.

This was secured from algal material collected at Sukey Lake, near Ann Arbor, Michigan. The vegetative growth, sporangial characters, and the formation and shape of gemmae do not differ in any particular from those present in *S. monoica*, *S. ferax*, or any other species of that group except *S. mixta*, which has weaker hyphae. The material was cultivated for nearly a year and a half on flies, in agar, and by transfer from a strong culture medium such as pea decoction or peptone, into haemoglobin, leucin, peptone, or other solution. During all this time no oogonia were produced. Toward the end of this time a series of tests was being made with several cultures by transferring vigorous mycelium to leucin to which various sugars and salts had been added. Among other combinations there was used leucin $\frac{M}{200}$ + levulose $\frac{M}{200}$, and in this a mycelium out of pea extract produced an abundance of oogonia. When these were examined they proved to be indistinguishable from the oogonia and antheridia of *S. monoica* Prings. Rarely an oogonium was found on which there was no antheridium, but in some solutions this may also be the case with *S. monoica*.

The fact that cultures of *S. monoica* were going on at the same time suggested the possibility of contamination. Check cultures were made, therefore, by taking mycelium from the dish in which the oogonia were formed and growing this on fly. Had the mycelium producing oogonia been that of *S. monoica* (no. 79c of my series), plenty of oogonia would have been produced. In fact, no oogonia were formed on the fly culture, but a fresh culture from this fly through pea decoction into leucin and levulose again produced oogonia as before.

We seem to have here, therefore, the remarkable case of a variety of *S. monoica* having lost sexuality, but recovering it under stimulus of this special combination, leucin and levulose in concentration $\frac{M}{200}$ each.

The gemmae of this form are perhaps a little more varied in shape than is the case with the species, but the shape of these organs is so variable in most species that they are of no value for systematic purposes.

Had time permitted, it would have been interesting to cultivate this form for many generations in leucin-levulose solutions to determine whether the vigorous production of oogonia which characterizes such forms of *S. monoica* as my 79c would be regained by this variety.

The forms described in this paper are remarkable examples of the intimate dependence of the members of this group on external conditions.

UNIVERSITY OF MICHIGAN

EXPLANATION OF PLATE XXI

FIGS. 1–4.—*Achlya Klebsiana.*

FIG. 1.—Portions of hyphae showing aborted sporangia; tip of one (at *a*) has also died and would shortly have been pushed aside; about ×85.

FIG. 2.—Short chain of gemmae showing how they break away and fall off; about ×85.

FIG. 3.—Oogonium with oospores; ×300.

FIG. 4.—Young oogonium showing much branched antheridial hyphae; ×300.

FIGS. 5–7.—*Saprolegnia Kaufmanniana.*

FIG. 5.—Two oogonia in series; ×300.

FIG. 6.—Intercalary oogonium; ×300.

FIG. 7.—Oogonium showing several antheridia; ×300.

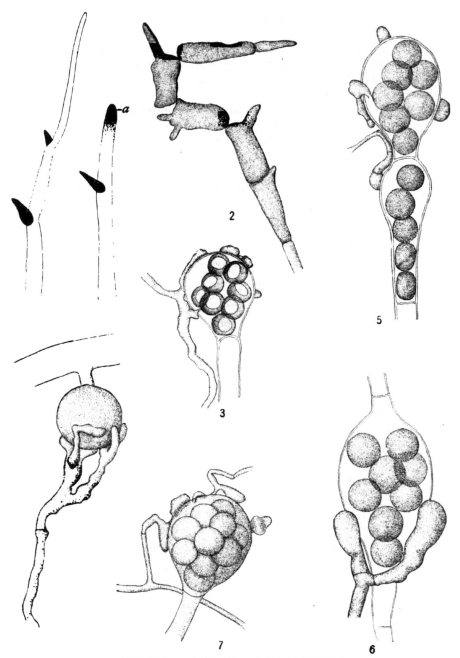

PIETERS on ACHLYA and SAPROLEGNIA

BRIEFER ARTICLES

DWARFING EFFECT OF TREES UPON NEIGHBORING PLANTS

The unfavorable effect of trees upon the growth of most plants rooted in the soil immediately about them is generally attributed to one or all of the following influences: (1) undue shade; (2) withdrawal of moisture from soil by the tree roots; (3) withdrawal of nutrient salts by the tree roots; (4) possible excretion of injurious substances into the soil by the tree roots.

Every observing farmer is well aware of the injury to most crops caused by the proximity of tree belts, hedges, or even of single large trees, and the loss caused in this way by trees is often considerable. Undoubtedly partial exclusion of light is an important factor. Cases in which fruit fails to mature may often depend upon this. The writer has found that the black raspberry (*Rubus occidentalis*) grew and flowered freely but failed to ripen fruit in a situation in which the bushes during the earlier half of the day received only one-twelfth to one-fifteenth of the total sunlight, although they were hardly at all shaded during the afternoon.

The present very rainy summer (1915) has afforded a suggestion as to the importance of the second of the factors previously mentioned, the withdrawal of moisture, in dwarfing plants growing under trees. The average rainfall in Boston for July is about 3 24 inches. This year the amount for July was in Boston 8 85 inches and in Cambridge 10 34 inches. The Boston record much exceeds that of any July precipitation during the 44 years for which the Weather Bureau has published a climatological summary of its observations.[1]

During the month of July of the present year the writer noticed that several species of perennial mesophytes which grew just north of and shaded by a belt of deciduous trees (wild cherries, ashes, and maples) were reaching unusual dimensions. Three most notable instances were

[1] No doubt the Cambridge rainfall for July of the present year is also the maximum for a long period. There is no readily accessible summary of total rainfall in Cambridge, month by month.

Aster novae-angliae, *Asclepias tuberosa*, and *Helianthus grosse-serratus*. None of these species was flourishing as well as did other individuals growing in open ground, but all were perhaps twice as tall as during an ordinary season and much more robust than usual. A plant of the moisture-loving *Chelone glabra*, which had for some years barely kept alive, grew luxuriantly and flowered freely.

On the other hand, the rather xerophytic *Sedum telephioides* and *Hedera Helix* showed no better growth than usual, and some other plants, such as *Saponaria officinalis* and *Oxalis corniculata*, showed little increase over their usual size. It would seem that the invariable dwarfing in ordinary seasons of the *Aster*, *Asclepias*, and *Helianthus* previously mentioned must be due mainly to abstraction of moisture from the soil by the roots of the trees. Doubtless many plants of agricultural importance are as sensitive to the effect of diminished water supply as are these three species.—J. Y. BERGEN, *Cambridge, Mass.*

STAMINATE FLOWERS IN ANEMONE

Anemone caroliniana is one of the most common of spring flowers in the vicinity of Grand Island, Nebraska. For several years I have noticed that in a large number of the flowers the pistils are lacking. In 1914 in one collection of 250 specimens, 190 were perfect, 50 had stamens only, and 10 had few or abortive pistils. There were none that had pistils only. The condition found is indicated in table I.

TABLE I

	STAMENS			PISTILS		
	Minimum	Maximum	Average	Minimum	Maximum	Average
190 normal flowers	20	45	28	25	60	35
50 staminate flowers .	7	55	28	....	....	....
10 with few pistils	10	52	40	10	20	16

In 1915 a collection of 133 specimens contained 55 staminate and 78 perfect flowers. A bouquet of especially fine large anemones was also examined. It contained 48 specimens, 46 of which were perfect, and in only 2 of which pistils were lacking. The average number of stamens in these was 86, the minimum and maximum being 72 and 100. The average number of pistils was 116, the minimum and maximum being 108 and 125. The fact that in these larger and more luxuriant plants the flowers were nearly all perfect and had a large number of both stamens and pistils would suggest that the absence of pistils is due to their degeneration through disease or some other cause; and the case of the 10 flowers with few pistils and an average of 40 stamens would suggest that pistils may be replaced by stamens. There was nothing in the vegetative parts of any of these plants to indicate disease.—CLARENCE J. ELMORE, *Grand Island College, Neb.*

CURRENT LITERATURE

Cretaceous plants

The second part of the British Museum *Catalogue of Cretaceous plants*, by STOPES,[1] treats of the flora of the so-called Lower Greensand, which is known on the Continent as the Aptian stage of the Lower Cretaceous. The beds of this age which have escaped erosion and are available for study are nearly all typically marine deposits, with only such traces of terrestrial vegetation as withstood the maceration and trituration of the sea. Consequently, the flora of the Lower Greensand has hitherto been supposed to have been essentially the same as that of the older and better known Wealden deposits. Thus, with the exception of the classic *Bennettites Gibsonianus* of CARRUTHERS and a few cones, the plant remains consist almost entirely of pieces of petrified wood. By a careful study of the latter, STOPES is enabled to list 45 plant forms, comprising 1 thallophyte, 2 Filicales (*Weichselia* and *Tempskya*), 9 cycadophytes, 27 conifers, and 5 angiosperms. Notable features are the presence of angiosperms, the preponderance of conifers, and the scarcity of ferns. Angiosperms have been sparingly represented by leaf impressions in deposits of this age or slightly older, but these have been few in number and vague in character, while these forms of the Lower Greensand, although vague in their affinities, are well characterized.

The proportions of the different groups represented, upon which STOPES lays considerable stress in emphasizing the differences between this flora and other floras of the Lower Cretaceous, is due almost entirely, I believe, to the methods of preservation; that is to say, to differences in the physical conditions of deposition of the sediments. Coniferous remains usually predominate and ferns are scarcely represented in coarse marine deposits. Thus, in the Trinity beds of Texas, which are about the same age as the Lower Greensand, there is but one fern, while there are 8 cycadophytes and 11 conifers. On the other hand, in the Lakota beds of the Black Hills, which are continental deposits partially synchronous with the marine Trinity, there are 13 ferns and only 8 conifers.

The two features of the present work which are of greatest botanical interest are the interpretation of the coniferous woods and the careful description of the wood structure of five indubitable angiosperms of a relatively high degree

---

[1] STOPES, MARIE C., Catalogue of the mesozoic plants in the British Museum. The cretaceous flora. Part II. London. 1915.

of organization. While they are the oldest angiosperms whose anatomical features are known, they are in no way primitive or pro-angiospermic. The forms, of which three received preliminary treatment in 1912,[2] are named *Cantia arborescens*, *Woburnia porosa*, *Sabulia Scottii*, *Hythia Elgari*, and *Aptiana radiata*. No attempt is made to indicate their family relationships, largely because of our sadly inadequate knowledge of the anatomy of existing dicotyledons. The Coniferae enumerated include, in addition to various forms of *Pinostrobus*, *Cedrostrobus*, and *Abietites*, anatomical material representing a *Sequoia* (Taxodieae), *Protopiceoxylon* (1 sp.), *Pityoxylon* (3 spp.), and *Cedroxylon* (2 spp.) in the Abietineae; four species of *Cupressinoxylon* in the Cupressineae; *Taxoxylon* (1 sp.) and *Podocarpoxylon* (4 spp.) in the Taxaceae. No remains of Araucariaceae are recognized. This list of conifers is of especial interest to American botanists in connection with JEFFREY'S rather sweeping conclusions from his studies of material from the Upper Cretaceous, from which STOPES dissents, apparently on the basis of rather satisfactory evidence.

The present contribution also shows that BUCKLAND'S genus *Cycadeoidea* offers anatomical differences from the American forms usually described under that name. A new cycadophyte trunk is made the type of a new genus, *Colymbetes*. It is found in the lower part of the Lower Greensand and may have been derived mechanically from the underlying Wealden sediments. The climate of the Aptian is considered to have been less warm than that of the preceding Wealden.

·The book on the whole is an exceedingly valuable contribution to our knowledge of fossil plants, particularly to that important part of the Lower Cretaceous represented in Britain by the Lower Greensand.—EDWARD W. BERRY.

Flora of New Zealand

Two quarto volumes illustrating the New Zealand flora have appeared under the editorship of T. F. CHEESEMAN, curator of the Auckland Museum, with the assistance of W. B. HEMSLEY.[3] The 250 plates are exceptionally good, having been drawn by Miss MATILDA SMITH of the Kew Herbarium, whose work in connection with the *Botanical Magazine* and *Icones Plantarum* has long been known. Accompanying each plate there is an account of the discovery and occurrence of the plant, as well as items of general interest. Since the technical descriptions are published in the *Manual of the New Zealand flora* (1906) they are not repeated in the *Illustrations*.

The problem of selecting approximately 265 plants to illustrate adequately such a flora as that of New Zealand can be appreciated. The main features of the flora, however, have been presented, and no important genus or group

---

[2] STOPES, M. C., Phil. Trans. Roy. Soc. London. B **203**:75–100. *pls. 6–8.* 1912.

[3] CHEESEMAN, T. F., Illustrations of the New Zealand flora. 2 vols. 4to. *pls.* 250. Published under the authority of the government of New Zealand. Wellington. 1914.

is without proper representation. The alpine flora has received full treatment, and special attention has been paid to the endemic genera. The editor says that he has also illustrated a number of plants of special interest "either on account of their economic value, or from biological or morphological reasons, or from their peculiar geographical distribution." He has not thought it necessary, however, except in a few special cases, to figure plants occurring in other countries as well as New Zealand.

It is interesting to note the result of the selection, as expressing in a general way the editor's opinion of what constitutes the main features of the New Zealand flora. Without going into detail, it may be stated that the families represented in this selection by more than ten species are as follows: Compositae (35), Gramineae (17), Filices (17), Orchidaceae (15), Umbelliferae (12), Scrophulariaceae (12).—J. M. C.

MINOR NOTICES

Pharmacognosy.—KRAEMER[4] has published a textbook of pharmacognosy which is encyclopedic in its wealth of information. He recognizes the synthetic character of the subject, especially in its combination of botanical and chemical aspects. Moreover, both of these aspects are shown to involve the ecological conditions under which plants grow, so that scientific pharmacognosy is a very complex subject. After an introduction dealing with the problems involved, the great plant groups are presented, the large majority naturally being families of angiosperms, 94 in number, from which drugs are obtained. As stated by the author, the book is intended for students of pharmacy, pharmacists, food and drug analysts, and pharmacologists. We might add to this list students of economic botany, for pharmocognosy is one of the great fields of applied science.—J. M. C.

Citrus fruits.—Corr[5] has published as a volume of BAILEY's "Rural Science Series" an account of the citrus fruit industry, with special reference to the requirements and practices for California. It seems that the citrus industry has not only reached a high state of development in California and Florida, but is still progressing rapidly. The present volume discusses the underlying principles in such a way as to emphasize the importance of certain fundamentals which must be kept in mind. The industry as it exists at present is described, and all current information that seems valuable is organized and made available. Of special interest to the botanist are the chapters dealing with the geography and climate of California, the gross structure and habits of growth of citrus plants, and the citrus diseases and their control.—J. M. C.

---

[4] KRAEMER, HENRY, Scientific and applied pharmacognosy. 8vo. pp. viii+857. *figs. 313.* Published by the author, 145 North 10th St., Philadelphia. 1915.

[5] CORR, J. E., Citrus fruits. 8vo. xx+520. *figs. 151.* New York: Macmillan. 1915.

Ferns of South Africa.—SIM[6] has prepared a second edition of his *Ferns of South Africa*, bringing together in this convenient form much widely scattered information. The preliminary chapters deal with the following topics: ferns, parts of ferns, reproduction and propagation, cultivation, identification and preservation, the ferns of South Africa, the natural home of ferns. The bulk of the volume naturally is concerned with the descriptions of species. Attention is called to the fact that the number of species of ferns in South Africa is remarkably small compared with the whole flora. The present volume contains 220 species of ferns and fern allies, an increase of 41 species over the first edition, published in 1892. The full descriptions and the numerous plates make the volume very complete for its purpose.—J. M. C.

Western wild flowers.—MARGARET ARMSTRONG,[7] in collaboration with Professor J. J. THORNBER of Arizona, has prepared a popular field book describing and illustrating the "common wild flowers" west of the Rocky Mountains. The book is "popular," not merely in the selection of plants for description, but also in the absence of technical terminology. The author says that "almost all technical botanical terms have been translated into ordinary English." The drawings for the numerous illustrations have all been made from life, and, in connection with the "ordinary English" of the text, should enable the "general public" to identify the conspicuous plants in which it may have a casual interest.—J. M. C.

Plant life.—This title has been selected by HALL[8] for a volume presenting the plant kingdom "to the amateur botanist and the lover of nature." As a consequence, the style is not technical, but appeals to general interest. The illustrations are numerous, and 50 of the 74 plates are colored. Some idea of the topics presented can be obtained from the chapter heads, which are in effect as follows: asexual plants, development of sex in plants and a study in evolution, seaweeds, fungi and lichens, archegoniates, phanerogamia, fossil plants, food of plants, perpetuation of the race, defenses of plants, ecology.—J. M. C.

NOTES FOR STUDENTS

Origin of monocotyledony.—In an address delivered at the twenty-fifth anniversary celebration of the Missouri Botanical Garden, COULTER,[9] because of continued studies on the origin of monocotyledony, chiefly in grasses,

---

[6] SIM, THOMAS R., The ferns of South Africa. 2d ed. 8vo. pp. ix+384. *pls.* 186. Cambridge University Press. 1915. 25s.

[7] ARMSTRONG, MARGARET, Field book of western wild flowers. 16mo. pp. xx+596. *col. pls. 48. figs. 500.* New York and London: Putnam, 1915. $2.00.

[8] HALL, CHARLES A., Plant life. 8vo. pp. xi+380. *pls. 74. figs. 80.* London: A. & C. Black. 1915.

[9] COULTER, J. M., The origin of monocotyledony. Ann. Mo. Bot. Gard. 2:175–183. *figs. 9.* 1915.

restated the conclusions previously arrived at from a study[10] of *Agapanthus umbellatus* and some other monocotyls as follows: "In the embryogeny of both monocotyledons and dicotyledons, a peripheral cotyledonary zone gives rise to two or more growing points, or primordia; this is followed by zonal development, resulting in a cotyledonary ring or sheath of varying length. If both growing points continue to develop equally, the dicotyledonous condition is attained; if one of the growing points ceases to develop, the continued growth of the whole cotyledonary zone is associated with that of the other growing point, and the monocotyledonous condition is attained. In like manner, polycotyledony is simply the appearance and continued development of more than two growing points on the cotyledonary ring. It follows that cotyledons are always lateral structures, arising from the peripheral zone developed at the top of a more or less massive proembryo. This reduces cotyledony in general to a common basis in origin, the number of cotyledons being a secondary feature. The constancy in the number of cotyledons in a great group is no more to be wondered at than the same constancy in the number of petals developed by the petaliferous zone."

To those whose mental processes require a "type" for everything, the embryogeny of the grasses is very puzzling, because it does not conform to the hitherto accepted monocotyl "type" of embryogeny. The structures have received the names "scutellum," "epiblast," and "coleoptile." Early in the history of the subject the scutellum was recognized as the cotyledon. The epiblast was recognized as a second and rudimentary cotyledon from its first discovery until the time of SCHLEIDEN, who in 1837 so clamorously dissented from this view that it was suppressed from the literature of the subject until 1897, when VAN TIEGHEM, studying the embryogeny of grasses, reaffirmed that the epiblast is a second cotyledon. In the meantime HANSTEIN, followed by FAMINTZIN, had made a study of *Alisma*, which has a filamentous and therefore highly specialized proembryo, and fixed what has been almost universally regarded as the monocotyl "type" of embryo. Because of these first studies, the massive proembryo, so prevalent in monocotyledons, has been called the "aberrant type," although it is undoubtedly the more primitive and generalized, while the filamentous proembryo is the more specialized. About two-thirds of the grasses show a well marked second cotyledon. This second cotyledon is quite pronounced in those grasses which have a relatively long internode ("mesocotyl" of English anatomists) between the cotyledons and the "coleoptile" or bud-scale leaves.

COULTER shows that there is a progressive reduction of the second cotyledon until *Zea Mays* is reached, where the only external sign of the second cotyledon is a small hump opposite the functional cotyledon; but even here a small procambium is present, being exactly opposite the procambium of the

---

[10] COULTER, JOHN M., and LAND, W. J. G., The origin of monocotyledony. BOT. GAZ. **57**:509–519. *pls. 28, 29*. 1914.

functional cotyledon. The absence of a vascular strand in the smaller cotyledon has been used as evidence to prove that it is not a second cotyledon. The reviewer has no doubt that when a relatively great quantity of material shall have been studied a procambium will occasionally be found in the second cotyledon. It is unfortunately the habit of some investigators to make sweeping conclusions from an examination of a very small quantity of material.

The facts shown by this reinvestigation of the grasses are as follows: "The terminal cell of the proembryo forms a group of cells; the peripheral cells of the group develop the cotyledonary ring or sheath, on which two growing points appear. One of these growing points soon ceases to be active, and the whole zone develops in connection with the other growing point; but at the base of the growing cotyledon a notch is left by the checking of the growing point. This notch is really the space between the two very unequal cotyledons which surround the real apex of the embryo. The apex of the embryo is at the bottom of the notch, and not at the tip of the large embryo. This apex soon begins to form leaves, and the so-called stem tip appears issuing from the bottom of the notch, in a relation apparently lateral only because the two cotyledons are so unequal. Furthermore, when the stem tip is examined, it is found not to be a stem tip, but a cluster of leaves, whose rapid development has aborted one of the growing points on the cotyledonary zone. All this is very obvious in grasses and is equally obvious in any massive proembryo, but it escaped the earlier observers of filamentous proembryos."

The general conclusion is that "monocotyledony is simply one expression of a process common to all cotyledony, gradually derived from dicotyledony, and involving no abrupt transfer of a lateral structure to a terminal origin."— W. J. G. LAND.

Anatomy of Helminthostachys.—In the third of his series of papers on the Ophioglossaceae, LANG[11] gives the results of a reinvestigation of the anatomy of the rhizome of *Helminthostachys*, together with the details of the vascular connections of two branching specimens. In rhizomes of young plants, the xylem forms a solid strand of two kinds of tracheids; those of the outer part having pitted walls, while the inner elements are smaller and are spirally thickened. By comparison with the bases of branch steles, where a similar condition exists, and by examination of the origin of leaf traces, where the protoxylem becomes evident, it is shown that the two kinds of elements in the juvenile type of stele are outer and inner metaxylem, respectively; and that therefore the stele of *Helminthostachys* is mesarch even in the juvenile condition. The transition to the adult condition begins by the appearance of parenchyma cells among the tracheids of the inner metaxylem, thus forming a mixed pith. In the adult condition the stele is greatly expanded and the pith is large.

[11] LANG, WILLIAM H., Studies in the morphology and anatomy of the Ophioglossaceae. III. On the anatomy and branching of the rhizome of *Helminthostachys zeylanica*. Ann. Botany 29:1–54. *pls. 1–3. figs. 7.* 1915.

All stages between the juvenile type with the solid stele and the adult type may occur; cases of reduction of steles of the adult type to the juvenile condition are described. These variations are held to be dependent upon physiological factors; the solid stele of the juvenile type is considered to be merely a physiological variation of the general type of the species.

The outer xylem of the leaf trace, after separation from the stele, curves round the protoxylem strand; at the same time, the trace, originally monarch, becomes diarch by the division of the protoxylem strand. These two processes result in the formation of a trace whose cross-section shows two protoxylem points, each surrounded by metaxylem. This "clepsydroid" stage of the leaf trace is characteristic of certain of the Zygopterideae and affords data for a comparison with that group.

The branches develop from the vestigial axillary buds. There is no vascular connection between the branch stele and the leaf trace immediately below, as in some species of *Botrychium*. Immediately below the point of origin of a branch, accessory xylem is developed by irregular divisions of the parenchyma within the phloem. While no cambium is present, this is considered to be a form of secondary thickening. In the base of the branch this accessory xylem surrounds an extension of the inner metaxylem of the main axis, thus forming a solid xylem strand which is in all respects similar to the steles of rhizomes of young plants. The further development of the branch stele is identical with that of steles of young plants.

The mesarch character of the stele, the clepsydroid stage of the leaf trace, the irregular secondary thickening of the stele, and the connection of the branch stele with the main axis rather than with the leaf trace, all emphasize the view of relationship of the Ophioglossaceae to the Zygopterideae. It is further pointed out that these features also afford a basis for comparison with certain of the Cycadofilicales.—L. C. PETRY.

Origin of dwarf plants.—STOUT[12] has described a dwarf form of *Hibiscus oculiroseus* which differs from the usual robust type in having more basal branches, shorter internodes, and smaller, somewhat crinkled and irregular leaves. The dwarfs are descendants of a single robust plant with some crinkled leaves and somewhat shortened upper internodes, a type known as "intermediate." Four other plants have given none but normal robust offspring, something over 100 in all. Selfed seed from the intermediate plant produced 45 plants, all dwarf. Open-pollinated seed from the same plant yielded one dwarf, 3 intermediates, and 11 robust plants. From selfed seed of one robust plant, there were grown one intermediate and 33 robust plants. Selfed seed of the one dwarf plant produced one robust, 8 intermediate, and 72 dwarf plants, the robust plant being apparently an accidental hybrid. These results are compared with the behavior of dwarfs of *Oenothera*, peas, and sweet

---

[12] STOUT, A. B., The origin of dwarf plants as shown in a sport of *Hibiscus oculiroseus*. Bull. Torr. Bot. Club 42:429–450. 1915.

peas. The author emphasizes the fact that in the dwarf *Hibiscus*, as well as in dwarfs of other plants, numerous characters besides stature are modified, and expresses doubt concerning the possibility of interpreting such phenomena as due to a change in a single hereditary unit. The reviewer is inclined to suggest that, until there are available adequate data derived from crosses between the new and the normal types, little is to be gained by the assumption of either single or plural genetic changes, or indeed by a discussion of any other hypothesis.—R. A. EMERSON.

Growth of sugar cane.—KUYPER[13] has investigated the growth of the leaf blade, leaf sheath, and stalk of sugar cane. His method was to make some holes (with a darning needle) through the young leaves and internodes. The distance between holes was made as uniform as possible; in practice the spacing could not be much less than a centimeter. After several days the leaves were removed, one after the other, and the distance between the holes measured. By comparing these measurements, the rate of growth of an area on different parts of the leaf, and on different leaves, can be determined. The results indicated that the region of most rapid growth moves basipetally over the blade, then over the sheath, and finally over the internode. The region near the base continues its growth after the regions above have completed their development. These conclusions were confirmed by measurements obtained from equally spaced lines of India ink, and also by cell measurements. Regarding an internode bearing a leaf at its summit as the unit of structure, it may be said that the blade first becomes fully grown, then the sheath follows, and finally the internode develops. KUYPER shows how these conclusions may be applied in studying the "top rot" disease of Java. The cause of the disease must be sought in temporary unfavorable conditions of growth, and this investigation furnishes a method of recognizing the period of growth influenced by these conditions.—J. M. C.

Stomata of sugar cane.—KUYPER[14] in connection with an investigation of the transpiration of sugar cane discovered a lack of knowledge of the structure of the stomata. Several methods were tried for measuring the width of the stomatal cleft. Direct measurements with the microscope proved impossible, not only because it is very difficult to make good preparations of the leaf for this purpose, but also because the variations in the opening are very small. Since, however, the application of the infiltration method of Miss E. STEIN showed that great variations really exist in the rapidity with which paraffin and kerosene penetrate the leaf tissue, it was clear that there must be something in the structure of the stoma which could explain this variation. The

---

[13] KUYPER, J., De groei van bladschyf, bladscheede, en stengel van het suikerriet. Archief voor de Suikerindustrie in Nederlandsch Indië 23:528–540. *pls. 2.* 1915.

[14] KUYPER, J., De bouw der huifmondjes van het suikerriet. Archief voor de Suikerindustrie in Nederlandsch Indië 22:1679–1707. *figs. 6.* 1914.

figures of transverse sections of stomata show that any noticeable widening of the slit in the ordinary way is impossible, because of the very thick cell walls. The guard cell, however, can move a little as a whole, because it is distinctly hinged with the adjacent cell. The general conclusions reached are as follows: a movement as described by SCHWENDENER for grass stomata is impossible; the guard cells can vary their distance from each other to a slight extent by changing their position among the surrounding epidermal cells; the position and structure of the neighboring cells makes it possible to change the length of the slit in the vertical direction.—J. M. C.

Internal temperatures of plants.—PEARSON[15] has made observations on internal temperatures of the stems of *Euphorbia virosa* and *Aloe dichotoma* at 4200 ft. altitude, on the western flanks of the great Karasberg Range in Great Namaqualand. *Euphorbia* reached its maximum at about 2:00 P.M., about the same hour as the shade maximum; while *Aloe* attained its maximum after sundown. In *Euphorbia* the highest internal temperature observed was $51°5$ C.; the greatest range of internal temperature was $27°5$ C.; and the excess of internal over shade temperature $15°35$ C., as against $38°5$, $16°$ C., and $5°$ C. respectively for *Aloe*. These observations were made in December. It is likely that in February and March both the internal maximum temperature and the excess over shade temperature is much greater. Xerophytes have much reduced transpiration, the greater of the two cooling methods in mesophytes, and depend largely upon thermal emissivity for cooling. Even in mesophytes as great difference between shade and internal leaf temperatures have been observed as are recorded here, and in xerophytes much greater differences. The author observes that wounding *Euphorbia* causes a considerable fall in the internal temperature. This is due partly to expansion of gases in the air chambers and partly to rapid evaporation from the wound.—WILLIAM CROCKER.

Javanese Pallavicinias.—CAMPBELL and Miss WILLIAMS[16] have recently completed a study of three species of *Pallavicinia* collected by Professor CAMPBELL in Java in 1906. The apical cell in the three species is dolabrate, being similar to the usual condition in *P. Lyellii*, which, however, the reviewer has observed rarely has a triangular pyramidal cell similar to the one in *P. decipiens*. The antheridium in development shows the usual situation among Jungermanniales. A delicate wall separates the pairs of spermatocytes. No "Nebenkörper," reported by IKENO for *Marchantia*, was found. The archegonia present few variations from the usual liverwort conditions. In *P.*

---

[15] PEARSON, H. H. W., Observations on the internal temperatures of *Euphorbia virosa* and *Aloe dichotoma*. Annals Bolus Herb. 1:41–66. 1914.

[16] CAMPBELL, D. H., and WILLIAMS, FLORENCE, A morphological study of some members of the genus *Pallavicinia*. Leland Stanford Jr. Univ. Pub. 7, Univ. Series, pp. 44. *figs. 23.* 1914.

radiculosa the primary cap cell may also add to the neck of the archegonium, but contributes nothing to the axial row. In this species the neck canal cells range from five to six. The embryo, while not different from that of other Anacrogynae, closely resembles *Aneura* in that a large haustorial, suspensor-like cell is produced. In the capsule the sterile cap is pronounced, and in consequence the dehiscence is by means of four longitudinal slits. The authors conclude that the evidence does not warrant the erection of the two families Aneuraceae and Blyttiaceae.—W. J. G. LAND.

Brown oak.—GROOM[17] has investigated the cause of what is known as "brown oak" or "red oak" in Great Britain. The phenomenon consists in the replacement of the ordinary heart wood of *Quercus Robur* by a firm, richer toned, often reddish brown wood, which varies in tint from dull brown to rusty brown. It is found to be due to the influence of a fungus which lives exclusively in the heart wood, which it infects through wounds. "Brown oak" usually occurs at the base of the trunk and in the adjoining root and extends more or less upward in the stem and downward in the root. If the fungus gains entrance to the upper parts of the tree, it produces in these regions masses of "brown oak." The color is due to the fact that the fungus produces a brown substance in the individual cells which is highly resistant to solvents. The source of the food of the fungus was not determined, although there are reasons to believe that tannin is one of the sources. The identity of the fungus is also left in doubt. It produces conidiophores resembling those of *Penicillium*, and on certain specimens basidiocarps appear, which were identified as a species of *Melanogaster*, but cultures did not establish any connection between the two phases.—J. M. C.

Morphology of Arisaema.—PICKETT[18] has made a careful study of *Arisaema triphyllum*, and has contributed materially to our knowledge of the morphology of the Araceae. The critical situations may be summarized as follows: There is a wide range in time in the development of the flowers, with a strong tendency to the earlier development of staminate flowers. The tapetal nuclei of the microsporangium are freed and "wander" among the developing microspores, as DUGGAR has described for *Symplocarpus*. The embryo sac is of the *Lilium* type, and a complex and permanent suspensor system is developed. The endosperm arises from one of the daughter cells (micropylar) of the primary endosperm nucleus, the other daughter cell not dividing. The primary roots are diarch, while the secondary roots of seedlings and all roots of mature plants are triarch to pentarch. The statement is made that the sex of mature plants is changeable, dependent upon the amount of available water.

[17] GROOM, PERCY, "Brown oak" and its origin. Ann. Botany 29:393–407. 1915.

[18] PICKETT, F. L., A contribution to our knowledge of *Arisaema triphyllum*. Mem. Torr. Bot. Club 16:1–55. *pls. 1–5. figs. 70.* 1915.

Occasionally flowers are also found showing a tendency to become bisporangiate.—J. M. C.

Physiology of parasitism.—Brown[19] has begun a much needed investigation of the physiological relation of host and parasite, his first paper dealing with *Botrytis cinerea*. From the germ tubes of this fungus he succeeded in obtaining a very powerful extract, whose action on cell walls results in the disintegration of tissue, and whose action on the protoplasts produces death. This extract loses its "lethal power" by heating, by mechanical agitation, and by neutralization with an alkali. Neither oxalic acid nor oxalates are accountable for the toxicity of the extract, which the author concludes must be due to the presence of a substance of colloidal nature. The only active substance discovered was an enzyme which was thought to be responsible for the lethal action of the extract. The multiplication of such investigations will result in some progress in knowledge as to the nature of immunity and susceptibility.—J. M. C.

Morphology of Ephedra helvetica.—In a thesis presented for the doctorate of science at the University of Geneva, Sigrianski[20] has reviewed and reinvestigated *Ephedra helvetica*. Two new facts are reported. The hypodermal archesporial cell does not divide periclinally and give rise to a primary wall cell. By division of the cells of the epidermis the hypodermal initial is placed deeply within the nucellus. The figures which illustrate this situation will not entirely satisfy a critical investigator. A second and most important fact is that the four megaspores are all functional, the wall of the megaspore mother cell being the embryo sac wall, as in *Lilium* and some other angiosperms. It would be most interesting to know with certainty whether *Welwitschia* and *Gnetum* have attained the *Lilium* level in this respect.—W. J. G. Land.

Jurassic wood.—Miss Holden[21] has described a new species of *Metacedroxylon* from the Jurassic of Scotland, under the name of *M. scoticum*. It is a good illustration of the merging of araucarian and abietinean characters during the Jurassic, since it is araucarian in the pitting of the tracheids, and abietinean in the pitting of the rays. It differs from *M. araucarioides* only in the absence of pits on the tangential walls of the tracheids and in the biseriate character of the rays.—J. M. C.

---

[19] Brown, William, Studies in the physiology of parasitism. Ann. Botany 29: 313–348. 1915.

[20] Sigrianski, Alexandre, Quelques observations sur l'*Ephedra helvetica* Mey. pp. 62. *figs. 74.* Geneva. 1913.

[21] Holden, Ruth, A Jurassic wood from Scotland. New Phytol. 14:205–209. *pl. 3.* 1915.

GENERAL INDEX

Classified entries will be found under Contributors and Reviewers. New names and names of new genera, species, and varieties are printed in **bold face** type; synonyms in *italic*.